“Memphis Belle”

It is a fearful thing to lead this great peaceful people into war...but the right is more precious than peace, and we shall fight for the things which we have always carried nearest our hearts - for democracy, for the right of those who submit to authority to have a voice in their own governments, for the rights and liberties of small nations, for a universal dominion of right by such a concert of free peoples as shall bring peace and safety to all nations and make the world itself at last free. To such a task we can dedicate our lives and our fortunes, everything that we are and everything that we have, with the pride of those who know that the day has come when America is priveleged to spend her blood and her might for the principles that gave her birth and the peace which she has treasured. God helping her, she can do no other.

Woodrow Wilson

Memphis Belle

Biography of a B-17 Flying Fortress

Brent W. Perkins,
Tennessee Colonel Aide de Campe

Schiffer Military History
Atglen, PA

Dedication

Thank you Michael, Steven, Jeremy, and Julia.
Dad didn't mean to be away so much.

For my mom and dad, Diana and Albert Perkins.
Thanks Mike & Kim, you've made so much possible.

For Christy, my Memphis Belle.

But especially for all those who left the vivid air,
signed with their honor.
To them, this work is dedicated

Book Design by Ian Robertson.

Library of Congress Catalog Number: 2001098658

Printed in China.
ISBN: 0-7643-1499-8

We are interested in hearing from authors with book ideas on related topics.

Published by Schiffer Publishing Ltd.
4880 Lower Valley Road
Atglen, PA 19310
Phone: (610) 593-1777
FAX: (610) 593-2002
E-mail: Schifferbk@aol.com.
Visit our web site at: www.schifferbooks.com
Please write for a free catalog.
This book may be purchased from the publisher.
Please include $3.95 postage.
Try your bookstore first.

In Europe, Schiffer books are distributed by:
Bushwood Books
6 Marksbury Avenue
Kew Gardens
Surrey TW9 4JF
England
Phone: 44 (0) 20 8392-8585
FAX: 44 (0) 20 8392-9876
E-mail: Bushwd@aol.com.
Free postage in the UK. Europe: air mail at cost.
Try your bookstore first.

Contents

Acknowledgments

To the combat crew of the "Memphis Belle" who have shared so many of their personal memories, as well as many of the other "Bassingbourn Boys," I wish to extend a very personal thank you for your contributions, then and now, to the creation of this record. You are the ones who are to be remembered in this very special and important time of American military aviation heritage.

In proud and grateful memory to those men of the United States Army Air Force, who from these friendly isles, flew their final flight and met their God. They knew not the hour, the day, nor the manner of their passing when far from home, they were called to join that heroic band of airmen who had gone before. - May they rest in peace.

American cemetery - Cambridge, England

Additional Picture Credits Unless Otherwise Noted:
The Commercial Appeal
James Webb, Sr.
Hamp Morrison
Harry Friedman
United States Air Force

With special thanks to:

Frank G. Donofrio - Founder of the Memphis Belle Memorial Association, and tireless Belle promoter and trusted friend.

Menno Duerksen - Author of the book "Memphis Belle - Home at Last." Your work is truly motivational.

Col. Bert W. Humphries - Squadron Operations Officer - 322 Squadron - 91st Bomb Group (H)
A tireless and devoted historian, and one of Bassingbourn's best!

Joe Harlick - WWII Photographic unit 91st Bomb Group (H)
Your work will never be forgotten.

John Harold Robinson - B-24 Flight Engineer and waist gunner. 703 Squadron, 446th BG (H) - Tibenham. Author of the book - "A Reason to Live." Thank you for the motivation and support.

James L. Fri Jr. - WWII United States Army Sergeant - 87th Infantry Division - "Golden Acorns" - Ardennes Campaign. One of the finest Americans I know.

David Tallichet - Mighty Eighth Air Force 100th Bombardment Group (H), Operator of the B-17 Flying Fortress N3703G aka "Memphis Belle." Blue skies Dave. Thanks for the wings!

For all the special individuals who helped with my research but are not named here, thank you. You know who you are.

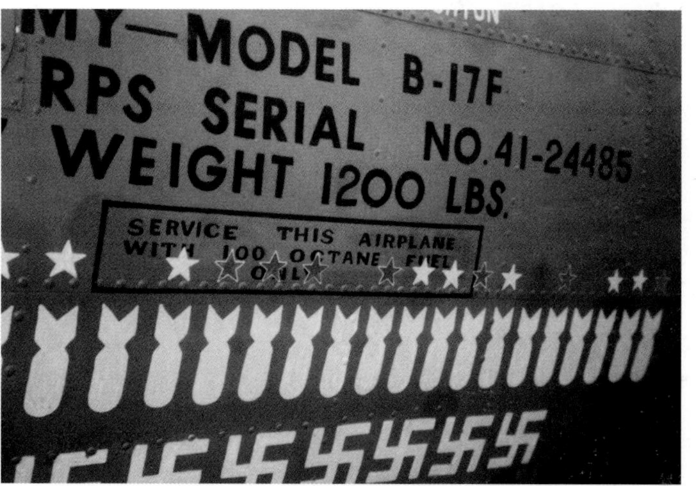

In the above photo, the mission symbols of the "Memphis Belle" are evident on the nose of this B-17. The yellow stars indicate that on that raid, the Belle was the lead aircraft for the 91st Bomb Group. The red stars indicate that on that particular mission the Belle led the entire bombing formation! (Perkins)

Preface:

I have met the generation gone past.

Before you begin this book, it is my hope that you will be able to feel the experience of the people involved. That you will be able to imagine the freezing high-altitude temperatures, the terror of aerial combat, and the tremendous emotion involved with the brutality of the type of war being waged in the skies of Europe. No one had ever done what these *kids* were doing. If you think about it, the simple majority of them should have been at home on their farms and in their cities preparing for college and starting their lives. The average age was between eighteen and twenty-five years. Their uniforms bely their ages. They were not old enough in many cases to even buy a beer, but they were old enough to crew a thirty-ton bomber and take it into the most hostile skies ever. They looked fresh and new, but their planes looked as if they were much older. Oil-stained, holed, patched, faded, dented, and frayed, these young men mastered the controls of these war machines and then wrote the book on how to bring an enemy to its knees from the sky. They left their homes not knowing if they were coming back. They joined in endless rows of other volunteers who also knew that on the other side of what was then a much larger world, someone else was putting on a different color uniform and training to kill them.

Today, I have been blessed to come to know hundreds of these important people. Most of them have unselfishly opened the dark pages of their memories and revealed to me, many for the first time since the war ended, their stories and their experiences. Some with laughter, though many with tears. Most of them today are well over the terrors that more than half a century has been able to erase. Their experiences gave them the ability to overcome many difficult obstacles. And most of them have gone on to lead fruitful and productive lives, but they, to this day, share a camaraderie that is beyond words.

Some of my veteran friends have endured the icy muddy winters of the Italian campaign and the Battle of the Buldge, living in a foxhole for weeks and finally running for their lives with wet and frozen feet when German Panzer tanks appeared over the horizon, cannons blazing murderous shells on them.

My own grandfather, William Kammer, endured the waters of the Pacific for months on end aboard a supply barge built out of concrete, never knowing when the Japanese were going to torpedo them. Still more endured the relentless baking sun of the south Pacific fighting the stalwart Japanese, who fought with unprecedented will and determination.

The purpose of this book is to share real and authentic experiences that happened not only to the "Memphis Belle," but to some of those important other B-17s that have been lost to memories through time. So significant are the contributions of these brave men, that I felt compelled to search for the true stories that surrounded this plane and the others of the 91st Bombardment Group (Heavy). At the "Memphis Belle" pavilion in downtown Memphis, Tennessee, I have answered the questions from literally thousands of tourists. I have labored to compile the questions that they ask most frequently and provide here, for the reader, those answers and even more of the dramatic events that they lived through. The result is more than five years of extensive research and study. We are still uncovering some details here and there. Sadly, however, it is becoming more and more difficult to find the men who were there then. Too many have already gone and taken their stories with them. It is my hope that you find a new and sincere interest in the people of the United States who did not question an order, who rode on their blazing trails of invincibility and took combat into the skies.

The "Memphis Belle" as she appears today in the pavilion on the downtown river park named Mud Island in Memphis. This has been the home of "Memphis Belle" since 1987 when, for want of a better place, the bomber was moved to this location. Although the environment is certainly an improvement over what had been some forty years of sitting uncovered and outdoors, this location is not enough, and the airframe has seen much corrosion and degradation. Efforts are under way to move the bomber into a climate-controlled structure. Chip Long

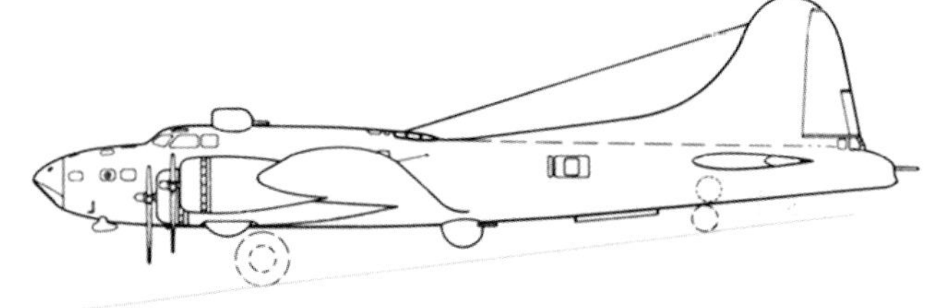

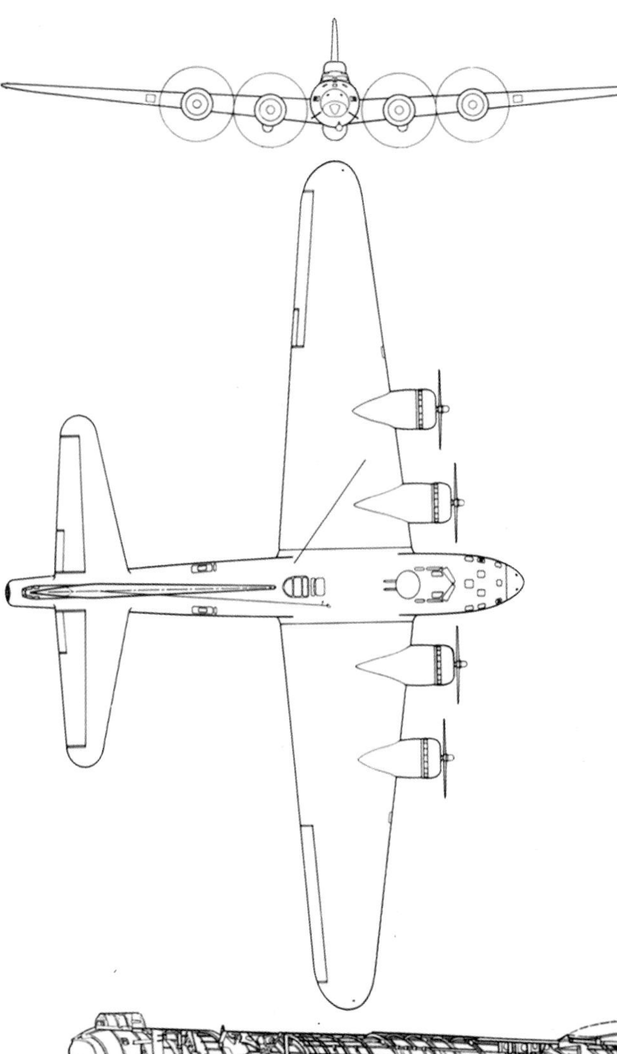

B-17F Statistics

Wing Span	103 ft. 9 3/8 in.
Length	74 ft. 8.9 in.
Height	19 ft. 2.44 in.
Engines	Wright Cyclone R-1820-97 1200 hp @takeoff, 1380 hp @ War Emergency Power 1,000 hp @ 25,000 ft.
Empty Weight	35,728 lbs.
Gross Weight	40,260 lbs.
Max. Takeoff Weight	74,900 lbs.
Cruising Speed	180 mph
Top Speed	325 mph
Ceiling	38,500 ft.
Range	4,420 miles
Crew	10
Bomb Load	8 X 1,000 lb bombs or any combination up to 24 X 100 lb weapons.
Armament	13 X .50 caliber machine guns.
Number Built (F models)	3,405 (All mfgs.)

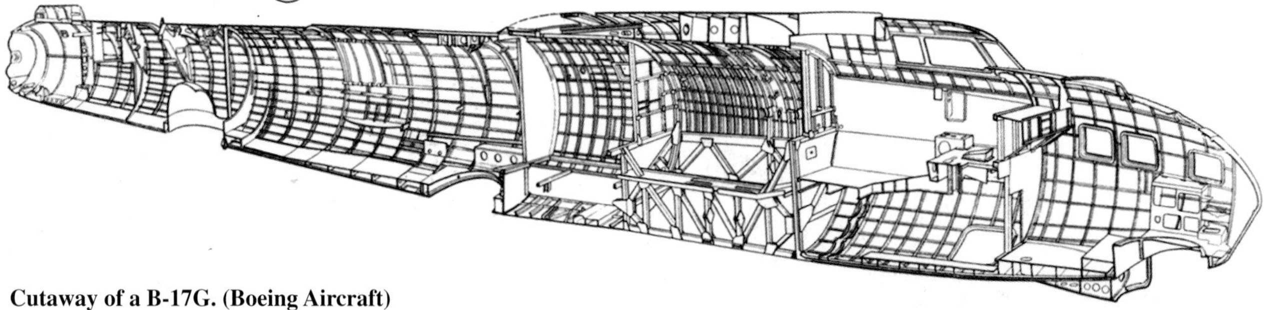

Cutaway of a B-17G. (Boeing Aircraft)

1

A Mighty Fortress

Ask anyone who has seen one, heard one, or flown one "Is the B-17 the best bomber ever built?" They will likely give you more explanations than you care to hear about this magnificent aircraft. To most warbird enthusiasts, there is just nothing like a B-17 Flying Fortress. With the classic lines of the early military, the Fort typifies what a heavy aircraft ought to look like. A large sweeping tail and broad wings, along with the angled windshield and distinctive nosepiece make the B-17 stand out among other designs.

Just sitting still, a B-17 gives the appearance of a warrior, while at the same time showing the graceful curves of a Hollywood movie star. Bristling with heavy machine guns in every direction and sitting tail low as if it is ready to leap into the air where, with its broad and deep wings, it finds its home. Dubbed "The Flying Fortress" by Richard Williams, newspaper reporter for the Seattle Times, the bomber was conceived under the Boeing proposal known as the model 299. It was April 1934, and a flyable prototype had to be ready in less than one year.

With the certainty of war on the horizon, military strategists were looking towards airpower to ensure total victory. At the time, there was no bomber in the U.S. arsenal that could meet the coming demands that aerial warfare would require. The United States Army Air Corps (the USAF was not born until 1947) needed more range, precision weapons delivery, and combat survivability than the aircraft in the inventories could handle at that time. The call went out. Design a bomber that can fly for ten hours at 25,000 feet, can climb to 10,000 feet in five minutes, and can maintain 7,000 feet with the designed useful load and one engine out.

Boeing had done it. After spending only $275,000.00, the 299 rolled out of the factory on 17 July 1935. The 299 could carry up to 4,800 pounds of weapons in the bomb bay just a little more than a year after going on the drafting tables. Little did they know that they had designed an aircraft that would set the standards for high altitude bombing through the end of the century.

Looking much like a B-17 on steroids, the B-15 lumbers slowly through the sky on a test flight. Many individuals believe that the giant Boeing B-15 was a predecessor to the B-17. It was not. It was developed alongside the Flying Fortress. Only a single B-15 was built, and it first flew in 1937. It survived the entire war as a cargo-converted bomber affectionately called "Grandpappy" by her crews. The B-15 had a wingspan of 149 feet, was 87 feet long, and was powered by Pratt & Whitney 850 hp engines. These powerplants barely made enough power to operate this huge aircraft safely. (Frank Donofrio)

Only one month and three days after roll out and accumulating some 14 hours and 5 minutes of test time, the 299 took off from Seattle, WA, bound for Wright-Patterson AFB near Dayton, OH. Nine hours and three minutes later it put the rubber on the runway in Ohio. Average ground speed was 233 mph. This flight stole the show from the up and coming Douglas B-18, which was really just a beefed up DC-2. It also overshadowed the much worked over Martin B-10.

On 17 January 1937, the Air Corps placed an order for thirteen YB-17s. Production for the plane would run a little more than eight years through August 1945, after some 12,731 Fortresses were built. It is believed that around 10,000 of these actually left the United States for assignments all over the world. Despite the Fort's uncanny ability to survive major battle damage, roughly 4,500 B-17s were lost in combat flying mainly from the European Theatre of Operations and the unsinkable aircraft carrier known as the United Kingdom. From this island, the mighty Eighth Air Force launched the greatest air armada ever—where the sky roared for a thousand days and the Boeing B-17 Flying Fortress marked its place in history.

What is the Best Bomber Ever?

Is the B-17 the very best bomber ever built? For years, the argument has continued between the B-17 and the Consolidated B-24 Liberator. Was the B-24 the better plane? This question has con-

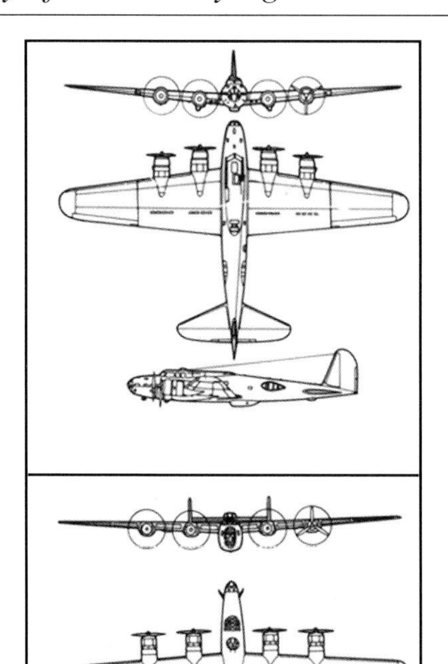

B-17C Flying Fortress

B-24D Liberator

Above, Right: The massive assembly line of the B-17 Flying Fortress. Right: A hybrid of a B-17 and B-24. (Frank Donofrio)

sumed far too many for far too many years. The Liberator carried more, was faster, and could fly higher than the B-17. Although they did fly next to each other in combat, their role was a little different. The Flying Fortress and the Liberator complimented each other and were never meant to be competitors. However, competition was inevitable, as they were both successful bombing platforms. The two bombers were simply the result of an exhaustive search by the War Department through aircraft builders for the new bomber designs that were so badly needed. Of the design submissions, the -17 and the -24 were the best. Their designs were simply the result of what was needed at the time. Consider the numbers. B-24 production ran up to 18,188 of the huge twin-tailed leviathans. However, by the end of the war, so many modifications had been made to the B-24 that it could no longer fly further or faster than the B-17.

Wartime Impact - The ability to penetrate hostile airspace, weather enemy threats, carry the correct bomb load, deliver the load successfully, and return home.

Defensive Characteristics - From single barrel machine guns through electronic counter measures, stealth, and even speed. What bomber defends itself better than any other?

Range - Could the bomber launch, ingress, loiter, deliver, and return?

Available Bomb Load - What could carry the most and how many different types of weapons could it haul?

Accuracy - The result of the bomb delivery system and the aircraft's stability and control predictability.

Survivability - Evading or withstanding enemy ground- fire and aerial attack. What bomber had the lowest loss to sortie ratio?

From the time that aerial bombing began through today, these aircraft stand out among the very best ever designed: the German Gotha bombers; the Heinkel 111; the B-17; B-24; Aichi D3A Val; the B-47 Stratojet; the B-52 Stratofortress; the F/B-111; the B-58 Hustler; B-1 Lancer; B-2 Spirit and more.

Most agree, given the above criteria, that the Boeing B-17 Flying Fortress is the best bomber ever built. Apparently, the U.S. military felt the same way. The baker one-seven was not fully retired from active service until the late 1960s!

The debate over which was the better plane began early in the War and continues even today. General James Doolittle wrote his studied comparisons, in which he partly stated that in an effort to improve the B-24's defensive characteristics, performance was compromised at the expense of the added weight of the armor and armament. One very unusual example of the Army Air Corps attempt to improve on the Liberator was to graft the entire front of a B-17G onto a B-24J. The results were terribly unsuitable, and only three tests were flown by this plane. The hybrid bomber could only climb to 18,000 feet and lacked both longitudinal and directional stability. One note from the study report stated that the installation increased the already excessive basic weight of the B-24J. But the Bombardier and Navigator reported that they favored the increased room in the forward compartment over the traditonal B-24 arrangement.

The Memphis Belle is seen here in an airbrush painting by Mickey Harris and Dru Blair. The painting is called "Lady of Grace - Freedom's Enforcer," and was finished in July 2000. The famous bomber is depicted on her solo return to the United States after finishing her 25 missions. A full color artist's print is available by contacting Dru Blair on the web at www.drublair.com.

2

The First 300 Fs

The "F" model B-17 Flying Fortress is considered by many to be the best looking B-17. With an elongated plastic nosepiece which was not hindered with metal bracing like her predecessors and the absence of the "chin" turret that would grace late "F"'s and all "G" models, she had a little higher speed but was not as well armed as her later offspring. With some 3,405 coming off the lines at Boeing, Vega, and Douglas, the initial 300 were among the very first to cut their teeth in combat.

Only two days after the final "E" model rolled out, the first finished "F"'s saw the light of day. Boeing boasted that over 400 improvements had been made to the airframe. The cost was $314,109.00, and the top speed was 325 mph at 25,000 feet. With the addition of the wider blade Hamilton Standard propellers, the giant could lumber up to 37,500 feet. These first 300 "F" model B-17s were at the edge of technology at the time they were built. They were delivered during a period of less than three months. Construction averaged about 100 aircraft per month between 30 May and 31 August 1942. Because of the astonishing attrition rate of these first F models, only 100 or so B-17s were listed as "operational" when Roosevelt, Stalin, and Churchill met in Casablanca in January 1943 and designed, among other things, the Combined Bomber Offensive. Production rates in American factories were soaring, and more bombers than ever were becoming available for combat. Planning for increased mission strength, new techniques were prioritizing the targeting of the enemy's U-Boat pens, the rail transportation system, oil refineries, and the German air industry. The "leader" bombing technique was devised with the idea being for all planes in the box formation to drop their bombs when the formation leader

B-17 #41-24340 - The very first of the "F"s. The bomber never saw combat and was assigned as a testing platform. She spent ten months testing mainly foward defensive firing positions. The airplane was even later fitted with a "chin" turret and bulged cheek windows.

"Memphis Belle" proudly showing off her forward defensive gun positions. The detail shows the unobstructed frame-free plastic nose piece. The flat teardrop panel was needed to provide a clear picture for the bombardier to use the super-secret Norden bombsight effectively. Technology did not allow for the manufacture of the large clear nose cone without some distortion, and it was not bullet proof. There were times when the nose was shattered from bursting flak, enemy bullets, and even spent shell casings expended from bombers flying ahead of other planes in the tight formations. Note the twenty-five bombs painted on the Belle's nose. This photo was evidently taken at the end of her missions.

Taking a test flight over the Washington mountains, these two factory-fresh forts hold a textbook formation for the photographer. Even though they both rolled out of the factory at almost the same time, both were soon assigned to different Bomb Groups and flew to opposite sides of the world to fight. At the top, B-17F 41-24520 "Fightin Swede" was assigned to the 403rd Bomb Squadron of the 43rd Bomb Group in the Pacific. She was flown by Lt. Folmer Sogaard, but was lost on 8 May 1943 with Captain Robert Keatts and his crew only 9 months after she rolled off the assembly line. The B-17 on the bottom carries the serial number 41-24523 and was named "snooks" by her crew. She was assigned to the 322nd Squadron of the 91st Bomb Group and flown by Lt. Yuravitch. The bomber would eventually be re-named "Lil Audrey" before becoming involved in a tragic and deadly accident involving two other B-17s on 31 August 1943. #523 collided with B-17 #42-29816 "Eager Beaver" on the way back to England, sending both bombers and their crews into the Atlantic. Debris from the collision struck another B-17 #42-29973 "Patty Gremlin Jr.," which limped back to Polegate where a devastating crash landing killed the entire crew. From that single incident only one crewman of thirty survived. #523 flew several combat missions in formation with the "Memphis Belle" (see formation diagram for the mission of 30 December 1942 later in this book).

A class photo at the Tulsa, OK, modification center for these B-17s of the 303rd and 305th Bomb Groups. Given the time and their assignments, most if not all would have flown a mission or more along with the "Memphis Belle" on the many missions the 91st BG was assigned to fly in the same combat wing with the 303rd and 305th. So new are these planes that they are yet to be named by their crews and no nose art can be seen. The photo was made in Aug. '42.
#612 would become *"The Devil Himself."*
#616 became *"Sam's Little Helper."*
#619 was actually an unnamed bomber, but her pilot referred to her as *"S for sugar."* (The fledgling reporter Walter Cronkite flew aboard this plane to file a news story about combat conditions on a Feb. 26, 1943, raid to Wilhelmshaven.) The B-17 went missing in action Jan 11, 1944.
#620 was named *"Snap, Crackle, Pop,"* missing in action Jan. 3, 1943. (See the story of this plane in the briefing of this raid explained later in this book.)
#609 would have *"Holy Mackeral"* painted on her nose.
#606 was "*Werewolf,*" (this plane was intentionally landed in a field beside a British hospital four miles from her airfield after suffering serious battle damage and two wounded on board!)
#605 was one of the elite B-17s. *"Knock Out Dropper"* became the first bomber to reach the fifty as well as the seventy-five mission mark! She accumulated some 675 hours in combat and delivered more than 316,000 pounds of bombs to the enemy. In all that time only one member of her various crews was wounded. The bomber was last seen after the War in Oklahoma City, OK, being prepped to be placed atop a gas station!
#607 became "*Jerry Jinx*" and did not return from a January 23, 1943, raid.
#608 was named "*Yehudi*" and failed to return on January 3, 1943.
#611 was known as "*Boomerang.*" She went missing in action on Feb 16, 1943.
#623 was an unnamed plane. She failed to return from a Feb. 26, 1943, raid on the enemy.

did. This CBO in particular called for "the progressive destruction and dislocation of the German military, industrial, and economic system, and the undermining of the morale of the German people, to a point whereby their capacity for armed resistance is fatally weakened."

At least one report on the original 300 Fs stated that among them, only sixteen of these Flying Fortresses ever returned to America after the War.

To extend the range of the B-17, Boeing designed and installed additional fuel tanks in the outer wing sections. Called "Tokyo Tanks," these units were said to enable the Flying Fortress to reach Japan. They had an intricate venting system which was being tested aboard this plane when a most dramatic event occurred. To create high-altitude pressure changes, Boeing test pilot Elliott Merril was putting the bomber through a series of rapid climbs and descents. During one thirty degree dive and at a speed of 300 mph, the fabric covered elevators were ripped away. The horizontal stabilizers broke free from the plane, which instantly lost all pitch control. Merril and the others aboard endured such high "G" loadings that they all blacked out.

With the plane in a slowing climb the crew aboard "woke up" and immediately prepared to bail out. The bomber approached a stall, slid back on its tail, and pointed nose down. The crew of six pilots and engineers started to jump from the plane, but Merrill was having some trouble getting out of the pilot seat. The B-17 entered another pitch-up, and as the plane slowed again to another wing-level stall, the nose fell forward again and the pilot was lifted by the zero gravity through the top turret into the bomb bay. The doors were down, and Merrill more or less fell out of the dying B-17.

From his perch beneath his parachute he watched the bomber fall to earth in a series of climbs, stalls, and dives until she hit the ground near Purdy, Washington.

The Boeing B-17 Flying Fortress

With accomodations for two and sometimes three men, the nose compartment of the B-17 is actually somewhat roomy compared to other parts of the aircraft. In this image, the Navigator's table is shown in the typical installation along the port side of the bomber. This is not the case with "Memphis Belle." The installation of this table in the Belle is reversed from the norm. Instead, the drift meter is situated on the port side and the Navigator's table is along the starboard side of the plane. It is thought by many that Boeing engineers were experimenting with crew comfort as this particular block of B-17s came off the line.

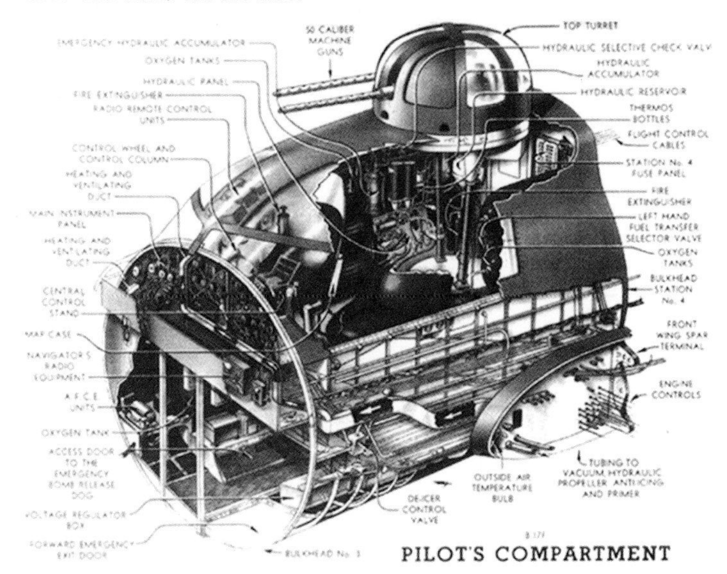

Reaching the flight deck through the nose hatch at the number two engine could be cumbersome, especially when one was wearing bulky flight gear. Once there, space seems precious, even though the instrument panel is nearly three feet in front of the pilots' faces. Immediately behind the two pilots is the top turret assembly. It is occupied by the Flight Engineer. He had various duties ranging from assisting in engine monitoring, defensive gunning, and even in-flight repairs. Getting off the flight deck during an emergency sometimes proved fatally challenging.

The bomb bay area of the B-17 is of typical truss type construction in the main members and supports. This area carries the bulk of the loaded aircraft. It is beautifully situated along the fuselage, providing a great center of gravity. Although small compared to today's bombers, the area could accomodate a maximum of 12,000

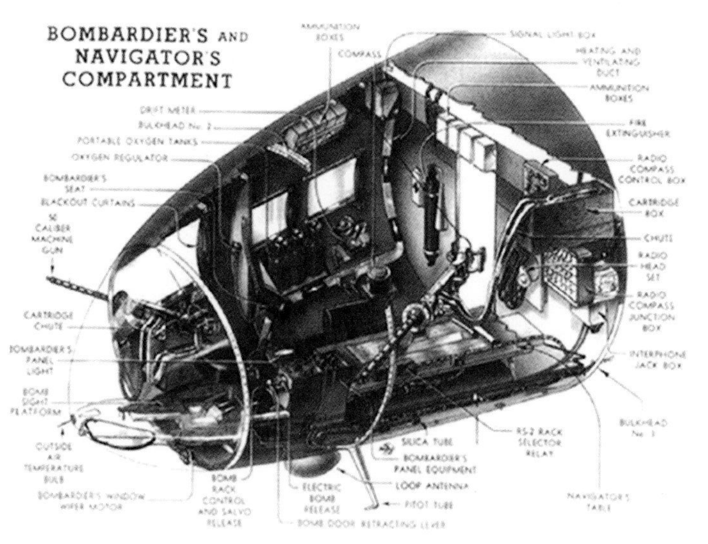

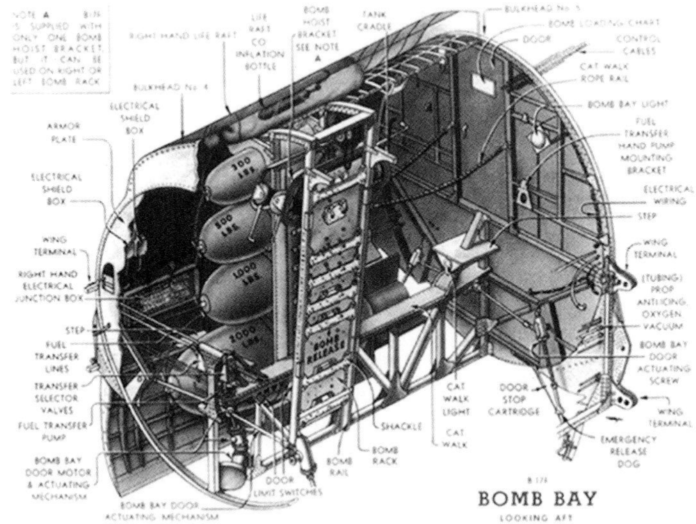

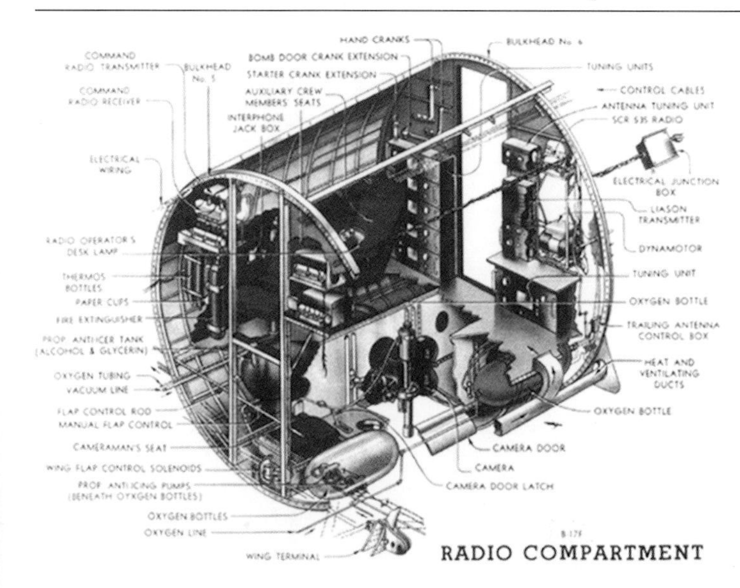

pounds of explosives. For long-range ferry flights, an additional fuel tank was installed here. Among the more popular memories of B-17 flight crews was traversing the narrow catwalk at altitude with the bomb bay doors open. The bomb bay was also a preferred place to exit a crippled B-17 when bailing out.

The radio room was considered to be one of the most comfortable areas of the ship. From here, all communications from plane to plane and from plane to base were handled. A pit for strike cameras was fitted beneath the floor. This was a very important part of each mission. If photographic proof of a target hit was not available then credit for a mission flown could not be given. The ROs were also supplied with a fifty caliber machine gun which was stowed (model F) when not in use in a sheathed area above the bomb bay. Normally, the top window hatch was removed and the weapon could be brought to bear by pulling it down and aft on two greased slides, then locking it into place. Because of the relative warmth of this room, wounded and injured aircrew were often brought here to receive medical treatment. As a result many of the photo department personnel were greeted by grisly scenes when they reached the plane to recover their film from the camera pit.

Probably the most notorious place on a WWII heavy bomber was the Sperry built ball turret. Operated both with electrical and hydraulic power sources, one had to be brave and small statured to fly here. Many believe that it was the most dangerous place to fly on a B-17. In reality, more crew injuries and casualties resulted from those who flew in the extreme nose and tail. Statistically speaking, the ball turret was the safest place to fly on the B-17 Flying Fortress. No one was permitted to be in the ball during take off or landing. Entry to the turret was done by hand-cranking the unit so that the guns were pointed straight down. This placed the hatch inside the plane and allowed the gunner to step down into his armored seat. Once this was done, another member of the crew "buttoned" him up, and the curled up position could be tolerated for hours. Also, the ball turret on a B-17 did not retract into the fuselage as it did on many B-24 Liberators.

The next part of the Fortress is called the waist. Here two gunners each manned a single caliber fifty machine gun. Their coverage provided a necessary cone of fire that radiated in both directions, protecting the flanks of the B-17.

In the "F" model B-17s, the windows were situated directly across from each other. In combat, it soon became evident that this arrangement created much bumping and jarring amongst the gunners. (It was often said that they were dancing cheek to cheek!) So in later models the windows were staggered, providing more ease of operating without interrupting the fellow gunner. Ammuniton was precious given the fact that the Browning fifty caliber fired at a rate of about 750 rounds per minute. With the load of only 800 or so bullets, this meant that for a mission of nine or ten hours, a gunner only had about one minute worth of ammunition!

To get to the tail emplacement on a B-17, you either entered through the small door beneath the right horizontal stabilizer, or during flight by crawling on your hands and knees through the number seven bulkhead and around the tail wheel fender through a small plywood door between the ammunition supply boxes, along their feed chutes, and to your seat. The seat was nothing more than a glorified bicycle seat, upon which more kneeling than sitting was done. An armored plate that you rested your chest on seperated you from your guns. Reaching around it, everything needed was within your grasp. Because of the large tail surface directly overhead, the tailgunner often felt as if he were sitting at the end of a pole in high wind. Although quite cramped, the comparment was considered comfortable by many. Later model Fortresses enlarged the space and added a computing gunsight over the old post and ring style. The computing sight required the gunner to know the wingspan of the attacking fighter and its distance. Once dialed in, the gunner could attempt to shoot the enemy plane down. Most tailgunners opted for the old manual ring and post, which popped up on demand from just outside the aft facing bullet resistant window.

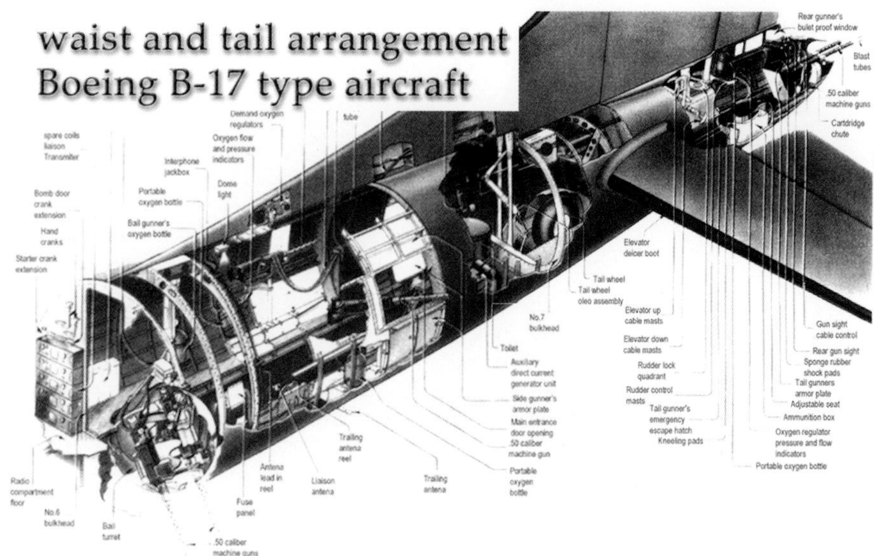

Opposite and this page: Boeing B-17 aircraft images.

3

A Book About the Belle?

I have been asked many times why I would write a book about the "Memphis Belle." To the average person, this is simply another old plane with round engines that flew before many of their fathers were even born. When some time is taken to study the important past of this particular airplane, then it quickly becomes clear that the Belle is not just another old airplane.

A recent poll revealed that the Boeing B-17 Flying Fortress is the most recognized airplane ever built, and the Belle is, of course, a B-17. That in itself makes her special. The Belle distinguished herself at her very first taste of combat when, by fate, she became the *first* bomber of the 91st Bomb Group over an enemy target. She is then, of course, famous for becoming the very *first* bomber of the Mighty Eighth Air Force to complete the required twenty-five missions and return to the United States. "Memphis Belle" was assigned to the *First* Air Division of the *First* Combat Wing. She is the very *first* bomber to have been assigned to a War bond and morale tour.

While finishing her twenty-five missions, "Memphis Belle" was the *first* American bomber to be reviewed by England's King George VI and Her Majesty Queen Elizabeth. After completing his missions requirements on the Belle her skipper, Bob Morgan, rode into more American military heritage, as he was the lead pilot for the *first* B-29 Superfortress raid on the city of Tokyo, Japan, in November 1944.

Of the 12,731 B-17s built during WWII, less than fifty remain today. Of those fifty fortress airframes, around a dozen or so are valiantly kept in airworthy condition. Finally, of those fifty remaining airframes, only three remain to this day that are confirmed to have a record of taking part in actual combat.

The "Memphis Belle" is currently the oldest B-17 on display in the world. She is the single survivor that flew her particular tour of duty. She is also the first and only B-17 to have two major Hollywood films produced about her.

With this in mind, it is easy to see why the Belle should be a subject of intense study. Out of all the bombers built for war, so many variables had to come together again and again to complete her amazing history.

It would seem unlikely that any single aircraft, through simple chance, would become the first at so many things. And realistically, these "firsts" pale in comparison to the honor and strength shown

Her skipper smiles as he taxies the "Memphis Belle" in from her 25th and final mission. Here Captain Robert K. Morgan became a major part of American military aviation fame and legend. A great many variables surround this particular aircraft which today make it one of the true treasures of flying heritage the world over. As he has sometimes said, "The joining of the Flying Fortress name to this plane made the Boeing aircraft special, but the joining of the 'Memphis Belle' name to this very plane made the Flying Fortress special."

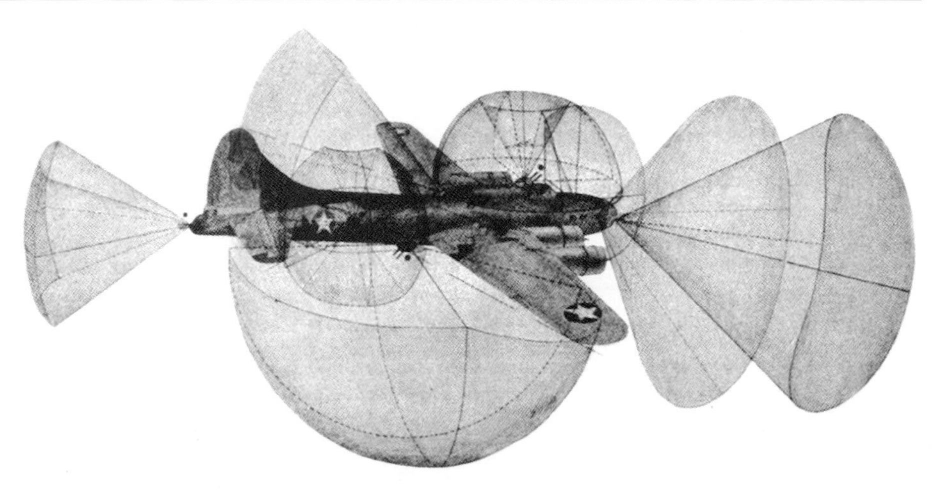

This Luftwaffe poster was titled "Viermotoriges Kampfflugzueg" (four engine fighter airplane). It was used to train German fighter pilots to attack a B-17F efficiently. Showing the radiating cones of fire from the many gun positions demonstrated where the B-17's guns could be trained on an attacking fighter. Once committed to an attack run, there simply were not many areas of the B-17 which could not be protected. Especially when a Squadron or Group offered the German pilots the chance to experience what the B-17s could do when they were mutually protecting each other. A lone B-17 could be considered somewhat vulnerable. However, when the Combat Wing consisted of hundreds of B-17s, the decision to fly into the American formation assemblies would have to be considered either brave or foolhardy. A thousand or more fifty caliber machine guns would be raining their protective curtains of high velocity lead.

by so many more bombers and their crews from that period of American military flight.

But her story really is quite unique and special. The "Memphis Belle" came along at a time when America badly needed her. She displayed to a thirsty Allied homefront that all-important American bravery and ingenuity which was needed to turn America from poverty to the largest production industry the world had ever seen. The Belle certainly was not solely responsible for anything so huge. But she played her part, slugging it out with the best the Germans could throw at her. Then she came home and showed millions of Americans her might as she roared from city to city. Perhaps most significant of all is that she showed the leaders of the Axis nations that Americans would never tolerate the stealing of human rights and freedoms.

The All-American incident: February 1943

B-17F #41-24406 "The All-American" was rammed by an FW-190. This photo shows the incredible damage. No one was killed when the fighter sliced through the bomber waist down through the tail wheel mounting stantion. The only thing holding the entire tail on was the bottom of the stressed skin fuselage! On the flight home, one of the crewmen reported that he had gone to the radio room to check the damage and when he looked back, he could see the tail assembly swaying back and forth in the slip stream. The gap created by the collision was about 2 feet wide. This B-17 landed and was re-built. "All-American" survived the entire war! (Frank Donofrio)

"Memphis Belle" waist gunners. Clarence E. "Bill" Winchell (top) and Casimer A. "Tony" Nastal (bottom) demonstrate why enemy fighters should stay away from their Flying Fortress. (Frank Donofrio)

4

Hall of Fame Fortresses

Among the B-17s that flew combat missions, these are certainly among the more well known. This list was created to highlight the exploits of these particular crews as an example of what was done on a daily basis with their planes, and not to dismiss the gallantry displayed by so many other air and ground crews. As the author of this book, I personally believe every Flying Fortress and crew belongs on this list, as they all made their contributions to the War. It is my hope that every individual that ever had anything to do with the B-17 would take pride in the accomplishments of these heroic crews. These pages are a tribute to the strength of the bomber design, as well as the dedication and determination of all those who put the Boeing B-17 Flying Fortress into the air and into history.

Alexander the Swoose, It flies, B-17D (40-3097)
This is the 129th B-17 built and the only B-17 to remain in active service throughout the entire war! Assigned to the 19th Bomb Group (Hickham field) on 14 May 1941, this bomber would log more than 4,000 hours before the pilot, Col. Frank Kurtz, could return to the United States. By looking at the "waist art" on the bomber, one can see that a "Swoose" is half swan half goose. This B-17 was the personal hack for General George Brett and later a V I P transport. Kurtz, a former olympic diving star, made speed records in this B-17 flying between San Francisco and Hawaii and on to Australia. He even named his daughter Swoosie after this famed aircraft. She is widely known on Broadway and has taken movie and television roles, starring in the film "Against All Odds" and on television in the series "Sisters." The B-17 is currently stored at the National Air & Space Museum in Washington, D.C. (Frank Donofrio)

Nine - O - Nine, B-17G (#42-31909)
This particular bomber flew with the 91st Bomb Group's 323rd Squadron and completed 140 missions! It is hard to say how many airmen crewed this particular B-17, but it is safe to project that it was somewhere in the hundreds during the time that she saw combat. During its service life the ground crew made twenty-one engine changes, eighteen "Tokyo" fuel tank changes, and fifteen main fuel tank changes. This bomber visited the German capital of Berlin eight times. The "909" did not abort until its 132nd mission—an Eighth Air Force record. She was scrapped in the United States on 7 December 1945. (Frank Donofrio)

Hell's Angels, B-17F (41-24577)
Assigned to the 358th Bomb Squadron of the 303rd Bomb Group, this is the first bomber of the Eighth Air Force to make twenty-five missions. The "Memphis Belle" is often mistakenly regarded to be the bomber to achieve this feat. Hell's Angels flew this milestone mission on the 14th of May 1943. The Belle was last flown in combat on the 17th of May—or was it the 19th? More about that later. This B-17 never had a mission ending mechanical failure and went on to complete forty-eight missions. (During the first twenty-five, Hell's Angels received just a single bullet hole and only a few flak holes.) Hell's Angels was dispatched to the U.S. to participate in a war bond and morale tour. The "Hell's Angels" B-17 was scrapped August 7, 1945. (Frank Donofrio)

Suzy Q, B-17F (41-2489)
The fightingest fortress in the world. This B-17 flew more long range missions against the Japanese than any other Fort. Suzy Q was never forced to abort a mission because of weather or enemy opposition. Between January 1, 1942, to January 1, 1943, the bomber flew around the world, completing a 35,000 mile journey which included crossing the equator four times. Twenty-six enemy fighters fell to the guns of the Suzy Q, and no one was ever killed aboard this aircraft. After one particular raid, the Suzy Q missed the field at Horn Island. Low on fuel and running on two engines, the pilot committed to a strip of land and saved the airplane by somehow landing in less than 1,000 feet! The field was peppered with huge holes that were more than five feet deep. Later, the holes were filled, fuel was trucked in, and a takeoff was made at only 80 mph. Suzy Q was scrapped July 15, 1946. (Frank Donofrio)

Knockout Dropper, B-17F (41-24605)
Assigned to the 359th Bomb Squadron of the 303rd Bomb Group, this B-17 was the very first bomber to make the fifty mission mark. This was accomplished on the 16th of November 1943, six months after "Memphis Belle" finished. By March 27, 1944, the bomber reached another milestone—seventy-five missions, and yet another first. Of the B-17s that arrived in England in September 1942, "The Knockout Dropper" remained as one of only two "F" model B-17s in service. This B-17 accumulated more than 675 hours over the enemy and delivered an unbelievable one million, five hundred thousand pounds of bombs on the German war machine. Through its combat days the gunners of the "Knockout Dropper" dispatched twelve German aircraft. Like the Memphis Belle, this B-17 was assigned to a war bond and morale tour and was autographed by everyone in the 303rd. She was scrapped July 19, 1945. (Frank Donofrio)

Ol' Gappy, B-17G (42-40003) of the 524th Bomb Squadron, 379th Bomb Group. This bomber made the list for completing an unprecented 157 missions. This was more than all other bombers in the Eighth Air Force. Ol' Gappy only aborted a single mission due to mechanical failure. "Topper" was going to be the first name of the B-17 after the popular Roland Young movies. The crew chief began to apply the nose art of a man in a top hat. The ship was damaged before he could finish the job, and only the top hat, gloves, and collar remained through the bomber's combat stay. After the original air crew rotated, one of the ground crew convinced the crew chief to apply the name "Ol' Gappy" to each side of the nose turret, because the new "G" model turret and its twin guns reminded them of a tech supply worker who had lost most of his teeth but retained two prominent "tombstone" teeth up front. The fast, well oiled, and patched up pride of the 524th was scrapped Nov 10, 1945. (Donofrio)

Cabin in the Sky, B-17F (42-230337) of the 571st Bomb Squadron, 390th Bomb Group. This bomber was under the command of pilot Captain Robert D. Brown. On October 10, 1943, "Cabin in the Sky" took part in a mission to Munster, Germany, and was the only Flying Fortress in the 571st Squadron not shot down. While defending their B-17, the gunners shot down an unbelievable eleven enemy fighters that were attacking them. This was an undisputed record for a single mission tally by any bomber in the Eighth Air Force. (Frank Donofrio)

"Snake Hips" B-17F (42-31713)

No other Flying Fortress took this much punishment.

Assigned to the 327th Squadron 92nd Bomb Group, this B-17 must be a contender for the most damaged B-17 ever. On 24 August 1944, with almost sixty missions behind, Snake Hips left Podington to hit the Germans at Merseburg. The formation approached the bomb run when they were bracketed by flak barrages. The bomb doors had just opened when a burst entered the bay and detonated. None of the ten 500lb bombs exploded, but three of the weapons had been blown through the side of the B-17. Two rolled down the right wing before falling off. Five bombs were knocked from their shackles and came to rest on top of two unreleased bombs.

The radio room and right wing trailing spar were destroyed and the ball turret gunner killed by thumbnail sized shrapnel that entered his turret. Below the formation, "Snake hips" was nearly destroyed by bombs from the formation above. The pilot and co-pilot watched in disbelief as a bomb just missed their nose, a second fell between the left trailing wing and the horizontal stabilizer, and a third fell just behind that. The supercharger controls were damaged, and the B-17 descended through 14,000 feet, where the engines found the air sufficient and roared to full power. Assymetrical power settings achieved a see-saw course for home; because of severed flight control cables, turns were hard if not impossible.

Many flight instruments were useless, the landing gear had extended, and when the engineer entered the bomb bay to manually retract it he was pushed aside by a circular wind that was created by the huge four foot hole in the side of the plane. This rushing wind was responsible for venting the bomb bay and kept it from exploding. The bomb compartment was literally soaked with fuel leaking from the main tanks. The gear had jammed in the down position, and the engineer actually bent the crank trying to bring it up.

The crew would have to take their chances at 12,000 feet over Germany. If a fighter showed up, hopefully the enemy pilot would take the sign of a gear down B-17 as a surrendering crew and leave it alone. The rudder pedals moved back and forth with tremendous violence, prompting the crew to keep their feet off for fear of a broken leg. The trim tabs could not be used. When the pilot turned the tab wheel, the cable just kept rolling out onto the flight deck.

When it came time to address the release of the bombs it was found the explosion severed all of the arming wires in the bomb bay. All of the bomb vanes were turning, and one weapon had reached fully-armed status. The catwalk in the bomb bay was severed. Unable to reach the back of the bomb bay, the crewman charged with dis-arming the bombs had to call the radio operator to address the tail fuses. After the bombs had been made "safe," the crewman re-entered the bomb bay minus a parachute and actually lifted one end of the bomb on top of the pile and gradually eased it backward until it slid out of the bomb bay. One by one, the five bombs were tossed out of the stricken B-17. Two remained in their shackles but would not release. A 20 inch screwdriver persuaded these two to fall, and the bomber, finding a new center of gravity, became increasingly difficult to handle.

As they passed the Dutch coast, the pilot asked the crew if they wanted to bail out rather than risk a North Sea ditching. They elected to stay and found a new problem. Number two engine failed owing to fuel starvation and the propeller would not feather. As they made the English coast a second engine reached fuel exhaustion, but an airfield was spotted. It was Woodbridge, a field built just for shot up airplanes. With an 8,000 foot runway it was perfect. The pilot instructed the crew to jump in their parachutes.

Because "Snake Hips" wanted to fly nose high the pilot made some check points and airport references for a skidding wide turn approach to the runway. The pilot and co-pilot smiled as they saw the runway come up from below their windscreen—they were lined up right on the center line! They let the bomber just sink onto the runway not touching a thing, since a very nice approach had been established. About 75 feet over the runway, air pressure changes brought the nose of the big bomber down through forty degrees of pitch before the men could completely recover. They felt a couple of bounces and a shudder and realized they were on the ground.

While they taxied in they found that there were no brakes and then were directed to a parking area by an RAF ramp worker. The poor soul had no idea that the B-17 could not stop, and as the crew pulled power and turned the mags to off, he fell to the ground as a spinning propeller churned just inches over his head.

Snake Hips was declared unrepairable the very next day. The story of this B-17 made bomber crews aware that the B-17 Flying Fortress could take far more than any designer could imagine on the drawing table. "Snake Hips" became a ruler to measure what a B-17 should be able to endure. She never flew again and was salvaged on 25 August 1944.

5

The "Memphis Belle"

Their wings of silver touched the passing clouds,
made soft white trails across the azure blue;
But not for them; this life we share on Earth;
they sacrificed that gift, for me and you.

A Memorial Inscription
398th BG(H) AAF Station 131
Nuthampstead, England

No other airplane from the Second World War has the notoriety or the visibility of the "Memphis Belle." While it is true that this bomber did not fly the longest, fastest, or most missions, today this airplane is recognized by millions around the world as the definitive bomber from the period. Without her timeless story, the Flying Fortress could easily have been just another of the hundreds of different types of aircraft from WWII. By itself, the B-17 is a graceful lady that cruises through the skies like a dancer who knows every step. At the same time, this beautiful lady can deliver precise and destructive blows that more than sixty years have been unable to erase. The "Memphis Belle" is one of the reasons the B-17 is so very well known.

When the name "Flying Fortress" was applied to this Boeing giant, military strategic bombing took on an unexpected heradlry. The joining of this name to this plane made the B-17 special, and the joining of this plane to this name made her story special.

The two complimented each other in a way no one could have predicted. "Memphis Belle" and the B-17 came together to create a story that America desperately needed. And her missions continue. While they are different now, they are carried out with dignity, as she now serves to provide the basis for reflections and memories of a time so long ago when, as it is sometimes simply stated, some very bad people were doing some very bad things, some young men stood up to stop them, and said that they would never happen again.

In the following pages you will read about the legendary "Memphis Belle" and why she became so well known. Does the "Memphis Belle" mean more to us than the other bombers from WWII? No. "Memphis Belle" represents them all. From the day this B-17 was built through today, she has proven herself a timeless example of the United States of America's stance on world freedom. She is a symbol of unity and strength, and exemplifies the term "Freedoms Enforcer."

(Chip Long)

The Time and the Place

America had been up to its neck for almost eight months in a world war. The second world war in a single generation to be thrust on the nations of the globe by the same country—Germany.

Just eight months before one special day, the Japanese had inflicted terrible devastation and death on the United States Navy at Pearl Harbor. And at this time the German Army had been perfecting "Blitzkrieg"—*Lightning War*—in the flat and fertile fields of Poland.

The German people were inflamed by perhaps the most successful madman of the 20th Century. A decade earlier Adolf Hitler, who had played on middle-class resentments, a fear of communism, and industrial ambitions, fanned the monarchists, political officers, and even the church. Hitler and his hodgepodge National Socialists (NAZI) party seized power and gained control of the Reichstag.

Years later on July 2, 1942, two days before America would celebrate her Independence Day, a significant event was taking place at the Boeing Aircraft Company in Seattle, Washington...but nobody knew it then.

Bread was a nickel per loaf, gasoline was about fifteen cents per gallon, a trip to the movies would cost you and your girl less than a buck, and a brand-new Boeing B-17F Flying Fortress ran about three hundred-thousand dollars.

On this day, one of a few brand new B-17s would see the light of day for the very first time. Built under the contract order number B-17F-10BO-3170, it had no graceful name painted on the nose, not even an army tail number. Olive drab paint was applied from nose to tail, and this bomber even smelled new.

Who could tell that only ten months later this very airplane would become the most recognized bomber to fly during the largest conflict man has ever known.

Army inspectors crawled through the plane looking for manufacturing flaws. Both company and military pilots climbed inside to run the engines and even take her around the field, just to see if she behaved like a B-17 should. Thirteen days later the Government took possession of her. She would now be known forever officially as Army 41-24485, and the Government was billed $314,109.00.

Army '485 was sent immediately to Wright field near Dayton, Ohio, to be outfitted for war. By the end of August, bristling with machine guns and with her bombing equipment checked and military radios installed, the B-17 was flown to a staging area for brand new combat ready Flying Fortresses at Dow field near Bangor, Maine.

Here is where the plane would meet the pilot. Dow field was the place where the newly formed Ninety-First Bomb Group (Heavy) was marshalling before departing the "Zone of the Interior"—the United States.

Who can say after all these years what his first impression was. After all, he had trained on the bigger and faster B-24 Liberator. He had seen crews lost in training and he was headed off to war. He did not even know if he would return home alive. But there must have been a moment where this tall lanky twenty-four year old Army Air Force Lieutenant stopped in awe when his eyes first met this gleaming B-17. He could not have known that he was set for destiny and fame when he first climbed aboard and took the Aircraft Commander's left seat. The date was September 3, 1942, and the gleaming bomber had only sixteen hours logged before Morgan took her up on their first flight together.

Lt. Morgan at the controls of his B-17. The compass card holder has the photo of his real "Memphis Belle." (Frank Donofrio)

Robert Knight Morgan, from Asheville, North Carolina, was the man assigned to this B-17, and he would grow to know it like no other. He would fly this bomber into the very center of the action taking place in the freezing skies of Nazi held Europe. He would command nine men who would be right there with him. And he, like hundreds of thousands of other young men, had to go. There was no not doing it.

Bob Morgan's crew was going to war in a bomber that they all agreed to christen with a name that rings of southern gentility, charm, and grace. A name that represented the girl he loved. No other name would do. By the time Morgan pointed the nose of this B-17 over the Atlantic Ocean towards war torn Europe, Army yellow lettering was applied to the nose of Army '485. She was now called *"Memphis Belle."*

Margaret Polk—The "Memphis Belle." (Frank Donofrio)

And why the name "Memphis Belle"? Bob wanted to name the plane in honor of his sweetheart, Miss Margaret Polk. She was just nineteen years old and had visited her sister who lived near Walla Walla, Washington. Margaret was in fact from Memphis, Tennessee.

There was a training base near Walla Walla where the 91st Bomb Group was in final training. Bob Morgan met Margaret Polk there. A relationship flamed immediately, and the two fell in love. He had a pet name for her. It almost became the name of the airplane—*"Little One."*

As the crew finished their training and went on to Bangor, Maine, before going overseas, Morgan and his Co-pilot, Lt. James A. Verinis, decided to stop one evening at the base theatre to see a film. Destiny was somehow again smiling on this story.

The movie was "Lady for a Night," a new Hollywood release. A romantic drama that starred Ray Middleton, Joan Blondell and John Wayne. The film was set in Memphis, Tennessee, in the late 1800s aboard a gambling riverboat where the starlet marries the wrong man in order to gain a coveted social position. At least twice in the eighty-eight minute feature the phrase *Memphis Belle* is spoken, referring to the boat as well as the "lady." The viewer can even see the name painted on the side of the riverboat in the beginning of the film.

In the movie the name was not meant as a compliment, but it was suddenly the only name that seemed fitting for Bob Morgan's B-17. "Little One" would have been a good name, but *"Memphis Belle"* was perfect! During a year 2000 interview with George Birdsong (who flew the B-17 "Delta Rebel no.2" with the 91st BG, and who was also in the theatre at the same time with Morgan and Verinis) Birdsong recalled Morgan's reaction the moment the words "Memphis Belle" were spoken in the film. Evidently Morgan immediately stood up and announced to everyone in the theatre "That's what I'm gonna name my plane!" An unknown jesting voice was heard in reply from the back of the theatre that said, "That's great Bob, but you'll never make Captain you S.O.B.!"

Just eighty-five days after the Belle rolled off the assembly line the Ninety-First Bomb Group left the United States for England. The date was the September 25, 1942.

The "Memphis Belle" was not alone. Thirty-six B-17s went along with her. They carried names like "Heavyweight Annihilators," "Mizpah," "The Jersey Bounce," and "The Great Speckled Bird." Thirty-six other crews. Three hundred-sixty men. These bombers made up the four squadrons of the group that was quickly becoming known as "Wray's Ragged Irregulars."

Col. Stanley Wray, the Commander of the Group, was well liked, admired, and respected by the men he led. Now he was in the air with them and leading the way to some of the most unforgettable combat the world has ever seen.

The course that lay ahead for these men and their war machines was terribly uncertain. They could not know it, but they would soon face combat horrors unlike anything they could have dreamed of. In the "Memphis Belle," the men were probably thinking of their girlfriends, swing music, and the adventures coming at them.

By May of the following year they had logged more than twenty thousand miles in combat, shot down eight German fighter planes, damaged twelve, and claimed three more as "probables." These men and the "Memphis Belle" would deliver sixty tons of bombs to enemy targets and lose nine engines, both wings, landing gear, and two tails on their brand new B-17. They were going to bring back a bomber filled with bullet and flak holes.

Margaret & Bob in Memphis during the war bond tour, June 1943. (Frank Donofrio)

By the next May, now only eight months away, only four of the bombers in this formation of thirty-six would be left.

Although there had been a great deal of emphasis on the seriousness of their training, there were always attempts by air crews to have some fun from time to time. This is how the Belle got her first "battle damage." It was during one of those early familiarization flights after the crew had picked up the Belle and were getting used to their new charge. At least two members of Morgan's crew later recalled the event, even if it is just a little different between them.

Apparently at some point during a long cross-country flight, the crew found themselves over a river and a small boat. When you are bored and just making some time, what do you have to do when a juicy opportunity like this comes up? You have to buzz the boat! The first pass apparently did not do enough to satisfy Morgan, because he turned the big bomber and headed back for another run. This time he would let the guys on the boat feel the heat from the superchargers!

The boaters evidently did not like it, because when the "Memphis Belle" returned to Dow Field, the crew found a couple of bullet holes in the skin. No one was hurt, the Belle was quietly patched, and the incident was never reported.

But now we are back in the planes of the 91st Bomb Group heading out over the Atlantic Ocean destined for a re-fueling and rest stop at Gander, Newfoundland. Here, some of the crews met

(Frank Donofrio)

the son of the President of the United States—Elliott Roosevelt. The air crews swarmed over their bombers and made them combat ready. After all, they could now theoretically be jumped by enemy fighters between here and their destination—Kimbolton, England.

They were not in training anymore. This was it. The pilots met in the control tower to quietly discuss the approach to Scotland and what they might expect. Their guns would carry ammunition, and they would be manned and charged. There would be a stop there before they would finally appear over English soil in flights of three bombers. A large formation of bombers could easily be mistaken for a German attack, and no one wanted to be shot down.

The first landing in England was at the base that was supposed to be the new home of the "Ragged Irregulars"—Kimbolton. The accomodations were not the best, but it would have to do. There was quite a bit of work to do, and the crews set about their days getting ready for the first operational mission of the Group.

Training was in order. The air crews would have to know the landmarks around the base. They began famaliarization flights right away. Fortunately for them the runways were found to be too thin, and in less than two practice missions they were cracking and in need of repair. The massive 30 ton weight of a B-17 was too much for them, and a new base would be needed.

Colonel Wray began the search for his Group. The kind of man who would "do it now and apologize for it later," Wray happened upon an RAF base near the town of Royston, just south of Cambridge, England, and ordered his trucks there immediately.

He had not asked for permission. The British had recently vacated this base and it was perfect, so the 91st just moved in and set up shop. Wray's Ragged Irregulars would now also be known as "The Bassingbourn Boys." By 14 October 1942 the complement of Ninety-first Bomb Group B-17s were re-located to Bassingbourn, where Wray eased concerns of the higher commands by stating that their stay there was only temporary. This temporary occupation lasted more than three years! Some three hundred-sixty missions from this base were flown by the 91st, which was also known officially as USAAF Station 121.

As a matter of fact, the British did not formally turn the base over to the Americans until April 1943—just three weeks before the "Memphis Belle" completed her required twenty-five missions.

This was a wonderful base, with accomodations far above what other American bases had. Brick buildings with fireplaces rather than cold and damp Nissen huts. Concrete sidewalks rather than muddy paths. And it was close to Royston, which meant a train station —*To London!*

ABOVE: 23 May 1998. For the first time ever, the pilot, the plane, and the Petty Girl are together. Morgan had never seen the original piece, which is housed with the entire Petty collection at the Spencer Museum of Art at the University of Kansas in Lawrence. Thank you Col. Terry Oldham! (Don Reber)

I'm the One with the Part in the Back

For as long as men have gone to the sea in ships, they have expressed a love for their vessels in the female metaphor. And in the mind of a combat airman, only a debutante of the species and her "class" can bring to mind a fitting ideal for comparison.

She was first seen in the pages of Esquire magazine. The publication was popular for printing the works of Alberto Varga and George Petty as gatefolds or centerfolds. Along with thousands of other servicemen, Lt. Morgan was an avid reader. At some time during 1941 Morgan noticed one of the girls in the magazine and knew immediately that it was the work that would grace the nose of whatever bomber would be assigned to him.

He wrote to the magazine and asked for permission to use the painting. They were all too happy to help and sent their approval, along with a copy of the print, which was immediately applied to Morgan's bomber. This was in the days just before his crew left for England from Dow Field at Bangor, Maine.

Morgan found a base worker there to apply the Petty Girl to his B-17, but it was not a very good job, and when he arrived in England, the Petty Girl was touched up and re-painted by none other than Corporal Tony Starcer. He was becoming known around the 91st BG as a pretty good painter, and actually ended up painting more than one hundred-fifty of the 91st BG B-17s!

The painting is often mistakenly referred to as "The Memphis Belle," but it is actually titled *"I'm the one with the part in the back."* The model for this piece was the sixteen year old daughter of the artist—Marjorie Petty-Macleod.

The story made its way around the bases in England during the war that German fighter pilots knew the B-17 with the leggy girl on the nose and vowed to shoot it down.

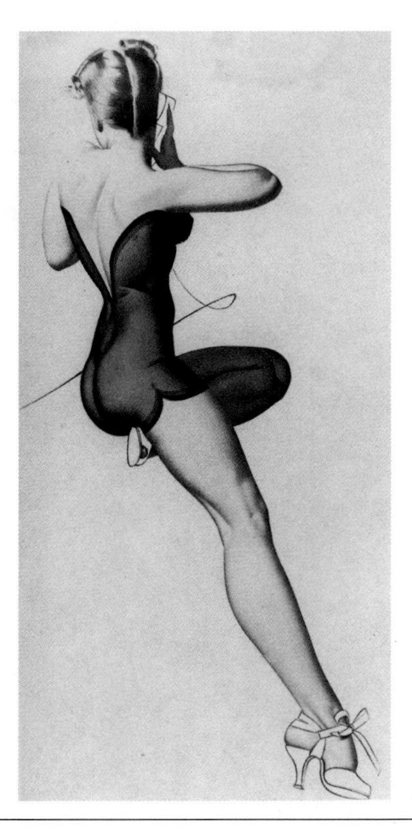

Col. Terry Oldham University of Kansas-Spencer Museum of Art. (The Petty Family.)

It was even said that during a German propaganda broadcast over the radio, a threat had been made. The announcer reportedly said "If the Memphis Belle comes back to Germany, she will never go home again."

At least eight German fighter pilots attempted to hold that reporter to his word and paid for it with their very lives.

The painting represents Miss Margaret Polk, the wartime sweetheart of pilot Robert Morgan. It is not, as many believe, a direct painting of Margaret herself.

So now we know some of the early history of the "Memphis Belle." We know that the plane was rolled out of the factory in July '42 and sent to England just eighty-five days later. We know why Morgan named his bomber the "Memphis Belle," and where the leggy pin up girl came from.

The missions that were flown by this bomber and her crew will all be profiled in detail throughout this record. There are many more intriguing things that occurred to the "Memphis Belle." You will know them well by the time you get to the end of this book.

The purpose of this chapter is to make you aware of the magnificence of the B-17. To share with you some of the strengths that made it a legendary aircraft, and why it is recognized today as the definitive bomber from the period. But it is also to introduce you to a particular B-17 that was very special during WWII—the "Memphis Belle."

Author Edward Jablonski summed up the baker one-seven perhaps better than anyone in his 1965 book *Flying Fortress* when he wrote the following:

> The job of the men flying in the Boeing B-17 was to strike at military targets. They did this with courage and skill and their stories are little known.
>
> They hated their job. They knew fear, dread, and death. Whatever their varied backgrounds they shared a hatred of war and a love for their plane. Many continue to recall the Flying Fortress as "the best plane of the War," capable of absorbing unbelievable punishment and still bring them home. With wings punctured and ablaze, tail surfaces shredded, with chunks of its graceful body gouged out by cannon fire, flak, or mid-air collision, the B-17 brought them home. With an almost human will to live, this great plane, shattered and torn beyond the limits of flyability, carried them to safety and, for some, to life itself.
>
> Watching one of these giant aircraft, like some living thing, clawing in the air in a vain attempt to remain aloft or at least in momentary level flight, was an awesome sight. The life and death struggle of so large a thing had its further poignance: there were ten men inside, some dead perhaps, some wounded, some not even so much as scratched, but in a moment all their lives had reached a crisis in that single plane, heaving and smoking in a freezing, hostile sky.

Without a doubt the B-17 is known to have been one of the most stable platforms in the air. These photos demonstrate this fact. This 463rd Bomb Group plane received a direct hit beneath the pilot's feet from what is believed to have been an 88mm shell, which entered the airplane and then detonated. The results are terribly obvious. The mortally wounded B-17 begins a level descent with all four engines still running. Everyone forward of the top turret was killed instantly.

The inherent stability of the airframe enabled the headless bomber to remain in loose formation with the Group for several minutes. Because of the smooth flight attitude those who survived the explosion from the radio room back were able to bail out safely.

Seeing friends lose their lives day after day created staunch and fatalistic outlooks among the crews. To make twenty-five missions was much more than a milestone for them. Everyone looked forward to flying that final raid, but many feared that they would never live to make it. The odds were simply stacked against the airmen of the Mighty Eighth Air Force at this time. They all knew that after their fifteenth mission, they were living on borrowed time. Superstitions ran wild; rabbits feet, horseshoes, ribbons from girlfriends, and other things all found their way onto many bombers. Many even refused to fly a thirteenth mission—they simply referred to that run as their mission number 12A.

These pictures were taken only moments after this fort was mortally wounded. As seen from the right waist of an accompanying B-17, she remained in formation for several minutes before falling to earth. (Frank Donofrio)

6

Why the B-17?

In the B-17 historical classrooms there are three dominant professors: Ed Jablonski, whose comments are referenced on the preceding page; Peter Bowers; and Roger Freeman. Bowers made his contribution in his spectacular 1976 book "Fortress in the Sky," now out of print but still very much sought after by Fort enthusiasts. The writing and references stand alone in the annals of B-17 lore. He wrote:

> The B-17 was more than metal, more than a machine, even more than a myth. An admirable aerial charger with a hundred million admirers, she will always remain a champion in our mind's eye, for not only did she triumph over our enemies, she vanquished the greatest ravager of all, time. The men who crewed her have long since joined an endless list of veterans from succeeding wars, but the B-17 lives on, her image burnished brighter than ever, enhanced by the selectivity of our memory and the passing years. High in what was then a hostile, alien sphere, she flies on, proud symbol of airpower incarnate, a haven for those within her, a Fortress In The Sky.

Roger Freeman takes a more objective look. His perspectives about the B-17 are harder to pin down because he has written so much about the aircraft that he grew to love. As a child growing up in war torn England, he watched huge formations overhead in Combat Wings that sometimes stretched more than fifty miles. He wrote:

The big tail B-17 was well-proportioned and looked good. And as so happens, a good looking airplane is a good flying airplane. From a pilots viewpoint, the B-17 was stable, responsive and dependable. The US news media paraded the Flying Fortress because it was known and accepted by the public.

Even though the B-24 series was built in far greater numbers (18,188 to 12,731 B-17s) the Liberator proved more versatile, and undoubtedly did more for the Allied cause than the Fortress, it was never to become as famous in the public's eyes. The public were presented with a picture of Fortresses fighting their way to the target and back again; a rugged battlewagon that could sustain extraordinary damage yet still bring its crews safely home. This undoubtedly caught the imagination of the masses, even if the true picture which had to be surpressed for secrecy, was far less glowing. In fact, by the Autumn of 1943, the bomber losses were so acute that it was obvious unescorted missions were sustaining prohibitive losses. Thus, the B-17 Fortress came to epitomise the fight-

During the 1960s Columbia Pictures released "The War Lover" with Robert Wagner and Steve McQueen. This low-level pass was executed by a pilot from Film Aviation Services in California after the B-17 was flown overseas to England for the filming. The plane seen here is actually B-17G #44-83563 and never saw actual combat. Today, the bomber flies the air show circuit around the U.S. with the National Warplane Museum in Elmira, New York, painted as "Fuddy Duddy." Note the markings of the 91st BG(H) on the vertical Stabilizer. (Frank Donofrio)

ing spirit of the Eighth Air Force and, aided by the media - in time of the entire U.S. Army Air Forces in World War Two. A constant and understandable irritant to those men who served in other warplane types and other theatres of war - but that's the way of legends.

Few man made machines have withstood the test of time like the Flying Fortress. In their circles, men can talk of powerful cars, ships, or any piece of equipment with engines and wheels. And even after almost seventy years, when the words "B-17" are spoken, a nearly magical silence prevails when yet another fantastic story about this legend comes to light.

Without a doubt, the Flying Fortress, through time, has solidly captured the love of aviation enthusiasts. She will continue in her Queen of the bombers role for the foreseeable future as even modern conventional bomber designs take a more pragmatic approach to their missions. The B-17 joined capability to beauty and strength. Many believe that this cannot be accomplished again despite even the best efforts of modern technology.

Three Air Divisions

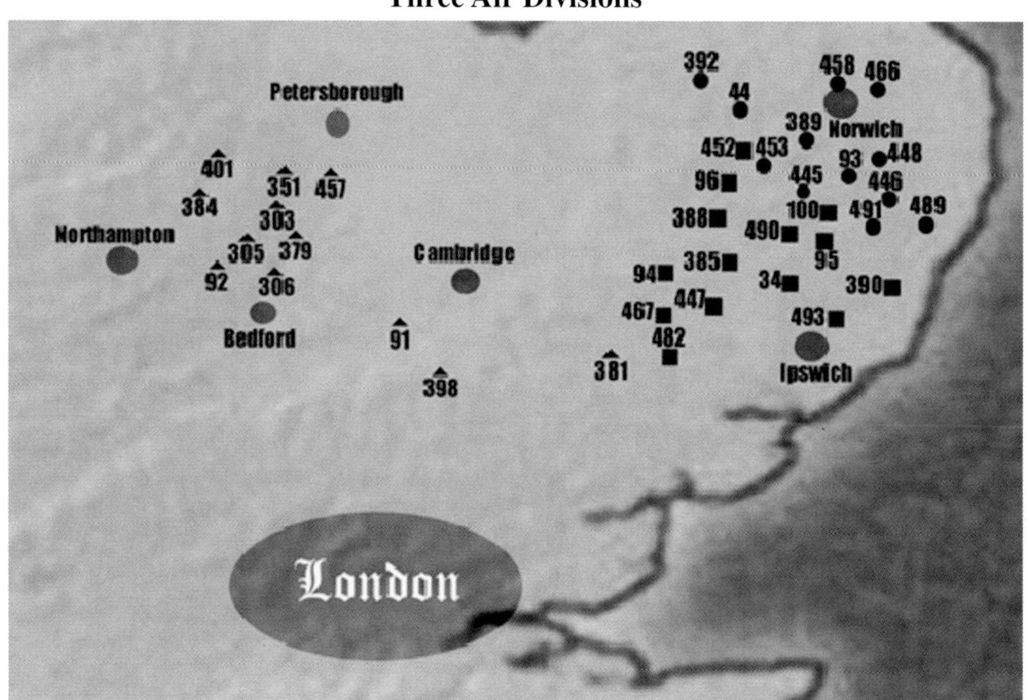

1ST

Group	Aircraft	Base	Tail marking
91st BG (H)	B-17s	Bassingbourn	A
92nd BG (H)	B-17s	Podington	B
303rd BG (H)	B-17s	Molesworth	C
305th BG (H)	B-17s	Chelveston	G
306th BG (H)	B-17s	Thurleigh	H
351st BG (H)	B-17s	Polebrook	J
379th BG (H)	B-17s	Kimbolton	K
381st BG (H)	B-17s	Ridgewell	L
384th BG (H)	B-17s	Grafton Underwood	P
398th BG (H)	B-17s	Nuthampstead	W
401st BG (H)	B-17s	Deenethorpe	S
457th BG (H)	B-17s	Glatton	U

2ND

Group	Aircraft	Base	Tail marking
44th BG (H)	B-24s	Shipdam	A
93rd BG (H)	B-24s	Hardwick	B
389th BG (H)	B-24s	Hethel	C
392nd BG (H)	B-24s	Wendling	D
445th BG (H)	B-24s	Tibenham	F
448th BG (H)	B-24s	Seething	I
453rd BG (H)	B-24s	Old Buckenham	J
458th BG (H)	B-24s	Horsham St. Faith	K
466th BG (H)	B-24s	Attelbridge	L
489st BG (H)	B-24s	Halesworth	W
491st BG (H)	B-24s	Metfield	Z

The symbols to the right of each Group listed reflect that Group's tail markings.

3RD

Group	Aircraft	Base	Tail marking
94th BG (H)	B-17s	Bury St. Edmonds	A
95th BG (H)	B-17s	Horham	B
96th BG (H)	B-17s	Snetterton Heath	C
100th BG (H)	B-17s	Thorpe Abbotts	D
385th BG (H)	B-17s	Great Ashford	X
388th BG (H)	B-17s	Knettishall	H
390th BG (H)	B-17s	Framlingham	J
447th BG (H)	B-17s	Rattlesden	K
452nd BG (H)	B-17s	Deopham Green	L
493rd BG (H)	B-24s/B-17s	Debach	M

Some historians claim that the number of 8th Air Force personnel in England during WWII was equal to the population of the State of Maryland!

7

Wray's Ragged Irregulars
aka
The Bassingbourn Boys

Bassingbourn, England

The airfield can be traced back to 1937 when pre-war fears of German Luftwaffe threats created the need for Royal Air Force expansion.

It is located in the parishes of Wendy and Bassingbourn just 4 miles north of the town of Royston in Cambridgeshire. The name comes from a Neolithic leader of some 1,200 years ago. *Bassa* would lead his people to settle near the bourn, or stream, of the area.

One of the first planes to ever land on the grass strip was an Avro Anson, which needed to use the field for a forced landing before the base was even operating. The British opened Bassingbourn on 27 March 1938. It is still in use today by the British Army. Although most of the runways are gone, much of the concrete is still there.

Bassingbourn was one of more than forty bomber bases used by the Americans in the Mighty Eighth Air Force. Tens of thousands of U.S. bombers flew missions from these stations scattered throughout much of East Anglia.

Bassingbourn was one of the first economized pre-war airfields, which made it somewhat different from what most of the U.S. Eighth Air Force Bomb Groups would become accustomed to. Arranged in an arc, four "C" type hangars defined the maintenance areas. These measured 300 by 152 feet each and three remain today. Many of the buildings on the base were constructed of reinforced con-

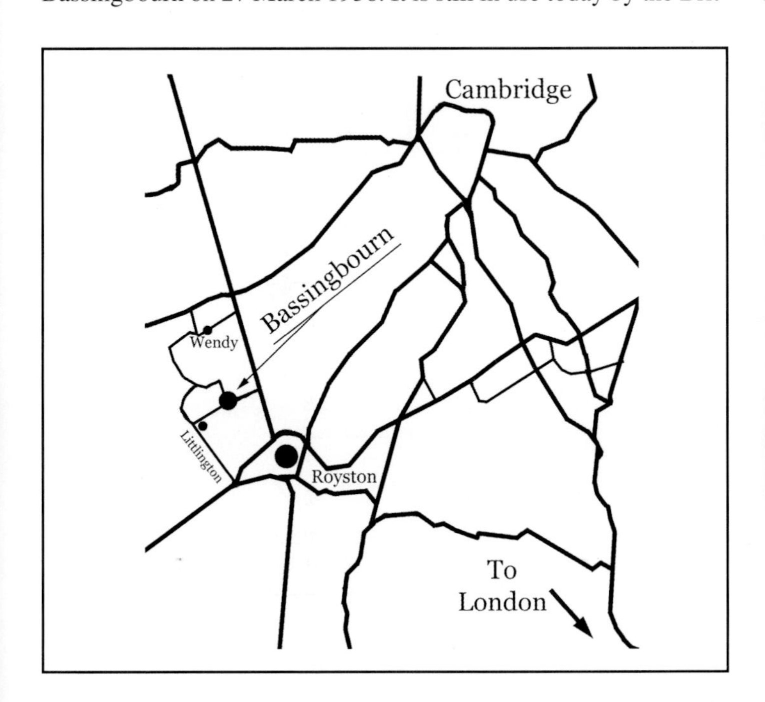

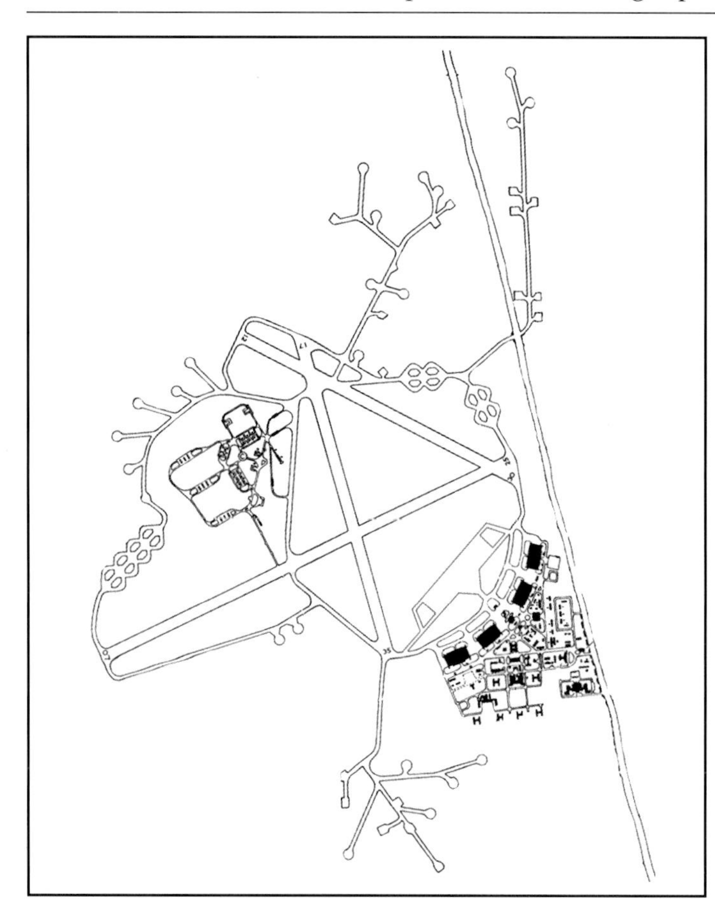

USAAF Station 121 at Bassingbourn circa 1944.

crete instead of lumber and brick. This is one of the reasons that the base was considered premium when the 91st BG went looking for a place to live.

It was said that Col. Stanley Wray was questioned by Eighth Air Force Commanders about the "procurement" and occupation of Bassingbourn, but claimed a misunderstanding of orders and also stated that he would continue to search for an alternate location.

Wray had been given command of this group just after it was born under Lt. Edward R. Akert on 14 April 1942 at Harding Field near Baton Rouge, Louisiana. The next month, the fledgling group arrived at MacDill Field near Tampa, Florida, to create the four squadrons, with Wray as their new Commanding Officer.

Initial training began in earnest—everything from navigating through actual bombing practice and crew consolidation would be drilled over and over to form the necessary cohesion that would glue these men together and form a fighting unit.

The 91st began moving across the country the very next month, and by 4 July 1942 everyone had arrived in Walla Walla, Washington, for the next phase of combat training. This training involved operating as a crew in the B-17E, which was close in appearance only to the new sleek "F" models that they had been hearing about. There were accidents, and some were killed. Learning to fly and fight the B-17 for a twenty-three year old was sometimes too much, and all the men did not stay.

It was here between early July and late August 1942 where fate would play into the lives of the men assigned to young Lt. Robert K. Morgan. None of them knew it, but all of their lives would be changed by a beautiful nineteen year old named Margaret from Memphis, Tennessee. While the men were busy learning about how to become a fighting team and adjustments were being made among the crews, they did what they could to entertain themselves, and this usually meant hot dates with local girls. Things were happening to them very quickly, and perhaps this was partly to blame for

In use by the British Army, the hangars are still in service every day. The picture on the right shows the hangar once used by the 322nd squadron during a recent training exercise. In the wartime photo on the left, it is the hangar 2nd from the top. (Frank Donofrio/Brent Perkins)

the final outcome of their failed relationship, but for now, Bob was deeply in love with Margaret and only a few weeks away from meeting the female which he would hold dear to his heart for the rest of his life—a B-17 numbered 41-24485.

So it was in the few short weeks that the 91st Bomb Group was training in Walla Walla that the "Memphis Belle" story was born. For if Margaret had not been visiting her sister there and met Robert, then who is to say what would have happened to the men and the B-17 they would fly in combat. It is very likely that they would have been just another crew on another B-17, and like the many thousands of young men they would have been treated as ordinary Eighth Air Force airmen with their names buried in the military records, known but to each other and the friends they made along the way.

However, everyone in the world was going to know about this crew. In just another nine months, they would be among the most famous airmen in all the world, flying in the most recognized B-17 ever built.

Station 121 Tower, Winter 1943. (Bert Humphries)

Col. Stanley T. Wray giving the "Rigid Digit." When an airman messed up or made a mistake, Wray would award him this, along with his adopted sign and motto *"My God am I Right"?* Wray himself was awarded his own *fickle finger of fate* by his men when he damaged two B-17s in two consecutive days in landing accidents! Note the B-17 behind him, "Hell's Angels." This is not the same "Hell's Angels" that first made twenty-five missions from the 303rd BG. The 91st BG had one that had this name, as well! Col. Wray flew the "Memphis Belle" twice in combat. Once over Lille, France, and once over Emden, Germany. (Frank Donofrio)

"Memphis Belle" wings her way over the English countryside and into history. (Frank Donofrio)

"Memphis Belle" crew left to right: Loch - TT/Eng; Scott - Ball Turret; Hanson - Radio Operator; Verinis - Co Pilot; Morgan - Pilot; Evans - Bombardier; Quinlan - Tail Gunner; Nastal - Right Waist Gunner; Leighton - Navigator; and Winchell - Left Waist Gunner. (War Department Photo)

Vivian Leigh and Laurence Olivier attend the christening of another 91st Bomb Group B-17 appropriately named "Stage Door Canteen."

General Eisenhower christens his namesake B-17 with a bottle of Mississippi River water.

James Cagney at the 91stBG(H) Headquarters.

Bing Crosby and Darlene Garner entertain the Bassingbourn Boys.

Bob Hope and Francis Langford visit Bassingbourn.

Glenn Miller and his orchestra at station 121 just 5 days before his disappearance.

Clark Gable and the B-17 crew that he flew with on "Delta Rebel no.2."

The Americans officially receive Bassingbourn from the British. Although the 91st BG(H) had been flying operational missions for months, official control of Station 121 was not handed over until just before the "Memphis Belle" finished her tour of duty. The date was 21 April 1943. (Frank Donofrio)

Among the more prominent visitors to the base were, of course, King George VI and Queen Elizabeth in May '43. The visit was arranged by Hollywood film director William Wyler, who had been working on his now-famous documentary "Memphis Belle." The entourage is seen here inspecting the base. They reviewed three B-17 and one B-24 crew during their visit. (Frank Donofrio)

King George VI and Queen Elizabeth get a close look at the Belle while they meet with her crew. The men were not aware that morning that the Royals were going to visit them, and some had to be rousted out of bed! (War Department Photo)

Going Home. "Memphis Belle" is center stage at a farewell ceremony prior to departing England for the Zone of the Interior (The United States). (Frank Donofrio)

Right: "Memphis Belle Crew Chief Joe Giambrone paints the 25th and final bombing symbol on the nose of the great fortress.

Bye Bye Bassingbourn. The 91st BG(H) officially hands the base back to the British 25 June '45. (Frank Donofrio)

The restored control tower as it appears today. The very building where so many commanders launched their missions is fittingly now a museum for the 91st BG(H). (Brent Perkins)

Just before leaving Bassingbourn for the United States, the brass assembled the famous "Memphis Belle" crew for a departing ceremony to bid them all farewell.

8

The Men

They were from all over the country. Just ordinary guys who were thrown together in extraordinary circumstances. No different really from all of the other bomber crews flying during WWII.

They had trained in various places, such as Charlotte, North Carolina, Jefferson Barracks, Missouri, and Shreveport, Louisiana. Some came from influential backgrounds, while others had very rough childhoods.

Among the similarities of this crew was that they all ended up being assigned to the brand new 91st Bombardment Group, and they were all young—very young.

Bob Morgan had first trained on B-24 bombers before WWII broke out. His co-pilot, Jim Verinis, was training to be a fighter pilot before he completed flight training as a first pilot aboard the Flying Fortress.

Vince Evans, their Bombardier, was operating a fleet of logging trucks while not quite an adult yet and enlisted in the Air Force shortly before the Japanese attacked Pearl Harbor. He too wanted to be a pilot. He did not qualify and became a Bombardier—one of the very best in the Eighth Air Force. Evans was responsible for quite a bit of Nazi real estate losses while he was aboard the "Memphis Belle."

Chuck Leighton was the fourth officer on the "Belle." Like the others he wanted to be a fighter pilot. but air sickness kept him from achieving that goal (An empty milk carton was part of his standard flying equipment.). When he switched to Navigation and was assigned to a B-17, the air sickness went away. Fortunate for the "Memphis Belle," as Leighton saved her bacon and that of her crew many times with skillfull navigation!

Then there was Dillon, Adkins, and Loch. "Memphis Belle's" Top Turret/Engineers. All from rural roots, and all with exceptional skills to understand the complex systems of the B-17: Dillon, from Virginia; Adkins, who spent some time growing up in a military environment in Georgia; and Loch up in Wisconsin, shooting ducks with a two-dollar shotgun so he could help feed his eleven brothers and sisters.

Bob Hanson (Radio Operator) grew up with his construction working father having to chase jobs around the country—he ended up staying with an uncle in Washington State working at his produce warehouses.

Cecil Scott had it worse in Pennsylvania before he ever got into the ball turret on the "Memphis Belle." So poor he did not even have a decent pair of pants to wear to his high school graduation. He was first introduced to death while still a youngster when he saw a friend struck by lightning while walking home from school.

The rest of the crew was made up of gunners. Winchell, who was born in Massachusetts, had to move with his family to Illinois. Miller, from West Virginia, grew up knowing all too well what hard work was on the family farm. And Nastal went from orphanage to foster home after foster home before he turned his knack of repairing washing machines into aerial gunnery.

And before he ever climbed into the tail gun position of a B-17, John Quinlan from Yonkers, New York, knew the value of life struggles. His dad had died very early on, and his mother took over teaching her son to endure terrific struggles and have devotion to family.

WWII brought these ordinary fellows together and placed them in extraordinary situations that changed them forever.

Foreword

The "Memphis Belle," a Boeing B-17, has been retired from active service in the European theatre after 25 successful bombing missions. With its distinguished crew, which has remained intact since its formation 10 months ago, the ship has been returned to the United States for another—and no less important—mission.

At my direction, Captain Robert K. Morgan, the pilot, his crew and his ship are making a tour of Army Air Forces training establishments in all parts of the United States. There are three principle reasons for this tour:

First, that combat crews in those units now being trained for the European theatre may be guided in their training to achieve combat skill, teamwork, mutual confidence, and fighting spirit.

Second, that student pilots, bombardiers, navigators and gunners may profit by the experience and knowledge of the men of the "Memphis Belle."

Third, that battle conditions now existing in Europe and the magnificent achievements of the United States Eighth Air Force in mastering these conditions may be represented properly to AAF personnel and the public.

I consider it important that the messages of these men be given maximum circulation. Therefore, I have had each member of the crew interviewed by an officer of the Army Air Forces and their stories, told in their own words, published in this booklet.

I commend the comments in this booklet to the thoughtful reading of all AAF personnel and especially those officers and enlisted men being prepared for duty in the European theatre.

Here are factual accounts of aerial warfare over Germany and the occupied countries. Here is an appraisal of the enemy we are fighting. Here is the advice of the men who have taken the war to the enemy. We must not fail to make the most of the experience of the men who have pioneered the American task in Europe.

H H Arnold

Commanding General
Army Air Force

Among these pages you will find several sidebar style interviews of some of the crewmen of the "Memphis Belle." These were prepared by the Army Air Force for inclusion in a booklet to be handed out to military personnel attending rallies around the famous plane as she was flown from city to city during the summer of 1943.

Given the intense need for success in what were very desperate days for the Eighth Air Force, Commanders approved the creation of the booklet. The pretense was that these were real-life adventures of this successful and famous bomber crew.

The Commanders knew what the newly formed crews in the states did not. That our losses were acute during the early days of high altitude bombing, and that if these messages flew back to the states and to these crews, that their confidence could be shaken. Then the war might not only take longer, it just might be lost all together.

And while these interviews included actual comments and events from the Belle's crew, liberties were taken with the writing before the booklet went into print for distribution. Literally tens of thousands of these booklets were handed out at every opportunity at bases they were visiting throughout the entire morale boosting tour.

The predominant message being boosted by some touches on the necessity of wartime propoganda. Military leaders had to instill confidence in their ranks, and the "Memphis Belle" and her story would help them accomplish this. The Belle was not the only bomber ordered to tour America upon fulfilling mission requirements, but she was the first.

The mission was for the Belle and crew to show the new ranks what they were capable of doing, and to leave them saying, "If they can do it, so can we."

Robert Knight Morgan

Born Robert Knight Morgan in Asheville, North Carolina, Bob grew up during the Great Depression. He had many things going for him from the very beginning. His curly-haired good looks often turned the heads of the pretty girls, and he also had a natural ability to always make things work out.

The Depression caught up with Bob Morgan, though, and he suffered through watching his father, David Morgan, go from being the president of a large furniture company all the way down to a night watchman for a mere $50.00 a month.

The Morgans had been part of a rather influential and well off segment of American society that many people can only dream of. While Bob's older brother David was learning the furniture trade from his father, their sister Peggy was off with their mother, Mabel Morgan, and none other than Cornelia Vanderbilt seeing the world. Bob lost his mother not long after that in what he refers to as "the biggest blow in my life." After quietly suffering the effects of the Depression, the radiant Mabel Morgan learned that she had cancer and decided that since there was no hope at all of any cure, that her only way to avoid the pain to her and her family was to take her own life with a .410 shotgun that young Robert had used as a child to hunt.

Sometimes referred to as a "wild" kid at Biltmore (he would even go on to live there for a while during the Depression), Bob was known to drive his car faster than anyone else in town, among many other things. This devil-may-care behavior would go on to serve him well as the pilot of the big new B-17 bomber that the U.S. Army was building.

His father would go on to rebuild the family's crushed company, and by wartime he had three furniture companies back up and running.

Bob Morgan went on to study business administration at the University of Pennsylvania, and his first job after graduating was with the Addressograph-Multigraph Corporation of Cleveland, OH, which made office equipment.

Robert Morgan joined the service in 1940 well before the Japanese would attack Pearl Harbor. "I could see the war coming, so I decided to get into the Air Corps." He nearly washed out because of some vision problems. The flight surgeon, who took a liking to him, helped him out. "He took me to a dark room and told me to hold some ice on my eye. After five minutes he gave me the eye test again and I passed with 20/20 vision."

Although most people would pick Bob Morgan to be the fighter plane type of pilot, they were surprised when he chose the big four engined bombers to fly. "I wanted some company up there" he said later. Throughout training Morgan would end up with some rear-end chewings, as he developed a reputation for buzzing things—mowing the lawn with the airplane! One Officer was having a lawn party down in Tampa when Bob nearly put a Lockheed Hudson into their punch bowl.

His training complete, Morgan was assigned to combat in a new "F" model B-17 that he would name the "Memphis Belle" in honor of his new sweetheart from Memphis, TN. Her name was Margaret Polk.

Bob Morgan would go on to command the very first crew to complete the required 25 missions over enemy territory and return home. Finishing 25 missions was thought by some to be impossible. The average life of a B-17 in combat was often less than 50 days! The "Memphis Belle" would also become the most recognized airplane of WWII, as dramatized in the extremely popular wartime documentary of the same name. Directed by William Wyler (who directed more than seventy feature films), this film describing the feats of Capt. Morgan, his crew, and their magnificent Flying Fortress was ordered into theaters around the nation by President Roosevelt himself.

After the European missions were finished, the crew of the "Memphis Belle" was ordered on a 3 month long war bond tour—the 26th mission. They would fly the "Memphis Belle" from city to city all over the U.S. promoting the sales of war bonds. Millions would end up seeing the B-17 flown wildly as the crew roared overhead in fighter plane fashion. Morgan even flew the giant bomber down one of the main streets in Asheville, North Carolina, and between the buildings there. With only 74 feet of airspace between two particular buildings, Morgan jinked the Memphis Belle between them, even though the wings of the great bomber spread out over 103 feet! He had a sixty degree bank angle on the plane when he did it.

He went on to volunteer for duty in the Pacific theater of operations. He would now fly the even bigger B-29 Superfortress. The name of this airplane was not the "Memphis Belle," and was instead called "Dauntless Dotty." His relationship with his Memphis sweetheart had dissolved, and the Dotty was named after his new wife! "Dauntless Dotty" would be the first bomber to lead a Superfortress formation over Tokyo. At the controls was none other than Robert K. Morgan.

Completing more than 50 total missions in both theatres of War, he would eventually retire from the Air Force Reserve with the rank of Colonel.

Currently married to his wife Linda, the couple remains active on the lecture circuit, airshow circuit, and in real estate. Bob and Linda were married in Memphis, TN, in 1991 under the wings of the grandest B-17 of them all—the "Memphis Belle."

His Own Words

War Department interview Pilot Bob Morgan - 1943

"If you want just one word how we were able to go through that hell of Europe 25 times and get back home without a casualty, I'll give it to you. The word is TEAMWORK. Until you have been over there, you can't know how essential that is. We had 10 men working together, each ready and able to help out anybody else who might need him.

Take just one example of what I'm talking about. If it weren't for the tailgunner using the interphone to keep me posted on the formation behind, the top gunner reporting to me what he can see, it would be almost impossible for me to fly the airplane in combat. I can't get up and look around. Those fellows are my eyes.

In the same connection, every man in a bomber crew should know something about every other man's job. Aside from the value of having someone to take over in case of emergency, understanding the other jobs and knowing each other's problems develops teamwork. It also promotes confidence in each other and enables us to anticipate each other's needs for help.

I would like to make one suggestion to improve the training of bomber pilots. Cut down on transition training—landings and take-offs—and emphasize high altitude and formation flying. I could have used a lot more of that. Now a few tips:

Before taking off on a mission, the pilot should check the airplane from one end to the other. I mean that literally. He should check even the smallest details and never be satisfied with anybody else's word that everything is OK.

He should consult with the navigator and get the course well in mind so that he can anticipate turns, can have some idea of where the heaviest flak is going to be, etc.

Heavy flying equipment is cumbersome, and a bomber pilot doesn't need it. It never gets very cold in the cockpit, and anyway, the pilot is so busy he could sweat at 35 below. I wear heavy underwear, a regular uniform and coveralls, and that's enough.

Keep your formation. I can't emphasize that too strongly. At first the idea seemed to be to get the bombs out and then go hell-bent for home. But we have learned how important it is that the formation be maintained. There have been cases when we turned a formation nearly around to pick up a man who was straggling. The Germans always try to break up the formation and then jump the stragglers. If we concentrate our fire power by keeping the formation, the only thing they can do is slug it out with us. They don't like that.

We like having fighter escort. We wanted P-38s but couldn't get them. The best escort we ever had was the P-47. If they'll give us fighter protection, they'll get a lot of stragglers back home that otherwise won't make it.

As for German tricks, here's one to watch for: they are painting Focke-Wulfs with white stripes like our P-47s. It's hard to tell them apart at a distance. Also the Germans have a camouflage color that is effective when they get up in the sun. If you aren't careful you won't see them, and they'll be on you before you know it.

The Germans are learning some things about the Americans. They used to say that we couldn't carry out daylight raids over Europe because our losses would be too great. They also said we couldn't hit pinpoint targets. We have proved them wrong on both counts."

James Angelo Verinis

Jim was born in New Haven, Connecticut, on October 23, 1916. His family was from Greece, and that was the language spoken in their home. His father ran a candy and ice cream business to put the four Verinis children through school.

In high school he was a basketball player, a sport Jim still enjoys very much today. Jim, or "Angie" as he was sometimes called, also played football. Jim was the Captain of the basketball team and was really pretty good at the sport. He still has many newspaper clippings from their championship games.

When he graduated from college he was prompted by a friend to join the Air Force. It was July 1941, and Verinis was headed to pilot training. The German invasion of Greece seemed to affect the young cadet.

Jim just about washed out of pilot training. He had some difficulty with his first attempt at landing solo in the "Vultee Vibrator"—the BT-13. But soon he had it mastered and went up to the P-39s and P-40s. These were some of the hottest fighters in the sky at the time and required quite a bit of attention from the pilot. He was going to stick it out no matter what. A failed engine, a blown tire, parachuting, and even a terrific ground loop after a fast, no-engine landing in a P-40. Jim spent several days thinking about the fuel running all over him after that incident while he lay trapped inside the flipped Warhawk, and he decided it was time for more engines!

After making the switch to B-17s, he was sent to the new 91st Bomb Group as a *first pilot*. There were not enough bombers for all the pilots, and it was then that fate stared at Jim when Bob Morgan approached him, asking if he would like to join his crew as co-pilot.

Verinis took the job, and consequently the move changed his life. He stood out right away. The kind of conservative fellow that makes you smile when you witness his quiet but resolute solutions to difficult situations. For instance, after a couple of missions where the "Memphis Belle" received her share of flak and bullet holes, including a couple that were very close to Jim's rear end, his response was to simply find some armor plate to enhance his seat in the cockpit. There were some laughs among the flight crews, but somehow a bunch of steel plates were finding their way onto B-17s all around Bassingbourn. When Command got wind of the activity, an order was issued banning all extra steel protection from combat B-17s.

Bob Morgan referred to Verinis as a conservative man who would not take part in some of the crazy stunts that other pilots would. Like the time later during the War bond tour stateside when Morgan rolled the "Memphis Belle" upside down over the Memphis airport. From the right seat Verinis was tending to his duties, and Morgan asked Jim if he thought they could roll the "Belle." "He didn't even give me time to answer," he said in a recent interview, "before I knew it we were rolling, and there was Bob with a big grin on his face!"

Morgan has often said that his first co-pilot was just the right guy for the job, and he well deserved his own Flying Fortress to command. He got it. After commanding the "Memphis Belle" on the raid to Lorient on 30 December 1942—a mission that Morgan could not make—Verinis received a B-17 which was taken from a weary bomber pilot. Jim's new charge was B-17 #42-2970. Thinking of his home, right away the name "Connecticut Yankee" was painted on the nose.

It was aboard this bomber that Verinis made Air Force history. On 13 May 1943, James Angelo Verinis became the very first Eighth Air Force airman to fly the required twenty-five combat missions—a fact that was not uncovered until 1999! Verinis showed terrific skill as a pilot. Once he was complimented by Major Putnam, Squadron C.O. for the 324th. The whole room cheered during the de-brief when Putnam announced that Jim could keep a string tied between two B-17s!

After the War bond tour Morgan wanted Jim to go to the Pacific with him on a B-29. Jim said no to that. He was now testing B-25s, B-26s, B-24s, and more. But after the War Verinis did join his first combat pilot in private business. Jim became a representative for several furniture companies that Bob Morgan was running. Jim did well with this, and now enjoys his retirement between his summer home in Connecticut and his winter home in Florida.

On a wall in his Connecticut living room, an original Robert Taylor painting of the "Memphis Belle" is hanging proudly. A print would not do. If Verinis' memories of his time aboard this special plane were going to be displayed in his home, then only the original painting would be appropriate for him.

His Own Words

War Department Interview with Co-pilot Jim Verinis - 1943

"To begin with, let me say to the fellows who go over as co-pilots that they probably won't be co-pilots long. The chances are that if they are good they will get their own ships. So don't get the idea that you are going to fight the war as a co-pilot. You should be prepared to take over your own ship and crew any minute you are called upon to do so.

But while you are co-pilot, you have very definite responsibilities. Captain Morgan and I had it pretty well worked out. As co-pilot you should do everything you possibly can do to relieve the pilot. The pilot has tremendous responsibilities, and I consider it the co-pilot's job to relieve him of all the worries that he can.

There is no question about the need for a co-pilot in a B-17. The strength of one man is not sufficient to kick it around in combat. Usually, the co-pilot will do at least half the flying, but the pilot takes over in actual combat.

But even in combat, when the pilot is flying the ship, the co-pilot should keep his hands on the wheel and his feet on the rudders. In that way he is ready, when needed, to apply pressure in the direction the pilot indicates.

If the co-pilot sees something that the pilot doesn't or can't see, he should notify the pilot or take the controls himself. It has been my observation that there is too much of a tendency to leave everything to the pilot. It's too much for one man. He can't see everything, and the co-pilot should help him watch.

I consider the training I had fairly thorough, but I did need more high altitude and formation flying. There is too much transition flying. There are fellows coming over with 500 to 600 hours in the B-17. A lot of that is wasted. After completing my tour, I have only about 450 hours.

If you are going over, talk to somebody who has had combat experience. The fellows who have been through it can tell you a lot about dodging fighters.

Here's something else important: learn to judge exactly when you are in position to be hit. And don't forget that pursuit planes have fixed guns and they've got to be pointed at you before they can hit you. You don't have to run all over the sky to dodge them if you'll just learn the positions that keep you out of their aim.

The Germans have some tricks they use on us, but they are all perfectly obvious if you keep on the alert. For instance, they will pretend to be shot down. They will shoot out a jet of black smoke and peel off. But if you watch, you'll see them come right back. They seldom fool anybody.

Now that the boys know that they have only 25 raids to go before they get a let-up, morale is good. As long as they have a goal, as long as they know there is a stopping place, they're OK. If they felt they had to go on indefinitely, the spirit wouldn't be so good.

Don't misunderstand me. It's no picnic over there. If anything, the missions are getting tougher. These daylight raids on the Ruhr valley must be hell. But the boys know that their equipment is good. They know that our bombers surpass anything that anybody else has. Our P-47 will tangle with a Focke-Wulf any day. And our pilots and crews are good.

Here's another thing that helps the spirit of the boys: The food and housing are excellent, and conditions generally outside combat are better than in any other theatre. There is no danger at the base. The tension is off when you get back from a mission. Just keep sending the B-17s over there, and our boys will be all right."

Vincent P. Evans

He was known as the spark plug of the crew, the "Kid Wonder," the unflappable Bombardier of the "Memphis Belle"—Vince Evans. There always seemed to be a romance, and it was always with a beautiful woman. When the crew badly needed a real egg for breakfast (a rare luxury during wartime) Evans was their man. He always had fresh hen fruit for the guys and even fresh milk. He somehow managed to persuade a local farm girl or two to donate the precious commodities to the American cause. And Vince never hoarded his booty—he shared these simple but sorely missed luxuries among his crewmen.

There was also a close "friend" down in London who was responsible for Evans' many excursions there. Her name was Kaye, and she was a nightclub singer. Evans would call back to the base to see if a mission was on for the next day. If there was no raid scheduled, he would simply stay—if the 91st was going to fly he would sometimes stay, but just long enough to allow him to get back to Bassingbourn in time for the briefing and the takeoff. Bob Morgan remembered that there was even a time when Evans had to run out to the "Memphis Belle" as they were taxiing out for the mission. Morgan simply said that if Evans was not going to make the mission then their Navigator would have to drop the bombs! Someone on the "Belle" saw Evans running up to the bomber, a door swung open, and Evans clambered up inside and off they went!

When on the job Evans was one of the best. The girls would leave his mind, and he would hunch over his bombsight and drop his bombs on the German war machine.

Vince was just one of those guys who seemed to be able to do everything very well. He was born in Fort Worth, Texas, in 1920 and was driving the family car at eleven years old. He was running his own logging company before he was twenty. After the War, he went to Hollywood and was married to Hollywood actress Jean Ames. He sparked around with the Hollywood elite and even wrote scripts for Humphrey Bogart. He was friend to June Allyson, Jimmy Stewart, and even Ronald Reagan, the future President of the United States! Vince Evans raced cars, he once owned the famed Anderson's Pea Soup Company, he operated a shrimp boat, and he was also a private pilot. Vince was everywhere and did everything.

Painted beneath the bombsight of the "Memphis Belle" is the name "Dinny Janie"—the woman to whom Evans was first married. They met while he was training near Walla Walla, Washington, and they married before he went overseas. Typical to most long range relationships, Evans was prone to stray from time to time. This was still an apparent trait of his when the "Memphis Belle" crew was needed in Hollywood to put voices onto the film that William Wyler was nearly finished with. In combat no sound could be recorded, so it had to be dubbed in during the editing process later in the states. *Note: it was good planning on the part of William Wyler to use the actual voices of the crew of the "Belle" when he could have just hired any group of actors to read scripts that were written by people that were never in combat. Just another example of the desire of Wyler to go the extra mile for his work.*

Out to California they went, and there were parties to go to. Actresses like Veronica Lake were on the town with Morgan and Evans. And there was Miss Ames. Evans was in love again, and the two were married, with one slight problem for Evans—he was still married to Dinny!

This was, of course, illegal, and Evans would have to find a way to get back overseas while things in the States cooled off and his divorce from Dinny up in Washington became final. When he married Jean Ames, he really believed that the divorce from Dinny was completed.

A call to his former skipper Robert Morgan would do the trick. Morgan remembered the phone call..."Bob, you know a lot of people. What can you do to get me out of the country?" Morgan was preparing to lead his new squadron of B-29s overseas to the Pacific. Knowing Evans' skill at the Norden Bombsight made Morgan all too glad to help Evans get assigned to a B-29 that was going out of the country. Some strings were pulled, and Evans ended up in the nose of another Bob Morgan bomber. Evans was with Morgan on the "Dauntless Dotty." He flew five missions on this bomber with Morgan. Girls were nonexistent on Saipan where these Pacific missions were being flown from, but they were still on the mind of Vince Evans. He woud quip during the missions, wondering what they were doing in all the Giesha houses in Japan.

Later, during the early eighties when Evans' business was growing so well, he became a pilot and bought a Piper twin Aztec to use between the cities he was developing in. Qualified on everything except instruments, Evans was considered a very good pilot. In April 1980 while on approach to a Santa Barbara airfield, he and an instrument rated pilot on board with him became disoriented in clouds and struck a mountain.

In an instant it was over for the brilliant businessman who had endured so much in the nose of the "Memphis Belle." A man who lived life to its fullest was killed with his wife and their daughter, along with the other pilot on board that day.

Vince Evans is survived only by his sister, Peggy, and his son Peter, who was not aboard the plane that day. Vince Evans was sixty years old.

His Own Words

War Department Interview Bombardier Vince Evans - 1943

"American bombing skill and equipment are the best the world has ever seen. There is no question about that. They all marvel at the results we get from high altitude bombing. And I am convinced that the German war machine will be destroyed by daylight, precision bombing from high altitudes.

When you go after one specific small target—not a whole plant, but one building in a plant—find your target, and put your bombs right on it, you are doing some real bombing. There's no guesswork about it. And that's what the Americans are doing day after day.

As for my job as bombardier, I wouldn't trade with anybody. Of course, I would like to be a pilot too, but if I had to do it over again, I would sign up for bombardier training. It's the greatest thrill in the world to see your bombs hit the target.

I would like to see our schools place more emphasis on high altitude bombing training. Also, our training should concentrate more on pilot-bombardier teams, teaching them to work together, and less on individual training. Bombing problems should be given—real problems, like bombing Denver from Seattle. That would be a great help to the boys in preparing them for combat.

Student bombardiers should be given more practice in photographic and map reading. I didn't get any of that in school. They should be given more pilotage navigation training. Every member of a B-17 crew should learn other duties besides his own. The four officers should be able to interchange.

A fellow who doesn't finish near the top of his class shouldn't get discouraged. The fact that a man leads his class in school doesn't mean that he will be the best bombardier or pilot. You just can't tell until you try him in combat.

The new Automatic Flight Control Equipment will be a means of shortening the bombing run. When we went over, they hadn't perfected AFCE. Bob (Captain Morgan) and I worked so well together that we were able to do allright manually.

Now that AFCE has been perfected, I certainly would recommend it to other crews. We have tried it out and found it to work very well.

Here's a tip that will help: When you first turn on the target, take out the big corrections by talking the pilot over, using the interphone, before clutching in. Then you will have only minor corrections to make.

Bombardiers would do well to spend a lot of time in the bombsight vault talking to enlisted men. I figure that I have learned thirty per cent of my knowledge from actual experience, and fifty per cent from enlisted men in the bombsight vault. These master sergeants can really put you wise to a lot of things.

I believe in dropping on the leader. It is the only way to keep the formation closed for maximum fire power. Of course, it puts the monkey on the lead bombardier's back. It makes me nervous to think about it.

Charles Leighton

Born in Anderson, Indiana, on May 22, 1919, Charles spent his youth in East Lansing, Michigan. The Leightons moved there when he was very young. He was one of three children in what was a very typical American family. His father worked as a machinist for Oldsmobile. While Charles excelled in high school football, his grades were considered average. "Lightning Leighton" even held the school record for the most touchdowns ever scored there, a record that stood for over fifty years!

He was not sure if he would actually go to college, but decided that it seemed to be the right thing to do. He enrolled at Michigan University and studied engineering. His grades remained average, unless he was studying something he really enjoyed—like navigating on a huge four-engined bomber!

Things happened quickly for Charles Leighton in early 1942. New to the Air Force, he had volunteered after the attack on Pearl Harbor. He laughed a little when he remarked that facing the Germans was easier than a college chemistry test he was facing, so the decision to join was pretty simple. Like so many other young men at the time Leighton wanted to be a pilot. There were so many pilot cadets and so few available training airplanes that Leighton found himself waiting.

When Charles was offered the chance to get right into the action by accepting some Navigator's wings, he jumped at it. Besides, the trainers he was flying in had introduced him to some unexpected airsickness. An empty milk carton even became part of his regular flying equipment.

He had no trouble on the bigger planes, though, and even found that he had a natural knack for navigating, often making calculations much faster than the other cadets. One of the lighter moments on the "Memphis Belle" for him occurred during a training flight. He found that nature was calling, and he made his way to the bomb bay, where there was a relief tube installed for this purpose. Before he was finished, the bomb doors suddenly swung open, and Leighton was nearly pulled out of the plane by the rushing air. He scrambled back into the nose, where Vince Evans was rolling in a ball laughing. Evans had opened the doors as a joke and nearly killed Charles. The harder Leighton punched Evans, the more he laughed.

He remembered the flight over the Atlantic to England. It was a night flight, and celestial navigation was necessary. He was busy shooting the stars and entering his calculations over and over. Fear arose, as he knew that he had the "Memphis Belle" somewhere over Canada instead of the Atlantic, where they were supposed to be. After a while he realized that he was entering the wrong base numbers for the take off time, and they were actually right on course. He got them to England right on the money.

Gunnery in the nose compartment was quite another thing. A very good shot with a rifle (he was in charge of the rifle range during high school), the fifty caliber turned out to be a very different weapon for him. He was given very little training on the gun and was expected to be able to service it with precision. He remembered that the nose guns on the first mission of the "Memphis Belle" were not working. Eventually armament specialists took care of the guns for the bombers. The guys that flew them would do just that—fly them.

Firing the machine gun was kind of an assault on the senses the first time he shot with it. With all of the bucking and action, Leighton's accuracy was the same as his high school grades. He often practiced by firing at wave tops.

Extra ammunition was sometimes brought on board the B-17s, and their metal boxes caused some unexpected trouble for the navigators. A very sensitive and delicate compass was mounted on the floor of the nose compartment; so sensitive, in fact, that any metal around it would throw it way off. Either the metal ammo boxes were going to move or the compass was. The compass moved and ended up in a station out on one of the wings, which could be remotely read from inside the plane.

Leighton's ability to calculate true readings and time resulted in him being one of the very best Navigators in the 91st Bomb Group. He led many raids for them, and because he was coupled with an exceptional bombardier, the "Memphis Belle" quickly became one of the stars of the Mighty Eighth Air Force.

He could tell Morgan in the cockpit which direction to head. Not many could tell Bob Morgan what to do—but Leighton could and did. Morgan trusted him. Especially after the notorious raid of 30 December (even though Jim Verinis commanded the Belle on that mission) when the returning strike force was lured over the Brest Peninsula by false radio beacons. Leighton kept the "Belle" out of trouble that day and on several other missions. He had a keen sense of awareness that Bob Morgan insists saved them time and again.

There were plenty of memories for the "Memphis Belle" navigator. Like the time that the flak was so bad that he and Evans, without saying a word, decided that it would be a good idea to put on a parachute. Nearly passing out for lack of oxygen. Shooting down a German fighter and how it felt. The strange sensation of a sudden hole punched through the side of the plane creating a cloud of insulation surrealistically floating around. How frightening it was to run out of bullets when the enemy fighters just kept coming and coming. And the worst part of it all, packing up the belongings of the friends who would not come back after a mission. Men he had breakfast with that would not be there for dinner. Even the toughest of the men hated that.

When the war was suddenly over for Chuck Leighton he took his opportunity to finish pilot training, eventually earning his pilot wings and a spot on B-17 training missions in the States. When WWII ended for everyone else, Charles Leighton became a civilian again and finished college with a degree in education, not engineering.

For almost twenty-five years he counseled and taught. Jane Leighton, his wife, was also a teacher. Charles Leighton died in Bellaire, Michigan, in 1991. He is survived by his two daughters.

His Own Words

War Department Interview Navigator- Charles Leighton - 1943

"We usually have three hours between briefing and take-off. During that time the navigator is busy as hell. He must make sure that he fully understands weather conditions, including any anticipated changes. He must check on any flak areas on or anywhere near the route so that he can avoid them. He must be thoroughly familiar with the formation that is going to be on the mission.

After briefing, he consults with the pilot. In fact, it is a very good idea to go over the route with the pilot and the whole crew before take-off. Otherwise, some of them won't be able to keep up with where they are, and that is a great handicap.

It is especially important for the navigator to talk to the radio operator and make sure that all the radio equipment is ready.

I think the navigator should brief the pilot on how to get back in the event something happens to the navigator. I have tried drawing a map showing the route on a small map for the pilot. I think it's a good idea, but every precaution should be taken not to let a map get out, and it must be destroyed if the ship goes down.

I should have had more gunnery training. The first moving target I fired at was a Focke-Wulf.

I could have used more practice in navigation. It would be good training to take a navigation student up, let him play checkers or something for a few minutes, and then have him try to figure out where he is. Without practice of that kind, you might have a hard time finding yourself after a fight.

There are two conditions that might cause a navigator trouble: First, poor weather when he can't see the ground. Second, a fight that takes him out to sea, causing him to lose all his landmarks.

This leads me to this important advice: Train yourself to write down your compass heading and to continue your navigation work even through a stiff fight.

I like being a navigator. Of course, I would like to get up and walk around a little sometimes, but I can't. I have to work like the devil all the time. But there is nothing like the satisfaction a navigator gets when he hits his ETA (estimated time of arrival) right on the head.

When you are the lead navigator, the whole formation depends on you. The responsibility is frightening sometimes. It really keeps you on the ball. Actually, though, you should be just as alert if you are the navigator of any of the other ships in the formation.

There are some things to watch for. The Germans will sometimes send out beams exactly like ours to throw you off. Another German trick is to stay just out of range and play around, trying to get us to use up our ammunition. I have also learned to keep my eyes off the flak. It helps a lot.

You get scared sometimes, but usually any feeling of fright or tenseness leaves you when you start mixing it up. I got scared the time I ran out of ammunition. We were away down past Paris. It doesn't bother me if I can shoot back. But looking down those barrels and not having anything to shoot is no fun."

Leviticus G. Dillon

He was born in Franklin County, Virginia, on 15 July 1919. Levi Dillon grew up in Roanoke, Virginia, in what was a very typical depression-era life. He had enlisted in the Air Corps on September 8, 1941, only months before Pearl Harbor.

A natural with his hands, Dillon was assigned right away to fly with Bob Morgan as flight engineer while the 91st Bomb Group was still in its infancy at McDill AFB, Florida.

While he is not recorded as a regular member of the crew of the "Memphis Belle," Dillon was there in the top turret from the very beginning and through the first five combat missions, of which he flew four.

Levi Dillon did not fly all twenty-five of his raids aboard the Belle—in fact, he was reassigned to the 306th Bomb Group. While returning to Bassingbourn after liberty there was a confusing disagreement at the gate, and Levi was wrongly accused of ripping an enlisted man's jacket during a scuffle. Dillon kept quiet and did not reveal the person who really was at fault. He took his punishment and went on.

Like the other members of the "Memphis Belle" crew, Levi had some stories that brought the occasional smile to his face. Because he was along with Morgan's crew from the start there was plenty of opportunity to make a few memories.

Once during training he and Charles Leighton were part of a crew that was carrying out a very long cross-country B-17 flight that took off from Walla Walla, Washington, bound for Dayton, Ohio. It was not on the "Memphis Belle," however (she did not exist yet). The pilot was Col. Stanley Wray, the Commanding Officer for the 91st Bomb Group.

Long range fuel tanks were installed in the bomb bay of the plane that day, and when they got to Dayton they found out that one of the main wheels refused to come down. When the crew went back to the manual extension socket in the bomb bay, they saw that the long range tank was in the way of the handle and they could not even begin to crank the wheel down. The tank had to go.

Bomb doors open, crew in position, and tank release pulled—nothing. It did not fall. Col. Wray ordered the lowest ranking man on the crew to crawl on top of the tank and try kicking it out! Leighton and the others tied a rope around crew chief Joe Gambrione's waist, and they held the other end while he jumped up and down on it! It did not move. It did not take Levi long to figure out that this was not really a good idea, and that applying brains over braun was the real solution to their dilemma. He quickly gathered some tools and worked on the release mechanism with Gambrione until the tank fell.

The newspaper headline the following day read "Air Force Bombs Chicken Coop." The tank destroyed some Ohio chickens, but the men got the wheel down and they landed safely.

Levi was the only member of the Belle's crew that dared to ride in the right seat during Morgan's "hot for a date" three-engined take off. He pulled the very first arming pins from the noses of the "Memphis Belle's" bombs on the very first mission. He was also injured during a combat mission, during which another unique story began.

It was the third mission for the Belle, 17 November 1942. During an enemy fighter attack a bullet came through the top plexiglass of the turret, somehow missed his head and struck him in the thigh. His flying suit caught fire and smoldered. Jim Verinis left the copilot's seat and beat out the flames, cut Levi's suit open, and put a bandage on him.

When the "Memphis Belle" landed at Bassingbourn, Dillon decided that his wound was not serious enough to keep him from the area pubs, so he jumped on the liberty bus and headed off for a night of relaxation and beer. Later that evening, someone noticed that there was some blood on his pants. Dillon headed for a Red Cross station and ended up being bandaged by Adele Astaire, sister to a famous Hollywood actor and dancer Fred Astaire.

Dillon did well in the Air Force, so well in fact that he stayed in until 1963 when he retired as a Senior Master Sergeant. He then ran a small marina in Virginia with his wife until he died in 1998.

Eugene Adkins

Gene flew seven missions total aboard the "Memphis Belle," and all but one raid was flown in the important position in the upper turret as Flight Engineer. His time on the "Belle" was during the first half of the missions, and then on the very last mission on the May 19th (that is, if the "Memphis Belle" flew on the 19th!). He was scheduled to crew the "Belle" that day, and he did indeed take part in the mission. Whether or not it actually took place on the "Belle" is not as significant as the fact that he was not flying as Flight Engineer, but was instead up in the nose dropping the bombs. It is not known if he was a Bombardier or a togglier. *(A togglier would watch the cues of the B-17s in front of them and release the bombs when they did.)*

Adkins had been through quite a bit during his short stay with the "Memphis Belle." Plenty of the typical shooting and flak, but he also fell victim to what was causing more problems than enemy bullets and flying steel fragments—frostbite.

On 4 February, the "Memphis Belle" was sitting back at Bassingbourn being repaired, and Morgan's crew was scheduled to fly a mission. They took B-17 #515 "The Jersey Bounce," and it was their first trip over German soil. Adkins' turret guns were giving him some trouble, and he removed his gloves so he could remove the covers. In less than two minutes the damage had been done. Even worse, since it was the beginning of the mission, Adkins had to stay at his guns through the entire raid. This was no "milk run" mission, either. The enemy was throwing some of their very best opposition at the Americans, and between attacks Gene was going back to the radio room to try to warm his hands. Every time an attack resumed he slid through the bouncing plane's freezing bomb bay, and with his frozen hands pulled himself back into the top turret to ward off the continuing attacks.

This would do it for Adkins. When the mission ended, he was taken to a hospital where doctors managed to save his hands. He was now officially off the "Memphis Belle" crew, but it would take a lot more to keep him out of the Air Force. Adkins eventually became well enough to win an officer's commissioning and a pair of 2nd Lt bars. He finished his twenty-five missions aboard different B-17s and took part in the infamous raid on Schweinfurt on 17 August 1943. He was even credited with shooting down an FW-190.

Adkins was part of the 91st Bomb Group from the beginning. He remembered that he had met Bob Morgan before the fledgling Group even left Florida. He had flown several training missions with Morgan's crew, but was assigned to fly on a B-17 named "Pandora's Box." *(This was actually the very first B-17 to take to the skies when the Group became operational on its first mission on 7 November 1942.)*

"Pandora's Box" was enduring some mechanical problems and some crew coordination troubles. It had aborted four missions and was among the first B-17s of the 91st that was shot down. That was on 23 November 1942 and, thankfully for him, Adkins was not assigned to fly that mission.

He became a replacement Flight Engineer, filling in where needed aboard different bombers, and when the regular "Memphis Belle" Flight Engineer (Levi Dillon) was transferred out of the 91st, Adkins took his place on the "Belle" immediately. Because of his time in the hospital, Adkins was unable to stay with the famous "Memphis Belle" crew when they were assigned to bring the bomber back to the States for the big War bond tour. He later reflected that he felt a little left out when they left England and he had to stay.

The first time he flew on the "Belle" was before it even left the United States. Crews were allowed to fly a training mission to a place of their choice, and Adkins heard that Bob Morgan was taking the "Memphis Belle" to Asheville, NC, where the pilot had grown up. Since it was not far from where Adkins lived, he hopped aboard for the flight and then took a bus to his home in Johnson City, TN. It was on this flight that Bob Morgan landed the "Memphis Belle" at Asheville on a runway that was too short for a B-17 and had to use the emergency brakes to stop the bomber. The brakes were burned out, and so was Bob Morgan's rear-end! His Commanders were not pleased that a crew had to be sent to North Carolina to fix the "Memphis Belle." (Adkins was actually the only regular "Memphis Belle" crewman from the Tennessee—none were from Memphis itself*!*)

Major Eugene Adkins finished his Air Force career training gunners and gunnery specialists. He flew on all the big bombers, including the massive B-36 Peacemaker. Thoughout his life there was the constant reminder of his days with the "Memphis Belle" whenever his hands would ache from the effects of the frostbite. Gene Adkins died in 1995 in Johnson City, Tennessee.

Harold P. Loch

Harold Loch was born November 29, 1919, in Denmark, Wisconsin. His father was a tavern keeper until prohibition erased his business. The elder Loch moved his family to Green Bay, where young Harold found a knack of shooting ducks on the many lakes near their home.

Much of this hunting was out of necessity, as he had eleven brothers and sisters—plenty of mouths to feed in those hard early depression years. With a rickety old shotgun Harold developed a knack for leading the ducks in cold high winds, a talent that would serve him well when he enlisted in the Air Corps after America entered WWII. He applied for aerial gunnery school and almost always scored the highest on the shotgun skeet training ranges. Back when he had been out on the lakes in his small boat, shooting at ducks, he developed a shooter's eye. He could not know it, but Loch had been training as an aerial gunner for years before he ever joined the service.

He breezed through radio training and engineer's studies before being assigned as a B-17 Second Engineer with Robert Morgan aboard the "Memphis Belle."

This was not a famous plane at the time. There was nothing special about this bomber or her crew. Just another B-17 with ten guys going to the other side of the world with so many other young fellows.

Many things stand out in his memories of that time. The terrible food and all the cabbage lice that infested their food. The terrible cold, and never enough coal to use for heating their quarters. Loch remembers sleeping in his flight suit to avoid what was known as "scabies itch" from the terrible British wool blankets.

One event during the latter part of their missions really left a mark on him—almost literally. While up on one of the "Belle's" wings checking the fuel tanks, tailgunner Johnny Quinlan climbed into the upper turret to fool around with it a little. He spun the twin fifties toward Harold and accidentally squeezed off a round that whizzed right by Loch's head (the gun still had a round chambered)!

Harold had begun his missions on "Memphis Belle" as the gunner in the right waist window, the Second Flight Engineer. This meant that he not only had to be good with his single barreled fifty caliber, but he was also required to know the complex mechanical systems of the Flying Fortress. After flying the first ten missions in the waist he was moved to the front of the bomber (so to speak) after Gene Adkins went to the hospital with frostbite. Harold finished his final fifteen combat missions in the "Belle's" upper turret.

Like so many veterans of air combat, missions begin to blend into one long single event, and the particulars become a little cloudy. But January 3, 1943, is still very clear to Harold Loch. The gunners were armed with over 1,000 rounds of ammunition for each gun. A terrible fight was on, and the enemy was not letting up. The "Memphis Belle" was being jolted all around the sky by the bursting flak. The German pilots were so determined to kill B-17s that day that they flew right through their own flak to hit the Americans. An Fw-190 pilot pointed his plane at the "Belle," and Loch's gun pointed right back. A steady stream from Loch's machine guns found its mark. The enemy fighter flipped on its back, started smoking, and went down.

There were a couple of occasions when the "Memphis Belle" was hit by shells that were supposed to explode when they hit a solid object—like a bomber. Loch remembers when a strut near the rear of the "Belle" was once nearly cut in half by a shell like this. If it had gone off, it would have killed Quinlan in the tail and blown the whole tail off. He also remembered another time when of these shells lodged itself into one of the main fuel tanks and did not explode.

Loch says that it was by the grace of God that they made it through. He stayed in after his stint with the "Belle" and did some more B-17s, flying on planes that were being rebuilt in the states after they were damaged.

After he married "that blonde girl" who stole his eyes at a Wisconsin bowling alley, he settled into a home-building business with his dad and brothers, then had eight children of his own. He spent nearly thirty years as a registrar and now dabbles in the real estate business. He revels at how his fifteen grandchildren sit and listen to his stories of being six miles high, half a world and half a century away. Much of his mature years are spent scouting about the Wisconsin woods—you guessed it—hunting ducks.

His Own Words

War Department Interview Top Turret Gunner / Engineer
Harold Loch - 1943

"The engineer of a B-17 doesn't have much ground work. I think he should have more, at least enough to keep his hands in. It is important that he know his engines, that he know every gauge, switch, and fuse. Fuses are especially important because they sometimes blow out, and I must know where they are.

Every man on a B-17 must know how to assemble his gun blindfolded. He ought to be able to fix it if it goes out in the air.

The biggest part of the attacks come from the nose. German fighters will come from away behind, slip up to the side just out of range, and gradually get closer. If you don't watch closely, they'll nose right into you before you know it.

Sometimes one plane will fly along in a line with you dipping his wings to attract attention while a lot of other planes sneak in on you from the other side. You have got to be on the alert all the time.

When you shoot at him and he peels off, don't worry about whether you got him. It doesn't pay to watch him. While you're doing that another one might sneak up on you. Just use common sense. That's all it takes.

The upper turret is a good position. You can see any plane that is in position to do damage to you. Also from there, you can let the ball turret gunner know when a plane is coming in and from what position so that he can take a crack at him.

We had good teamwork on our ship. I think that is the reason we were able to complete our 25 missions without a casualty. It doesn't pay any dividends to have trouble in the crew. All of ours were good boys, and we worked together and had confidence in each other.

Before going over, bomber crews should get used to high altitude flying. It gets cold up there, and gunners should get accustomed to it. The first time most of the fellows see an electric suit is after they get over there.

We had a lot of excitement. I'll never forget our March raid on Rouen. We flew over the French coast, feinted, and flew back across the channel. We knew the Jerries would get wise sometime, and they did. They jumped us over the channel, 30 or 40 of them. They attacked from every position. Then just after we dropped our bombs, more fighters came from out of nowhere. Our tail got hit. We weren't bothered much more until we got almost to the channel. Then six of them jumped us, circled around our tail from seven o'clock to five o'clock, and went to work on us. Shells were bursting everywhere. Finally, the foremost fighter began to smoke. He turned away and the rest followed him.

We hear they get an Iron Cross when they shoot down a B-17. They are a pretty determined bunch.

When you cross into enemy territory, you have a tense, expectant feeling. You never know just what you're getting into. But get busy and then you're OK."

Robert Hanson

Bob Hanson was born 25 May 1920 in Helena, Montana. His childhood was spent in the state of Washington, where he lived with his uncle and grandmother. His father was a construction worker who had to go on the road to find work. The Hanson family was moving from job to job, and this quickly became too much for Bob's mother. She decided that the family could not tolerate all the different homes and schools and sent young Bob up to Washington state, where he went to work warehousing produce and grain as soon as he was old enough.

Right: The bullet-ridden log book of "Memphis Belle" Radio Operator Robert Hanson. Note that it is Royal Air Force issue!

High school was pretty typical for a kid growing up in a town of only 500, and Bob excelled in baseball and football. Many others would have said that he had a rough childhood, but he took it all in stride and developed a very positive outlook. From running around with a .22 rifle shooting squirrels for a penny a piece to hoisting a wagon on top of the high school as a prank, Bob seemed to make the most of each situation that came his way. This was a trait that would serve him well in his coming days in the military.

He wanted to volunteer for service months before Pearl harbor was attacked, but waited until he was drafted so that he could then re-enlist for a benefit of $9.00 extra per month. He was sent to Jefferson Barracks in South St. Louis County, Missouri, for basic training, and it was here that his Commanders found out about his attempt to squeeze a few extra dollars out of Uncle Sam. It nearly cost him a court-martial, but he ended up on what were endless weeks of punishment. He was back in the produce business—peeling tons of potatos!

He was sent to Radio school just across the Mississippi River at Scott Field, Illinois. Here he breezed through the instruction, well ahead of the others in the class. The codes he was learning were being sent at a rate of only five to six words per minute. He could handle more than thirty per minute and sometimes read the newspaper during the training.

From there Hanson was sent to the brand new 91st Bomb Group at McDill, Florida, near Tampa. He was assigned to the crew of Robert Morgan, and again was ahead of many of the other Radio Operators in the Group.

When the 91st moved to Walla Walla, Washington, for more training Bob had a chance to re-acquaint himself with the father that he had not seen in many years. The base there was still being built, and Bob's dad was there doing some of that work. Somehow his father found out that young Bob was coming, and even found out which B-17 he was aboard. When Hanson's pilot landed (it was not Bob Morgan, because Bob was serving a punishment for an earlier buzz job and was ordered to take a troop train across the country), a work truck came rolling up to them, followed quickly by some Military Police. From the Radio room, the only thing Bob saw was some guy being hauled off for coming too close to the big bomber. He did not know it then, but that was his dad. The two got together after Mr. Hanson explained that he did not know that civilians were not allowed out next to the planes, even if a son was on one of them.

When he is asked about his memories of his time on the "Belle," Hanson is a little reluctant to discuss particulars: "All I'll say is that the good Lord had a big hand in what happened to us. We were no different from anyone else flying those missions, and it couldn't have been luck, either." After some time, the stories begin to unfold and the true level of risk and danger becomes very evident. True to his nature, Hanson plays these dangers down.

He endured the terrible moments when the "Memphis Belle" was perforated with enemy gunfire, holes opening up in her body all around him. He remembered wondering about the critical systems on the bomber and if they were destroyed during the attacks. His thoughts went to the time when some of the control cables were severed and the "Memphis Belle" entered a steep and uncontrollable dive. Thinking the bomber was going down, he struggled above the increasing "G" forces and made his way past the two waist gunners, who were pinned to the floor in the waist beneath the weight of their flak jackets. On that mission, Morgan was able to pull the "Memphis Belle" out of the terrific dive just before Hanson released the rear door. Hanson remembers that he was determined to jump from the plane, and had he been a little faster, he would have.

Another time during a raid, he was leaning over his small table in the waist of the "Belle" making entries into his radio log book. The constant roar of the engines was interrupted by a sudden noise, and when he looked down at his table, there was a pretty good-sized hole in it *and* his log book. Had he not leaned away to gather some thoughts, there would have been a good-sized hole in his head as well.

Today, the eternal "39 year old" Hanson enjoys retirement at his home in Mesa, Arizona, with his wife Irene, whose name is painted below the radio room window where he used to work on the "Memphis Belle." He remembers most of his days then with some slight humor, but with a deep belief that God was there with them all the time.

His Own Words

War Department Interview Radio Operator Bob Hanson - 1943

"If you are in a new combat crew, you would do yourself a favor to sit down and have a good bull session with men who have been through it. Talking to them, hearing what they did and how they did it, you can pick up things it would take you a long time to learn for yourself.

Also, you should get accustomed to talking over your interphone. Learn not to talk in an excited, high-pitched voice. A little noise on the interphone going over always helps, because everybody is nervous. When you are in combat, use the interphone to keep the rest of the crew informed about what you can see. In a fight, the interphone is one of the most important things on your ship.

You will find that Jerry is fond of putting out false signals and false beams to confuse you. You have to be careful.

There should be others on the crew besides the radio operator who can take code. There is always a chance that somebody will be hit, and it may be the radio operator. The ball turret gunner or the tail gunner should be able to take messages by blinker code if the radio operator can't see them.

Practice wearing your helmet before you go over. If it doesn't fit, get it fixed. You probably won't be able to get it fixed over there. At best, our helmets aren't satisfactory. The wind whistles in, and if you pull them tight, they hurt your eardrums. The British helmet is far superior to ours.

The radio operator's position is a good one, but it's a rough place to ride. In the Lorient raid, when we got the tail shot off, Captain Morgan put the ship into a terrific dive and we dropped two or three thousand feet. It pretty nearly threw me out of the airplane. I hit the roof. I thought we were going down and wondered if I should bail out. Then he pulled up again and I landed on my back. I had an ammunition box and a frequency meter on top of me. I didn't know what was going on. Captain Morgan didn't have time to tell us, and I wouldn't have heard him if he had.

American morale in England is very high, and the boys are really on the ball. The fellows would enjoy more current magazines. About the only one we got was Reader's Digest. By the way, the Red Cross does a swell job. If it weren't for them, we couldn't have gone to London because we wouldn't have had any place to stay.

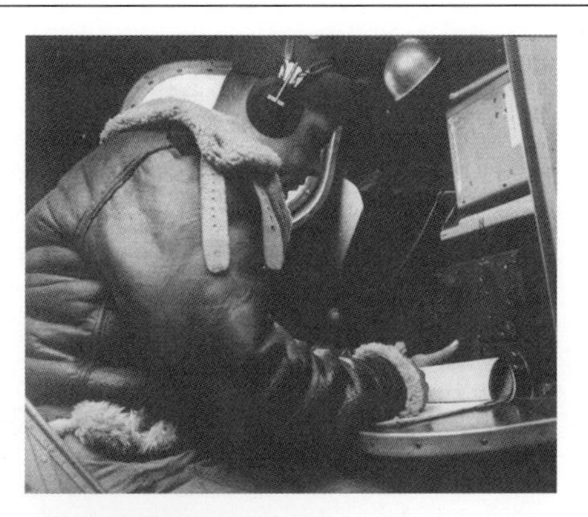

Hanson adjusts the frequency on a radio in the "Belle." It was from here that he performed his vital role of keeping in touch with other planes and England.

Cecil Harmon Scott

Cecil Scott was born in Hollidaysburg, Pennsylvania. The Scott family was not very different at all from so many others that had hard times during the depression. Cecil's father operated a logging company that went belly up, as he used the company's last dollars to pay his workers instead of his creditors.

There were seven children in all for the Scotts to feed and clothe. Young Cecil, along with his brothers and sisters, made their way picking berries, trapping animals for their furs, and even making their own maple syrup, which they would pour over clean snow. They called it poor folk's ice cream.

Those who knew Cecil Scott say that there are only a few people that know the art of survival like him. So poor that an entire summer's work yielded just eighty dollars, almost any kind of work would do, as long as he could make a buck or two. Fighting brush fires with a wet gunny sack, hoeing crops for farmers, even grinding engine valves on his first car by himself. The only real luxury was a large bear rug which gave some warmth in front of the fire.

Adulthood thrust itself too early on the youngster when he witnessed first hand the death of a childhood friend. Walking home in a thunderstorm and climbing fences to shorten the distance, his friend was climbing over some barbed wire when a bolt of lightning struck. In one terrifying moment, Cecil turned to see his lifeless friend hanging there. Fearing electrocution himself, Cecil wrapped his coat around his friend's body to pull him from the fence. Only then did he run for help.

When it came time for high school graduation, Cecil's mother Tena shed tears over the fact that her son did not even have a decent pair of pants to wear. He simply told her not to worry, because he would be wearing a cap and gown and no one would see his tattered clothes underneath.

Through the rough times the industrious Scotts found ways to enjoy themselves. The girls played with dolls filled with sawdust, and the boys found holes in fences to watch ball games through. An old swimming hole was a favorite hangout for them as well.

Cecil Scott was among the throngs of young men that showed up to enlist in the service just after the Japanese attack on Pearl Harbor. He was rejected at first, as he was too short and underweight. By hanging from a bar every day to stretch his height and eating everything he could, he was eventually accepted. It seemed that now the Air Force wanted small guys that they could jam down in the tiny ball turrets that hung under the bombers.

Flying below the thirty ton B-17 in the tiny turret was considered by some to be impossible. Just looking at the small space was sometimes enough to make even the most stalwart man cringe. Scotty, as the crew called him, took it very well. He flew all of his raids on the "Memphis Belle" in that position. During this time he did not receive any credit for shooting down an enemy fighter, but was credited with damaging at least one.

From his vantage point in the plane he saw quite a bit of action. During one terrible attack, he swore that his guns brought down two German fighters, but he did not have enough witnesses to support his claim. His job was to protect the entire belly of the "Memphis Belle." Among the enemy fighter tactics was one effective maneuver. The German pilot would hang his plane from its propeller, climbing straight up at the bomber. It was thought among the crews that they were trying to hit the bomb bay of the B-17s and cause the bombs to blow up inside. It can't be known just how many of these attacks Scott endured hanging below the Belle, but it is known that the famous plane was never seriously hit in this manner. Evidently Scotty was down there really working to keep the Luftwaffe off of this particular Flying Fortress.

The enemy pilots found a weak spot in the defensive firepower of the B-17, and they wasted no time in exploiting this weakness. A frontal attack was becoming the preferred way to knock down the bomber. Closing speeds were mostly above six hundred miles per hour. But if flown well, the Germans could bring an effective cone of machine gun fire to bear in a very quick and concentrated pattern that often crippled the American machine.

The "Memphis Belle" was attacked in this manner several times, and Bob Morgan usually put the plane into a dive to get away from the fighter. On one run this was not possible because other B-17s were positioned below the Belle, but Morgan did something that the German probably never expected. He hauled back on the yoke and pointed the nose of the Belle right up at him. This put Scotty down below in a direct line of sight with the German, and he let loose with both of his guns. Enemy bullets began gouging holes in the Belle all beneath the plane. Down in the ball, Scotty did not know that much of the tail was shredded and useless. Then the Belle went into a very steep dive. All Scotty knew was that it must have been over for them. He rotated the ball turret to put his hatch inside the waist of the plane, and as the Belle fell from the sky the forces pinned him into the turret. Wondering how he would ever get out of this mess, find his chute, and then a hole to jump from, he eventually felt the Belle leveling off and heading back to base.

Norma was the woman he met and fell in love with. They married in 1947 and had three children. He found a good job at the Ford Motor Company and spent thirty years with them. A typical veteran who took part in some of the worst aerial battles ever, Cecil Scott suffered a massive heart attack and died on the 6 July 1979, just nine months after he retired.

His Own Words

War Department Interview Ball Turret - Cecil Scott - 1943

"The ball turret is the best position on the airplane. You see a lot of action in that position, you know what's going on, and you are always busy. If the plane catches on fire you know it first because you can see all four engines, and you can get out as quickly as anybody else.

It isn't too uncomfortable. Of course, a big man shouldn't have the ball turret. I'm small and I get along all right. I was in it seven hours one time, and didn't get very tired.

You should get as much practice in the ball turret as possible. Practice using the sight, operating the turret, and getting in and out.

The Germans have some tricks that you'll soon know if you are alert. When they attack and come under the ball turret, they turn sideways or clear upside down. They go into a slow roll and they are awfully hard to hit. Sometimes they'll shoot out smoke to make you think that they are hit. I have seen ME-109s come out of the clouds and hang on the prop under our ship, probably to try to hit our bombs.

Before the attack, you are usually scared, but when the planes start coming up and attacking you are all right.

I have known fighters to follow us for fifteen minutes before attacking. They seem to be looking us over while they circle around, trying to decide where to attack. When you see them start to peel off, you'd better start shooting.

If you can't get a good shot, you might be able to figure out a way to change the position enough to make it possible. For instance, if you get a plane at the right wing where you can't hit it, you can ask the pilot to lift the wing.

On the Romilly-Sur-Seine raid, about 300 fighters attacked us in relays. The fight lasted a couple of hours, the longest one we'd been in. They attacked us as soon as we crossed the coast and circled around us like indians. Then they started attacking us from all directions at once. I thought that I got two, but I didn't get credit for them. We kept plugging away at them and somehow got by. Practically all our ammunition was gone when we got back. As many times as they shot at us, we didn't get a single bullet hole in the plane.

Last winter American morale in England got pretty close to the breaking point because we weren't getting any reinforcements. It's good now.

Clarence E. "Bill" Winchell

They called him "Winch." His full name was Clarence E. Winchell, and he was born November 4, 1916, in Cambridge, Massachusetts. No one really knows why many called him Bill. Whatever his name, Clarence E. "Bill/Winch" Winchell was very likely the one member of the crew of the "Memphis Belle" who was more proud than any other to have played an important role in the story of this B-17.

Bill became a resident of the state of Illinois before the age of ten. His father, who was working for the YMCA, moved the family to Oak Park near Chicago to take a "Y" assignment there. He quickly fit in and filled his childhood with all of the special things that make fond memories in the minds of young gunners on bombers when they fly in combat: canoeing; camping; and swimming (he made the diving team). Mostly the Winchell family was the kind that some said could have posed for a Norman Rockwell painting. Until October 29, 1940, and a thing called the draft.

Congress, fearing that the U.S. would eventually be pulled into the war already raging in Europe, passed a mandatory draft for young men throughout the nation. Millions of names were to be drawn lottery style. Just about the entire United States was tuned to radio stations from coast to coast to hear whose names were going to be pulled first for service to their country, and Clarence Winchell was number 52!

Volunteering for duty was still an option for draftees, and Winchell decided that because of his love for horses, he would enlist in the calvary. When he showed up to join, they told him that they had all the personnel they needed. So he joined the infantry, and shortly thereafter they stopped using horses. After going through a rough time there, he was up for promotion and told his Captain that he wanted to go into the Air Force. Even though it meant that he would have to start at the lowly rank of Private, he said he would take his chances.

Next stop was Jefferson Barracks on the South side of St. Louis, Missouri. The cold, wet, and muddy midwestern winter would have been almost unbearable if it had not been for a friend in supply that slipped some extra blankets to Winchell. Then he was on to McDill Army Air Force Base near Tampa, Florida, and assignment with the 91st Bombardment Group. Sweeping hangars was not a career that appealed to young Winchell, but the flying programs they offered were. The 91st was a bomber outfit, and the planes needed guys like him to crew them.

Something else tried to stand in his way. Since birth, Bill had a slight vision problem and was told he only had 20/50 vision. This would have stood in his way had he not memorized all of the eye exam charts. Now a gunner with less than perfect vision could theoretically place all of the men and the plane at risk if he could not shoot straight. "One-eyed" Winchell, as some called him, always said that he did not put his crew in danger and pointed out that no less than two enemy fighters went spinning to earth from the sting of his fifty caliber. There were plenty of gunners with perfect sight that never shot down a single plane.

Before being assigned to Morgan's crew there was still plenty of traveling for Bill. He volunteered for aerial gunnery school at Las Vegas, and he had his wings in six weeks. Back in McDill, Florida, "Winch" tried out for radio school but found that it did not suit him very well. Then on to Bombardier school. He did very well here and was soon flying around the U.S. in the nose of bombers with the top-secret Norden bomb sight under his care. At the time the Air Force allowed enlisted men to take part as Bombardiers. There was even a time when he had to draw his pistol at an officer who wanted to have a look at his bombsight. Having taken the Bombardier's oath, Winchell was prepared to protect the sensitive piece of equipment and its secrets—even if he had to shoot an unqualified officer to do it.

After the 91st Bomb Group moved to Walla Walla, Washington, for more training, Winchell was told that he was not going to be a Bombardier after all. The brass decided that this was the job of an officer. But since he knew so much about the bombsight, he could stay with the 91st fixing and repairing the sights, so off to Minneapolis he went for more training. It looked like he was going to be on the ground. When he found out that the B-17 crew was being changed from just one waist gunner to two, he knew it was his chance. If he was going to war, he wanted to fly. They took him with no questions and assigned him to the crew of Lt. Robert K. Morgan.

After the war was over Winchell carried with him the memories of seeing a B-17 blown to bits from a direct hit. He watched his friends being shot down, not knowing if they were dead or taken prisoner. He talked of the times when he remembered staring down the barrels of an attacking German fighter trying to kill him as bullets tore through the "Memphis Belle" all around him, and remembered feeling lucky that no person or vital part of the plane was hit. He thought of the time when he was so busy on his gun that he did not notice his oxygen mask had frozen closed from the condensation in his breath. It was 28 March 1943 over France, and the other waist gunner had already passed out. Winchell needed to shove him out of the way and for a time manned both guns. Too busy to notice, he too succumbed to a lack of oxygen. Had it not been for Bob Hanson in the radio room coming back to revive them, they would have died.

After the "Memphis Belle" returned to the states in 1943, General Arnold himself asked the men what they would like to do next. Both Winchell and Radio Operator Bob Hanson said they wanted to become officers. Next stop, Officer's Candidate School in Florida. Bill received his Lieutenant's bars and a wife, Laura, there. He was hoping for assignment to a B-29 outfit when doctors discovered his bad eye. When asked how he managed while crewing the "Belle" he told them that he must have damaged his sight while peering into the sun looking for German fighters. He was finally mustered out in October 1945.

Bill Winchell, one of the most proud members of the "Memphis Belle" crew, claimed two enemy fighter kills from the left waist position of the "Belle." He spent the rest of his career as a chemical engineer and raised his daughter Jacqueline. He died suddenly at his home near Chicago in 1994.

His Own Words

War Department Interview Left Waist Gunner
Bill Winchell - 1943

"You can see that Germany is getting desperate. A good example of the things they are trying is dropping bombs on our planes from above. It isn't effective, but it's something that bomber crews should watch for.

They seem to be pulling their defenses back into Germany itself. St. Nazaire used to be tough, but it's not so bad now.

Although some of those places on the occupied countries are easier than they used to be, they are still no snap. It's good advice never to sell a target short, because you may get a surprise.

You still find stiff fighter opposition. The Germans seem to be trying to feel us out, trying to learn more about us and our ships. They have gone from mass attacks to single attacks and back to mass attacks.

Our crews should have more high altitude training. Bombers and fighters should be in the air at the same time on these training flights so that they can get accustomed to each other and learn what to expect.

The interphone is the most valuable piece of equipment on the ship. Most of the fellows would rather go over with half their guns out than with the interphone out.

Bomber crews must not relax just because they are out of enemy territory. They jumped us once in the middle of the North Sea and damn near shot us down.

There is one incident that stands out in my memory. It was on the Emden raid. We had bombed our target and the fighters were after us. I never saw such crazy flying as they were doing. One Focke-Wulf came in at 9 o'clock and seemed to concentrate on me personally. I was looking down the barrel of a 20 mm. He went over our left wing ship and under us. I don't know yet how he managed to slip through. I was petrified.

The Memphis Belle had no better crew than a hell of a lot of other B-17s. If there was anything remarkable about our taking all we did without a casualty, it was a combination of things. We had some luck. We had a good crew, and what's just as important, we had absolute confidence in each other."

Emerson Scott Miller

The important right waist position of the "Memphis Belle" was first manned by Harold Loch. He was also the bomber's 2nd Engineer. When Loch was moved up to the top turret as First Engineer, the left waist position was vacated and the "Memphis Belle" received a new crewman.

Emerson Scott Miller first flew on the "Belle" on the mission to Emden, Germany, on 14 February 1943. The available records show that he completed sixteen missions mostly in the right waist, but also in the tail gunner's place on the 19 May 1943 raid.

The West Virginia farm boy was born in 1918 into a life of hard work. He was, like so many others, just an ordinary guy placed into extraordinary circumstances. Originally trained as an autopilot mechanic, he was assigned to be a member of a different B-17 crew in the fledgling 91st Bomb Group. While he was sent to school away from the base, his pilot, Capt. Richard Hill, took his B-17 and his crew up on a training flight. They crashed into a mountain and all aboard were killed. John Quinlan, the "Memphis Belle" tailgunner, was also supposed to be on that B-17 but was not. Fate played into the story of the "Belle" even before she was built. Who knows what might have happened to the "Belle" if Quinlan and Miller had been on that crashed B-17!

Miller was not very fond of his job repairing bomber autopilots in England, but since he had also received training as an aerial gunner, he decided that he wanted to fly on the B-17. He approached the 324th Squadron Commander Col. Wray, who was all too happy to put Miller on a crew. It just happened to be the "Belle," and again Miller was flying with John Quinlan.

During his time on the "Belle," Miller received one confirmed German fighter shot down. This was during one of their raids to a target in Belgium several weeks after joining the crew. He was very proficient at his position with his stuttering fifty caliber machine gun. This would not do him any good, however, when he found out that the rest of the crew would be taking the "Memphis Belle" back to the states for the War bond and morale tour. He did not have his quota of twenty-five missions in and would have to stay behind in England.

The "Belle" left England, and Miller finished his missions aboard different 91st Bomb Group bombers. He then became a gunnery instructor in Scotland before he, too, could return stateside. The blast from the fifty caliber guns was causing him some pain in his ears, so the Army cut him orders and he spent the remainder of his military life at bases in Florida.

He returned to his quiet life in Albright, West Virginia, and took a position with that state's Alcohol and Beverage Commission. Then he held a job with the West Virginia Department of Ag-

Left image courtesy of: The Miller Family. Right: Frank Donofrio

riculture from which he would retire. E. Scott Miller and his wife, Louise, raised two daughters, Mary Elizabeth and Martha Sue. During this time the crew would reunite in Memphis around the "Belle," but Miller was always missing.

No one could find him. He just quietly slipped away and did not keep in contact with any of the men he flew with. Through extensive records searches and a bit of luck from an old newspaper article, the Memphis Belle Memorial Association was able to locate Miller, who was busy just raising some cattle and looking after his wife.

He had not been located at the time the pavilion was dedicated for the "Memphis Belle" in 1987, and the rest of the crew mentioned how they would like to have seen him there. In 1989, Miller came to Memphis for a book signing to promote the book *Memphis Belle - Home at Last* written by Menno Duerksen.

This was the first time he had seen the famous B-17 that had carried him through hell and back in more than forty years. Emotions ran deep, and Miller posed for some pictures in his old position in the right waist window. It was also the last time he would ever see the "Belle.," as E. Scott Miller passed away at his home soon afterward.

Casimer A. "Tony" Nastal

Casimer A. "Tony" Nastal was born in Detroit, Michigan, on October 7, 1923. There were four children in all in the Nastal home: Walter; Ted; Tony; and Ann. All of them endured the kind of childhood that prepared them for tough times as adults. When their mother died, the youngsters were moved from foster home to foster home. Their father was a drinker and could not take proper care of the children. They were sometimes split up, and sometimes they were able to stay together. Mostly, the Nastals found caring adults in the homes they were sent to that needed young kids to work.

By the time he got into high school Tony's life had become somewhat normal, and he was even able to join the football team and was part of a couple conference championships. He was also living with a family that had a small coin-operated washing machine business, and he found that he could fix the things pretty well. It was about that time that America was thrown into WWII, and Tony decided that he was going to go. His brother Ted, who was a little older than him, volunteered for the Air Force. Knowing that she could not stop him, his older sister Ann signed the papers allowing the 17 year old Tony to enlist. He followed Ted into the Air Force and was even assigned to the same base as a Sergeant's jeep driver.

Fate stepped in one day during lunch in the chow hall. Some officers walked in and started picking out short fellas for tail gunner positions. Since Ted was not tall, he was in. His next stop was gunnery training in Nevada. Not wanting to be far away from his brother, Tony volunteered for gunnery instruction and was sent to Nevada as well.

Ted was very upset when Tony showed up and chewed him out. Ted was sent right into the 97th Bomb Group and shipped overseas. Tony would end up in the 91st Bomb Group and was shipped to McDill.

To Tony, the worst part of gunnery training was in the beginning when he was required to stand up in the rear of a T-6 trainer fitted with a small machine gun that he would fire at target sleeves towed from another plane. The only thing that held him in was a belt snugged tight around his waist. He eventually got used to it and became pretty good at putting holes in the sleeves. Tony was among the very first men of the 91st, but was not assigned to the crew of Bob Morgan. His regular crew was commanded by Lt. Paul fisher, and the plane they normally flew was the "Jersey Bounce."

In the days of early daylight bombing, Tony was subjected to the terrifying rigors of combat flying, and as was typical, flew aboard various B-17s when the "Bounce" was unfit to fly. His normal position on the crew was tail gunner, but he flew as ball turret gunner and waist gunner as well. He knew perhaps better than most why the B-17 was called a Flying Fortress.

While flying in the tail of "Sad Sack" on 23 January 1943, Tony got a chance to feel real fear. This particular raid saw fierce opposition from the Luftwaffe, and "Sad Sack" was under heavy fighter attacks. Tony saw the tail emplacement around him opening up from enemy bullets. Several hot rounds struck his ammunition feed boxes and they lit up. His own bullets began firing right out of the box. By some miracle he was never hit as the rounds flew in all directions. Burning, acrid smoke filled his compartment, and something began to trickle down his forehead. Knowing he was hit badly, Tony reached to wipe what he was sure was blood from his head and pulled back a black sooty hand. Unspent gunpowder from ruptured casings was floating everywhere in the tiny compartment. Somehow, it never exploded. Much of it was clinging to the sweat on his head and face and began to run down. Who ever said that a man could not sweat at forty below zero?

Since he was originally assigned to a crew other than the "Memphis Belle," Nastal had flown only a single mission on the famous B-17. He also flew only a single raid with Bob Morgan. Crews were constantly shifting from plane to plane, and as a result Nastal was assigned as tail gunner on one mission with Bob Morgan on 27 February 1943. Morgan was making up a raid he had missed and was ordered to fly the "Jersey Bounce" with most of its regular crew. Then, on 13 May 1943, the "Jersey Bounce" crew was ordered to fly. Since the "Bounce" was grounded, they were assigned to fly the "Belle" while her regular crew stood down for rest. So Tony flew once with Morgan and then once more on the Belle with a different pilot.

When the "Belle" was ordered to return to the United States, the regular right waist gunner, E. Scott Miller, did not have his twenty-five missions in and was unable to return with the regular crew. Tony had completed his mission quota aboard other B-17s and was sent home to the states with them. In the end it did not matter much at all. Tony Nastal had done his part. He shot down two enemy fighters and wasn't even 19 years old. Back in the states, he was told that he could have his choice of assignments. He chose gunnery instruction, but did not take to the strict officers there. It seemed that they did not want to listen to some kid who was much younger than they were—even if he had flown tweny-five grueling raids against some of the Luftwaffe's best fighter pilots.

Tony did what he felt was natural and volunteered to return to England as a gunner and give Uncle Sam his money's worth. This time he was assigned to the 209th SQ of the 447th BG on the B-17 "Dixie Marie." He flew twice on D-Day supporting the invasion. By the time Nastal finished combat, he had accrued more missions than any other regular member of the "Memphis Belle" crew with fifty-five completed raids.

That was it for him, and when the war was over he found a wife in Chicago where he was engaged in the appliance business. He and Doris raised two sons, Frank and Joe. Tony retired from a career as a grocery store manager at Jewel foods near Phoenix, Arizona—where it was much quieter than the freezing skies over Nazi Europe sixty years ago.

His Own Words

War Department Interview Waist Gunner - Tony Nastal 1943

"My first advice to gunners is to take good care of their guns. This is important. They shouldn't depend on anybody else to do it for them. Every gunner should see that his oil buffer is set right. He should check his electrical equipment before taking off, because it gets cold up there, and if his equipment isn't right he'll suffer.

The Germans will try to fool you. They'll come in and attack, and as they pass they'll let out a streak of smoke as if they're going down. Then they'll come back. Also, they try to imitate our escorts. They get where you can hardly see them and unless you watch closely you might think they are your own fighters. Then suddenly they'll break in as fast as hell and start shooting. If you have escort you're likely to be less alert than when you don't have them. But if you'll keep alert you'll be OK.

Always watch the other gunners if you can. If you are a waist gunner, watch the other waist gunner. If he needs help give it to him. He may have attacks coming in and be short of ammunition. If he is, give him some of yours.

Combat crews should never go into combat with the idea that they are not coming back. Those who have that in their minds are the least likely to get through.

It's always a great thrill to get a fighter in your sights and let him have it. I don't know how many I have hit, but I have two confirmed. I'll never forget the day when one came in shooting from five o'clock. I let him have it and saw my tracers go into his gas tank. He went down. I didn't see the pilot get out.

The Germans are a wild bunch sometimes. On our Bremen raid, the fighters came in bunches of 20 or 30. At the target, the flak started. It was bursting outside the waist windows. I could have reached out and grabbed it. I kept thinking, "Let's get the hell out of here." I saw two or three fighters hit by their own flak. It was so thick, you could hardly see the ground. The Focke-Wulfs were even bursting through our formation.

You can see the effects of our missions. In England, they used to call St. Nazaire "Flak City." The raids have softened it up.

At the waist gun position, you can see what's going on just about anywhere. It's cold, but when you're up in flak you warm up. You don't have time to think about cold. You don't have time to think about being scared either. You might be scared on the way over. You think of all the things that could happen. A lot of funny things run through your mind.

I want to go back as a pilot. I put in for fighter pilot, but if they give me a B-17 I'll take it. I guess it just gets in your blood."

John Patrick Quinlan

He was perhaps the most colorful man who ever flew aboard the "Memphis Belle." The other guys liked to call him J.P., and Bob Morgan liked to call him "Our Lucky Horseshoe." Like sitting on the end of a flagpole in a high breeze, Johnny sat in the extreme tail end of the "Memphis Belle" in a very important position. It is often thought that the only job of a gunner is to shoot at enemy fighter planes, but the gunners were also the eyes of the flight crew. Up in front, they had a restricted view of the dynamic changes in the formation of planes all around them. Gunners kept the flight crew up front aware of how the other planes were faring, and also when enemy fighters might sneak up on the tail of a bomber. Eyes constantly straining against the sun, they could not let their minds stray from the near constant threat that an enemy fighter might blast through the bomber stream. A good tail gunner would also tell the pilots when the flak gunners on the ground had their range, speed, and bearing. Bursts could be counted, and when a straight pattern began to zero in on the planes, the tail gunner often had the best view of it. A warning over the intercom and a slight change in course would force the Germans to spend precious time recalculating their guns. Many of the men said that Johnny Quinlan was the best tail gunner the "Memphis Belle" ever had, and he lived up to it.

He was born June 13, 1919, into a strict Irish-Catholic family in Yonkers, New York. His father worked there for the sanitation department, but died when Johnny was very young. Things were rough, and the Quinlans sometimes had to get by on the chickens that their mother raised. He was good with a gun even before he was a teenager. He would hunt squirrels with a BB gun before his mother decided that if he was going to hunt them, he had to do it properly with a .22 rifle, which she bought for him.

It was after he finished high school when the Japanese attacked Pearl Harbor, and John Quinlan volunteered the next day. Johnny was the only son to Mrs. Quinlan, and she did not want him to go. Despite her pleading, he went anyway. "I wanted to join any damned thing" he said years later. "I didn't really choose the Air Force, they chose it for me."

Basic training at Jefferson Barracks was horrible. There was cold rain and mud, and everyone was sick with the flu, meningitis, or something else. It was not long after that when he was relocated to a new training base in Florida where even harsh basic training seemed like paradise in the sun compared to what he had seen in St. Louis, Missouri. Gunnery training was next, and Quinlan was a natural at filling sleeve targets with .50 caliber holes. He mastered the large machine gun quickly and impressed his instructors.

Like some of the others, he was not originally assigned to fly as a regular member of the crew of the "Memphis Belle." After the 91st Bomb Group moved across the country to Walla Walla, Washington, he was informed that he would fly as tailgunner on the B-17 commanded by Lt. Richard Hill. Johnny liked him. Hill was running a good ship, and the tail gunner was enjoying his compatriots.

A night navigation training flight was scheduled, and Hill told Quinlan that he did not need to go along if he didn't want to. He asked for a pass, and Hill wrote one out immediately. Quinlan thanked him and headed to town. That was the last time the two men spoke. Hill flew his B-17 into a mountain and all aboard the plane were killed. Another "Memphis Belle" gunner was scheduled to fly that night on Hill's plane—E. Scott Miller. Again fate looked at the men of the "Belle."

Quinlan was upset about the loss of his team. Then he was told that he would join the crew commanded by Lt. Bob Morgan. "I knew about him. I remembered that he had this red convertible that he would tear through town in. He had a reputation for getting into trouble, and I thought he would get us killed, so I didn't really want to fly on his plane," he said. The Army wrote an order, and it was final; Quinlan was going to fly on B-17 #41-24485 "Memphis Belle." Quinlan would say years later in an interview, "Say what you want about Bob Morgan, but he always got us back."

It takes several chats with Johnny Quinlan to get him to open up about his experiences. When he does, a harsh air of reality sets in and the listener is given an in-depth feeling, almost as if he is there in the tail with Johnny. He delivers an almost monotone description of some of the worst kind of war man could ever dream up. As if it was nothing, Quinlan talks about the missions in a steady macabre sort of way, sometimes motioning with his hands to help you understand what he is saying:

Long after the War, after I was married and had children, I had nightmares, bad dreams. Something would bring it back. I would be up in an airplane. Something would be burning, something you remember from a plane on fire. You wake up at night. You feel cold. Or you hear a noise, maybe a plane flying over your house. I would remember going on a mission, a hell of a lot of ammunition shot. We'd shoot 1,400 rounds easy.

When we got back, we were losing altitude. We were coming in, and actually it was the coast of England, but I thought we were landing in enemy territory. I thought, *Gee, we're going to surrender the plane. The pilot said the landing gear was down.* I said, 'Hell, we're going to surrender. I want to bail out and get captured because I don't want to surrender the plane.'

Then I recognized the landing strip on the ground and we were landing, but I did not remember a damned thing about the mission.

It was one hell of a dogfight. You were half frozen because it was twenty or thirty degrees below zero. Your hands were numb. You were shaking scared. We had electrically heated gloves, but half the time they didn't work. You would be kneeling down in the firing position, and the electric suit would burn the back of your legs. You couldn't trust the old suits. You had to take off your gloves to get your oxygen mask off, as it was freezing up all the time.

Sure we were scared. But most of the guys don't want to be called chicken, so you put on a pretty good face so nobody will know you're scared. Then, when a mission is over, you go to town and drink a few beers and you become one of the knights of the air, telling everyone who you are.

Then it came down to those mornings when you were out there with your flight gear on, walking to the plane, and you knew it was going to be a bad one. You had seen that in the briefing session. We knew about the losses we had been taking. Some said it was a hundred percent, counting replacements. An awful lot of losses. You knew it would eventually happen to you because a lot of your friends had got it already. We were like brothers, all the way from McDill field. You knew all their stories and all about their girlfriends and their mothers and sisters. They were like brothers. And then they would disappear. You move their beds out and then replacements come in, and you're afraid to be friendly with them because you don't want to get that feeling again of losing them.

If you were one of the old timers that had lasted a few missions, you had a little moss on your shell, you had ribbons on your chest. The new ones, the replacements, would figure they would try to get chummy with you. They think you are a man who made it. Maybe you can tell them something that will help them make it. They need somebody to cling to. But you never let them get close. The longer you flew, the more distant you became. You'd stick around with the guys who had survived.

When you knew you were going on a mission you got a little scared. You told yourself you might be able to figure out a way to keep from going. Going over to the plane I would try to figure a good way out. Maybe I could manage it so that one of the trucks that hauled the bombs out to the plane would hit me in the back, just bad enough to put me in the hospital, just bad enough so I would not have to go that day.

Even after you get on the plane and it starts up you say to yourself 'Maybe not all the engines would start and we won't get off the ground. Maybe the plane will slip off the runway and get stuck in the mud.'

Then you hear the priest up in the tower, blessing you. Even after the plane begins climbing you think *Maybe something will happen so we can abort.* But we'd soon gain altitude and be up there, and you'd tell yourself: 'Well, here we go.'

Then after you got up there you would stop thinking about those things. You lose your fear. Then you want to make sure nothing happens to the plane because of you. You don't want to do something dumb so the plane gets shot down or someone killed. You feel responsible. You want to make sure you do what you are supposed to do and do it right. That was the one thing that kept you going.

One of the worst things was to see one of your other planes get shot up, an engine out, trailing smoke, dropping back and out of the formation. We always wanted to drop back with them and help them, but our orders were to stay with the formation. If you dropped back you would lose more planes. You would get your ass shot off.

Back in the tail, I was the one who got to see most of that. I watched our friends fall back, hit, and the German fighters waiting to pounce on them like a pack of wolves. They always ganged up on a crippled plane. Those guys in the bombers were our friends, and it made you feel bad because you could not help them.

After that, when they come at you, you want to shoot them. You want to kill them because they killed your friends. You got frustrated because you shot at them and couldn't stop them from coming in. I was shooting and doing everything right. I was leading them right and firing the guns just right, but they just kept coming... kept coming.

John was not accustomed to all the attention thrown at him and his compatriots during the War bond tour. The newspaper and radio reporters would call them heroes, and to this day Quinlan will tell you that they were just in the right place at the right time. Crews who were just as good with just as much experience were lost in sudden violent flak explosions. Bursts that went off in blue sky that the "Belle" had just flown through. He never liked being referred to as a hero, preferring to be called a patriotic American who simply wanted to do all he could to stop the enemy.

That's just what he did. When told that he could have his choice of assignments, he elected to go to the Pacific. He trained aboard the much bigger and brand new B-29 Superfortress and tried unsuccessfully to be assigned to the crew of Robert Morgan. Instead he was sent to the primitive conditions of the China, Burma, India theatre of operations, hitting Japanese targets from American bases in China.

He proved himself to still be a very competent tail gunner and claimed three official Zeros killed at his guns. His official credit from the "Memphis Belle" was two German fighters—a number that he disagreed with for years. "It was more like five German planes," he says. And even though numbers are really not as important to Quinlan as his devotion to duty, his records officially list a total of five enemy fighters downed by his guns.

In the Pacific, he flew aboard the B-29 named "The Marietta Misfit." It was during their fifth mission that he would finally face the worst he could imagine—jumping from a burning bomber. The number three engine was hit and burning, and as the flames ate their way into the wing, the intercom came alive with orders to jump. For some reason the reality did not set in right away and Quinlan made his way forward into the waist of the Superfortress and found that he was alone on the B-29. He strapped on his chute and leaped from the plane somewhere over Manchuria. But his troubles were just beginning.

As he fell away from his plane, he pulled on his rip cord and the cable release came out but the chute did not pop. The pack did not open, and he was now falling from thousands of feet up. The others from the bomber, already in full chutes, witnessed this and saw him fall with no parachute into a cloud deck below them. They all landed in friendly territory, while he landed in Japanese held Manchuria. John Quinlan was reported as Missing in Action, and a telegram was sent to his distraught mother back in New York.

What no one realized was that as he fell, he began to tear at the closures on his chest parachute pack and ripped them open so he could pull the silk out by hand. Unbelievably, his parachute bloomed into a full canopy. He was captured and taken prisoner. At the first opportunity, he effected an escape and fell into the hands of sympathetic Chinese guerillas. One of their officers who spoke some English came up to him and tried to determine who he was. Quinlan was happy that the Officer seemed to take particular interest in the fact that he had flown, only months before, as the "Memphis Belle" tailgunner. After a few moments of discussion, the Chinese Officer arranged for a new uniform and a rifle for Johnny, telling him that he knew of the "Memphis Belle" and also that he had gone to school in the United States at Ole' Miss!

Before Quinlan could get back to an American airfield he would have to stick with the band of guerillas. Several hectic skirmishes were fought, and right alongside the Chinese was Quinlan, firing a rifle at Japanese ground soldiers. Weeks went by walking, fighting, and hiding from patrols. Among his memories of his time there was how young some of the Chinese slodiers were and how quickly and efficiently they were killing. Terribly dirty and hungry, they sometimes ate dogs to survive.

Eventually, the chance he was waiting for came and he was put aboard an American B-25 Mitchell and flown back to safety.

His Own Words

War Department interview Tailgunner John Quinlan - 1943

"I like being a tailgunner. It's my own private little office back there. I sit down all the time, and when I get a chance, I relax. I get a lot of good shots, too.

The tailgunner is in a good spot to help the pilot by telling him over the interphone what's coming up from behind. But he should be careful to call out only the ones that are attacking. If he calls out everything he sees—one at five o'clock, one at seven o'clock, one at six o'clock—he'll get the pilot so confused he won't know what's going on. He should call out only the ones that are after him.

Don't be afraid to use your ammunition, but don't waste it. That's the best advice I can give.

You've got to be alert all the time. You never can tell what will happen. The time they shot my guns out and hit my leg, I hadn't expected any trouble at all. I thought that mission would be a cinch. It was a short raid and we were going to dip in and pop out again. Just after bombs away, I thought I saw flak. It wasn't, it was fighters.

A fighter will climb until he thinks he can give it to you, then he'll dive on you. That's what this one did. I looked up just in time to see his belly. It always gives you a funny sensation to see those big black crosses on the wings. I could hardly miss him. I got him. He burst into flames. I guess I was gloating over the one I got. Then I saw the other one. It looked like he had four blow torches in his wings. All of a sudden, it sounded like someone hit the tail with a sledge hammer. It got my guns and me.

But the one I got the biggest bang out of was the Lorient raid. Captain Morgan went up, then down. I lost equilibrium. I didn't know whether to jump or stay there. I didn't know what was going on. That was the time the horizontal stabilizer got on fire. I guess it was the wind that put it out.

Johnny Quinlan will play down his part in the Second World War. He won't be comfortable if you call him a hero. Perhaps the most heroic thing he ever did was raise six children and a nephew with his wife Julie.

Always good with his hands, he made a career in construction driving trucks and operating backhoes. When asked, he might get up from his chair and limp slightly to a wall and point to where his medals and ribbons hang unceremoniously in his house. There, the sounds of Wright Cyclones fill his head and the sight of burning airplanes comes to mind, and he will tell you that he was just an average guy who wanted to do his part. That he was no more special than all the others whose luck ran out.

During a phone call to Quinlan at his New York home in November 2000, Johnny relayed to this writer one of the more shining moments of his military career. It was during the British Royal review of the "Memphis Belle" and crew just before they left England. The King and Queen had approached him while they walked down the crew lined up in front of the Belle. Her Majesty asked Johnny (the poor boy from Yonkers) just what he thought of all the attention. He could not believe what his response was, and he said it really before he thought about it. "My Irish ancestors are rolling in their graves, Maam!" Less than one month later, Quinlan was gone. He died very quickly at a Veteran's Hospital in New York State. Somehow, the Belle lost a little color that day.

Tailgunner Johnny Quinlan perched in the tail position of the "Memphis Belle." From this position Quinlan (like all tailgunners) was primarily responsible for defending the rear of the bomber, but also for keeping the pilot informed on the status of the formation behind them and the tracking of the enemie's flak bursts. (Frank Donofrio)

Close up shot of some damage received in the tail area of the "Memphis Belle." While at first it does not look very extensive, everyone realized that the relatively thin skin of the B-17 did very little to slow down or stop a bullet or shrapnel from flak. The lights located below the gun mounts were used to communicate with other bombers in the formation during the missions. In this way, radio silence could be maintained and prevent the enemy from listening in on the bombers' conversations. (Frank Donofrio)

(Frank Donofrio)

Joseph Giambrone

Boeing technical orders stated that an engine change on a B-17 would require twenty-five hours. Joe Giambrone and his ten man team used to do the job in only four. As a matter of fact, they changed nine engines on the plane he was assigned to care for. He was the Crew Chief for the bomber that would become known the world over as the "Memphis Belle." It was really his plane. The aircrew just took it out from time to time. Giambrone was there every time they took off. Straining his ears to hear if the engines were tuned just right. Listening to the brakes squeal, looking for oil leaks and anything else that might not be just right. Known as a guy who just would not put the tools down, he knew that it was the job of his team to keep this bird flying perfectly. If something went wrong during a mission because he did not use the right part or fasten a piece properly, it was his fault, and he took special pride in the meticulous maintenance of his B-17.

Like most other people, Joe Giambrone and his family suffered the effects of the depression. His family ran a bar, and Joe worked there doing everything, including repairing slot machines. They were illegal, but no one really cared. The important thing was that it gave him some sense of pride that he was able to figure the complicated mechanisms and make them work well. Of course he could not know that this would lead him into becoming one of the best Ground Crew Chiefs in the Eighth Air Force.

Having enlisted long before the war began, Giambrone enlisted in the National Guard in an infantry unit. He quickly realized that it really wasn't for him and took advantage of volunteering in the rapidly expanding Air Corps. Just several weeks later, he was at Chanute field south of Chicago working his way through aircraft mechanics school.

Considering the training to be natural for him, he quickly ended up in the brand-new 91st Bombardment Group (Heavy) that was being formed near Tampa, Florida. He was among the original members of the soon to be famous Group.

The Group was relocated to Walla Walla, Washington, where the aircrews were in the final flight phase of their training. None of the new "F" models had arrived yet, and the crews were wearing out their older "E" model bombers. The first "F" model B-17 was scheduled to come to them from Dayton, OH, and Col. Wray flew over on another bomber to pick it up. Coincidentally, Charles Leighton was the navigator and Joe Giambrone was asked by Col. Wray to go along as Crew Chief in the event a problem or two arose. Col. Wray was right to bring Giambrone along, because they were about to become involved in a terrific event that they would remember for the rest of their lives.

The trip to Wright field was uneventful. The plane was readied for the long range flight to Washington state, and no fuel stops were planned. This was possible because two additional fuel tanks were fitted inside the bomb bay. The gleaming new B-17 fired up and roared down the runway. Col. Wray hit the switch to raise the gear and nothing happened. They were not going to be able to fly all the way across the country with the wheels down, and immediately Giambrone headed into the bomb bay to manually crank the wheels up. The long range fuel tanks fitted so tightly that there was no way he could fit the crank into their sockets. The officials at Dayton gave them permission to salvo the full tanks in an uninhabited area, and when they got there, only one of the tanks dropped. Despite their best efforts, including erratic flying, hard turns, proposing the bomber, and pulling positive Gs, nothing seemed to be working. Col. Wray ordered the lowest ranking man aboard into the bomb bay to attempt to kick the stubborn tank from its mounts! A rope was tied around Giambrone's waist, and he held onto some of the trusswork inside the bay and kicked and kicked for all he could. Circle after circle over the empty field, and finally, not realizing exactly where they were, the full tank fell right onto a farmer's chicken coop, completely destroying it. The "bombing" even made the local newspaper headlines the next day.

Like all other ground crews, when the 91st was flying missions every day, the dedicated crews worked throughout the night prepping the bombers for their raids. In these early days of the air war, bullet holes were patched with glued fabric over the aluminum and sometimes caulking. Later pop rivets were developed. Then repairs could be made with aircraft metal, and many of those metal skin repairs remain on the "Memphis Belle" today. Things like replacement oil coolers were nonexistent, so if one was damaged in battle, the plane was grounded. Scavenging derelict B-17s for parts became an art form. This is how the entire right wing of the "Memphis Belle" was replaced. The inflatable de-icer boots were constant problems. One in particular on the Belle was causing so much trouble that Morgan complained to Giambrone about it. Morgan rarely used them anyway, so he didn't know just how Giambrone fixed them. (The practice of removing them entirely became common, as they could create quite a problem if they came off the bomber in combat. This created a lot of drag and sometimes would damage other bombers if they were ripped off in flight.) In this instance, during a crew reunion some forty years after the war, Morgan brought up the story again and said that he never figured out how Giambrone repaired the troubled boot. "I fixed it good chief" Joe said, " I just went down it with an ice pick!"

The grounded crews "sweated out" the missions every time one was flown. There wasn't much to do while the planes were off the base. While their planes were gone, the men back at the base would spend the anxious hours playing ball, gambling, or whatever they could do to distract them from thinking about what their compatriots were going through. When the return to base time approached, the games would stop and the bicycles would appear with their riders as the throngs would sort of wander out towards the runways. Up in the tower, the Commanders would begin to scan the horizon with binoculars and listen for radio traffic. Then someone would spot the planes and the counting would begin. Hopefully it would add up to the same number that left earlier in the day. It was almost certain that some would spout red flares asking for priority to land. These bombers had wounded on board, and the ambulance drivers prepared to chase the plane down the taxiway once they landed. The fire crews, dressed in their asbestos suits, stood high on their vehicles trying to see if any of them had battle damage and may not be able to land safely. Some had parts of the plane gone. Some had parts of the plane on fire. Many times B-17s with no landing gear had to commit to landing alongside the runway on their bellies, sliding to a stop in the grass.

Joe and his crew did a remarkable job keeping the "Memphis Belle" flying. For an airplane that only accumulated 148 hours and 50 minutes in combat, they replaced 9 engines, 5 superchargers, both wings, both main landing gear, and two tail assemblies. Hundreds of bullet and flak holes were patched, and untold hours were spent inside the Belle testing, repairing, and inspecting all of the equipment. He cared for other B-17s after the Belle returned to the states. One in particular was also well known—"Pistol Packin Mama." (Frank Donofrio)

Giambrone spent so much time repairing the "Memphis Belle" that he began to wonder what combat was really like. He began to ask Morgan if he could go along on one mission, and finally got the pilot to agree to it. Even though it was strictly against regulations to have him along, Morgan (who was known to bend a rule from time to time) had his crew chief standing behind the co-pilot for the entire raid. It was supposed to be a milk run mission, so Morgan did not really have a big problem taking an extra passenger along. (Some records indicate that this was the mission that was flown to Meaulte, France. It would appear unlikely, as the "Memphis Belle" was flown that day by a replacement crew under the command of C.L. Anderson. Bob Morgan and other members of the "Memphis Belle" crew were in London taking part in a recorded radio broadcast back to the United States.) Nonetheless, Joe Giambrone did fly a mission, and while the flight into the target was relatively calm, the flight out was filled with enemy fighters and a terrific air battle. This had Giambrone wishing to a degree that he had stayed back at the base. Perhaps the Commanders should have allowed the ground crews to fly from time to time. Joe came back after the raid with a very clear understanding of the difficulties faced by the brave men that flew these missions day after day.

As a Technical Sergeant, Giambrone was among the finest Crew Chiefs. He impressed the Commanders so much that soon he found he was receiving a promotion to Master Sergeant and would have a larger responsibility as Squadron Inspector—guiding all of the ground crews at Bassingbourn. When his "Memphis Belle" was ordered back to America for their war bond and good will tour of the country, Morgan wanted his crew chief along. After all, Joe Giambrone was one of the main reasons behind the success of the plane. He would have to stay. Giambrone remained in England until after the Nazis surrendered.

When he got back to the states, he again started on the bottom rung of a corporate ladder for a construction company. He started out with shovels and ended up with a pen when he eventually became their office manager. Through that time there was a marriage to Mary Jane and a son—which they, of course, named Joseph. Joe Giambrone, the best crew chief the "Memphis Belle" ever had, died in 1992.

The bomb bay of the Belle as it appears today.

9

Miss Margaret

"I was in love with love," she used to say, as Margaret Polk would sit in her living room in her modest East Memphis home just off of Poplar Avenue. Reflecting was easy for her, and she retained her dignified southern grace even while she remembered some of the sadder moments of her life. She was born December 15, 1922, into a hard working but rather well off family. As a matter of fact, she was a direct descendant of the 11th President of the United States—James K. Polk! Margaret's parents would see to it that she would have the best education available. Her father, Oscar Polk, was an entrepeneur and involved in various businesses which ranged from lumbering to cotton. His hard work paid off, and his family did not want for much. She would attend the elite Miss Hutchison's school for girls, and in the summer a lot of time would be spent at the family's vacation cottage in Hickory Valley, Tennessee. Picking strawberries and skinny dipping in the cow's pond kept her busy, and her joyful spirit captivated all who were fortunate enough to know her.

By the time she entered college, she had blossomed into a beautiful nineteen year-old who could pick just about anyone she wanted to date. Her older sister, Elizabeth, was married to a doctor who was in the Air Force and assigned to the 91st Bomb Group at McDill, Florida. When the 91st was sent to Walla Walla, Washington, naturally the doctor was ordered along. While he took the train, Elizabeth would have to drive their car across the country. A stop in Memphis was planned, and after she arrived, she complained to Margaret and their mother about the long lonely drive. Mrs. Polk suggested that Margaret go along with her sister to Washington and make a summer adventure of it. She had no way of knowing how this trip would change the rest of her life.

Naturally, a doctor in a bomb group gets to know the pilots. Robert Morgan, one of those pilots, had offered a time or two to give the good doctor a lift to his home off the base. Morgan always seemed to have a car when no one else did. Neither Bob Morgan nor Margaret remember their first meeting, but feel strongly that it

occurred when Edward "Mack" McCarthy (the doctor) invited him inside after one of those lifts home.

Even though it was not the typical love at first sight sort of meeting, Margaret was evidently interested in Bob. What happened was that Margaret became involved in an argument with her sister and brother-in-law after Bob invited her to his birthday party. She already had a date lined up for that day, and it would not be proper to cancel at the last moment. The hotter the argument became, the more she wanted to go to Bob's birthday party. It more or less threw her into his arms, and they began dating fairly often.

Bob was determined to impress Margaret in the best possible way he could. Not many girls could awaken to the roar of 4,800 Wright Cyclone horses blowing the shingles off her roof! Bob took to the notion to really show off whenever he could by taking off early for practice missions and buzzing the heck out of his girl. Most young girls were happy to be able to date a guy that had a car, but Margaret's guy had a B-17! She remembered years later "Here Bob would come, around four or five in the morning. He was so low and the plane was so loud that you would have thought that he was going to fly right through the window! The whole house shook, and it was so exciting."

The summer was coming to a close, and Bob was being transferred from Walla Walla with the rest of the 91st Bomb Group. Margaret had to go back to college anyway, so she headed back to Memphis alone. The dating had been exciting, but the couple had not made any real commitment and Margaret even renewed a romance with a young Park Ranger whose photo she kept for the rest of her life. But she did love Bob, and decided to become more serious with him. The pilot had her heart.

The stream of letters from Bob began in earnest. He called her "My Dearest Polky" and "My Little One" while he pledged to come to her after the War if they could not be together then. He kept telling her that he was flying just for her. At a time when a long distance phone call was a luxury, Bob made it a point to call her often.

When the 91st Bomb Group was declared ready for combat, the flight crews were ordered to Gowen Field near Boise, Idaho, to turn in their old B-17Es and pick up their shiny new "F" model bombers before going overseas. Only a few of the new planes were there, and the Group Commander Col. Wray grabbed one of the first ones. Morgan managed to hitch a ride across country with Wray after learning that a stop was planned in Jackson, Mississippi—which was close to Memphis and Margaret! A telegram was on its way to her. Then she was on her way to Jackson. They met up at the Hotel Heidelberg there and the romance went into full bloom. The do not disturb sign was hung on the door of room 931. Margaret and Bob only had a few hours together, and they would make the most of it.

Bob went to Bangor and Margaret went home. He was assigned his B-17, which would soon be called "Memphis Belle," and part of their check out was to fly the bomber for 100 hours to test the systems and fuel consumption before going overseas. As a result, the crews were given their choice of destinations during their cross-country flights, and there was only one place Bob wanted to go—Memphis, Tennessee. Another telegram was sent to Margaret, "Land airport (Memphis) about ten your time Saturday."

On 12 September 1942, B-17F #41-24485 appeared over the Memphis horizon and headed directly for the airport. There was no name on the nose yet; the Belle was still that new. But Bob was going to show his girl the big plane that the Air Force trusted to him. He was proud of it and of her. Bob obtained special permission for Margaret to enter the military section of the airport, and unknown to anyone there, a little history had been made when this B-17 made its first landing in Memphis, Tennessee.

That night, Bob took Margaret to a dance at the Peabody Hotel where they later shared room number 782. He gave her a love knot ring, which was only the precursor to the package she would receive a week later. A diamond engagement ring—Bob was asking her to marry him!

The letters, phone calls, and telegrams flew with urgent fury, and just fifteen days later, Bob was gone. The training complete, the 91st Bomb Group took Bob overseas to England and the War. It would be nine months of hell before the two could once again be in each other's arms.

When Morgan flew the Belle to Memphis in June of '43 there were plenty of mixers for him to attend with Margaret. They are seen here with Vince Evans and his escort.

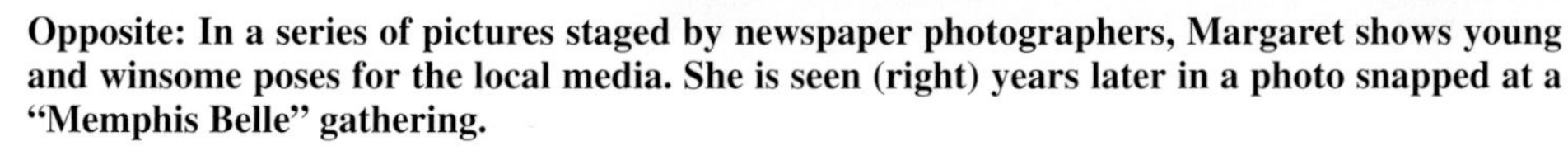

Opposite: In a series of pictures staged by newspaper photographers, Margaret shows young and winsome poses for the local media. She is seen (right) years later in a photo snapped at a "Memphis Belle" gathering.

10

The Missions

In this chapter you will fly along on every mission of the "Memphis Belle." You will get to know some of the crewmen and what it took for them to survive. You will read excerpts from their diaries and, in some cases, the actual records of the 91st Bomb Group (H).

To the right you can see one hundred ninety-seven B-17s in a symbolic formation, which represents the losses suffered by the "Boys of Bassingbourn." This is a very graphic way to feel a little of what it must have been like to live on an Eighth Air Force base at that time. While the 91st Bomb Group stayed in England for nearly 36 months, they had four hundred twenty-eight B-17s assigned to them. Nearly half of those were lost in combat and training.

The men grew used to coming back from a raid and seeing their roommates' personal effects gone. His mattress would be rolled up on his cot, and someone was getting ready to ship his things home. He was not coming back.

You will begin to understand why some of these men became very superstitous. How could you possibly explain the irrationality of war? Who will be killed? Who will be disfigured permanently? And which of your friends will finish the war in a prison camp?

You will know why, for instance, a man will get up on a mission morning, roll out of his cot on one particular side and put on

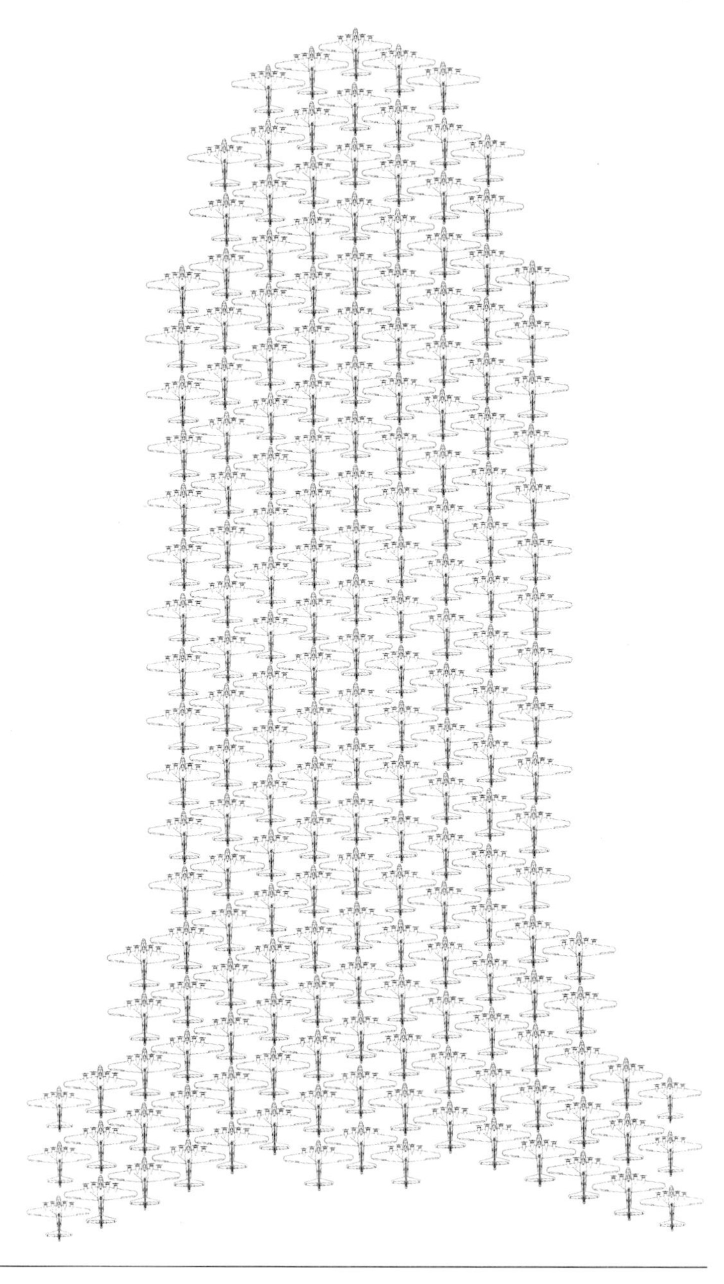

(Frank Donofrio)

his clothes in a certain order. Maybe he will eat some fruit cocktail and have some coffee before his briefing and then fly and survive this mission. Well from now on, he will get out of bed on the same side, put on the same flying suit until it can stand by itself, and eat the same food before he flies.

The map and table on this page work together to show several things. The base from which the "Memphis Belle" flew her missions is shown as a star just to the north of London.

The targets which were hit are indicated by letter and correspond to the graph below in the map column.

The last two columns in the graph indicate how many times the "Memphis Belle" or her crew visited these areas, and then on which raids.

The French cities of St. Nazaire and Lorient saw more of the "Belle" than any other during her missions. Those cities were bombed five times each by raids in which the Belle took part.

Of course, it is known that the Belle made 25 raids. The reason for the listing of 30 missions is that there were 5 times that the Belle was flown by different crews, and five missions that Morgan and his team flew different B-17s.

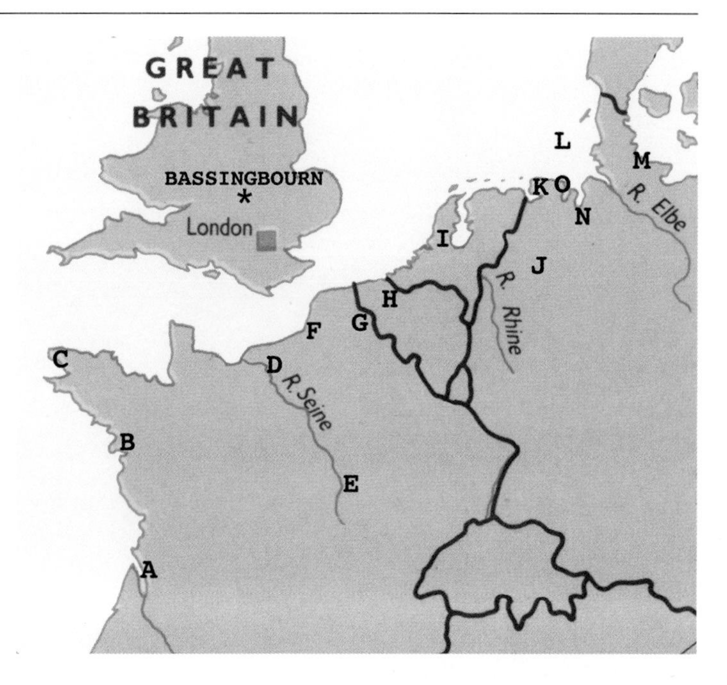

Raid #	Date	City	Target	Map	B-17 flown	# visits	Raid #
1.	7 NOV'42	Brest	Sub pens	"C"	Mem Belle	2 raids -	No's 1,14
2.	9 Nov'42	St. Nazaire	Sub pens	"A"	Mem Belle	5 raids -	No's 2,3,7,12,24
3.	17 Nov'42	St. Nazaire	Sub pens	"A"	Mem Belle	5 raids -	No's 2,3,7,12,24
4.	6 Dec'42	Lille	Train works	"G"	Mem Belle	2 raids -	No's 4,8
5.	20 Dec'42	RomSurSeine	Air Field	"E"	Mem Belle	1 raid -	No 5
6.	30 Dec'42	Lorient	Sub pens	"B"	Mem Belle	5 raids -	No's 6,9,15,22,29
7.	3 Jan'43	St. Nazaire	Sub pens	"A"	Mem Belle	5 raids -	No's 2,3,7,12,24
8.	13 Jan'43	Lille	Train Works	"G"	Mem Belle	2 raids -	No's 4,8
9.	23 Jan'43	Lorient	Sub pens	"B"	Mem Belle	5 raids -	No's 6,9,15,22,29
10.	4 Feb'43	Emden	Sub pens	"K"	*Jersey Bounce*	1 raid -	No 10
11.	14 Feb'43	Hamm	Rail Center	"J"	Mem Belle	1 raid -	No 11
12.	16 Feb'43	St. Nazaire	Sub pens	"A"	Mem Belle	5 raids -	No's 2,3,7,12,24
13.	26 Feb'43	Wilhelmshaven	Naval Base	"O"	*Jersey Bounce*	2 raids -	No's 13,18
14.	27 Feb'43	Brest	Sub pens	"C"	*Jersey Bounce*	2 raids -	No's 1,14
15.	6 Mar'43	Lorient	Sub pens	"B"	Mem Belle	5 raids -	No's 6,9,15,22,29
16.	12 Mar'43	Rouen	Rail Yards	"F"	Mem Belle	2 raids -	No's 16,19
17.	13 Mar'43	Abbeville	Air Field	"D"	Mem Belle	1 raid -	No 17
18.	22 Mar'43	Wilhelmshaven	Naval Base	"O"	Mem Belle	2 raids -	No's 13,18
19.	28 Mar'43	Rouen	Rail Center	"F"	Mem Belle	2 raids -	No's 16,19
20.	31 Mar'43	Rotterdam	Ship Yards	"I"	Mem Belle	1 raid -	No 20
21.	5 Apr'43	Antwerp	Plane Engine wrk	"H"	*Bad Penny*	2 raids -	No's 21,25
22.	16 Apr'43	Lorient	Sub pens	"B"	Mem Belle	5 raids -	No's 6,9,15,22,29
23.	17 Apr'43	Bremen	Plane Factory	"N"	Mem Belle	1 raid -	No 23
24.	1 May'43	St. Nazaire	Sub pens	"A"	Mem Belle	5 raids -	No's 2,3,7,12,24
25.	4 May'43	Antwerp	Plane Engine wrk	"H"	*Gr.Spkld.Bird*	2 raids -	No's 21,25
26.	13 May'43	Meaulte	Plane Repair	"P"	Mem Belle	1 raid -	No 26
27.	14 May'43	Kiel	Shipyards	"M"	Mem Belle	2 raids -	No's 27,30
28.	15 May'43	Heligoland	Naval Yards	"L"	Mem Belle	1 raid -	No 28
29.	17 May'43	Lorient	Sub pens	"B"	Mem Belle	5 raids -	No's 6,9,15,22,29
30.	19 May'43	Kiel	Ship Yards	"M"	Mem Belle	2 raids -	No's 27,30

Bombs away! (Frank Donofrio)

Statistics of the 91st BG(H)

The men and machines of the 91st flew 360 combat missions between 7 November 1942 and 25 April 1945. While doing this, they compiled one of the finest records of any Bomb Group in the Eighth Air Force. It was not without penalty.

197 B-17s were lost and more than 1,150 men became prisoners of war, while some 600 more were killed in action. In all, some 428 B-17 Flying Fortresses were assigned to fly in the 91st Bombardment Group (H) divided between her four squadrons.

The sortie tally was nothing short of incredible. 9,591 individual sorties were flown by these amazing men. While American losses were hard to endure, Axis losses were harder for the enemy to swallow. The gunners of the 91st BG shot down 420 enemy fighter planes—almost one enemy fighter for every B-17 assigned to the "Boys of Bassingbourn."

To the right are the three other B-17s that the crew of "Memphis Belle" flew while their B-17 was grounded. The "Great Speckled Bird," which was lost to enemy action on 17 August '43 (Schweinfurt), 10 POWs. "The Bad Penny" was utilized as a VIP transport, then salvaged on 19 October '44. And the plane that is believed to be the "Jersey Bounce" of the 91st BG, lost to enemy action on 21 May '43 on a raid to Wilhemshaven. The "Bounce" crashed in the North Sea, killing every member of her crew. (Frank Donofrio)

Below: A very early picture of some of the 401st's B-17s preparing to depart for a mission. The first bomber is called "Hellsapoppin." She would be lost on a future mission over Bremen on April 17, 1943. Five of her crew would be killed, while the other five would become prisoners of war. The Memphis Belle was on that mission in a different squadron. The next B-17 is called "Bomb Boogie," and would be lost on a mission on September 9, 1943, to Stuttgart. Four of her crew evaded, the other six became prisoners. In these early days of the war the losses were very high. Some historians say as much as a staggering eighty-two percent! (Joe Harlick)

"Down the Glory Road"

7 November 1942
Target - Submarine Pens at Brest, France
Alert no.2 - Mission no.1 for the 91st BG(H)
10:30 AM. The B-17s of the 91st position themselves for the launch of their very first operational mission. In the cold morning rain, the lead B-17 #41-24503 "Pandora's Box" is piloted by none other than the Group Commanding Officer, Col. Stanley Wray. The B-17 immediately behind him is none other than the "Memphis Belle." This would be the day that the very first bomb symbol would be painted by "Memphis Belle" crew chief Joe Giambrone.

B-17F #41-24485	"Memphis Belle"
Pilot	Robert K. Morgan
Co-Pilot	James A. Verinis
Bombardier	Vince Evans
Navigator	Charles Leighton
TT / Engineer	Leviticus Dillon
Radio Operator	Robert Hanson
Waist Gunner	Clarence Winchell
Waist Gunner	Harold Loch
Ball Turret	Cecil Scott
Tail Gunner	John P. Quinlan

Much of the information included here was taken directly from the extensive writings of Col. Bert Humphries, 322 Sqd B-17F 41-24482 "Heavyweight Annihilators." His material provided a great deal of information never known before. Most of his writings were made the day of the missions.

On board the "Memphis Belle," her pilot Lt. Robert K. Morgan made final adjustments, and his co-pilot Lt. James A. Verinis checked the gauges on the flight instrument panel. Just at their feet and slightly in front of them navigator Chuck Leighton and bombardier Vince Evans were busy checking and re-checking all of their equipment. Nothing was going to go wrong today if they could help it. Levi Dillon, the flight engineer, was standing in his takeoff position between the two pilots. His job was to assist the co-pilot and observe the critical engine readings to insure that no harm would come to them during this short pause before the takeoff. Dillon would also keep a sharp eye on these instruments throughout the climb and formation assembly.

In the Belle's radio room, Robert Hanson was on the radio listening to the tower's commands, while the gunners crowded around him for the takeoff. Gunners Bill Winchell and Harold Loch (Loch was also the assistant flight engineer, a position that would see him become the important flight engineer/top turret gunner on later missions) crowded into the radio room, and on the floor just behind the radio room were ball turret gunner Cecil Scott and tail gunner Johnny Quinlan. The gunners were not allowed to be at their respective stations on the bomber during takeoff. These positions were important should something go wrong and the big bomber lose control and crash. There was much greater protection in the center of the B-17.

Two of the four squadrons of the 91st would fly this raid. the 322nd would assemble in trail behind the 324th squadron, which included the "Memphis Belle." Each squadron would put up six B-17s and a spare, making up a formation of fourteen bombers on this morning. On these pages the author has included some of the thoughts of a man who flew missions alongside the "Memphis Belle," Col. Bert Humphries of Southern California. Humphries gives an interesting perspective from another cockpit. In this, his very first taste of combat, he is gazing at the bombers ahead of him, which include the Belle. It is a very unique insight into the occurrences of this historic day.

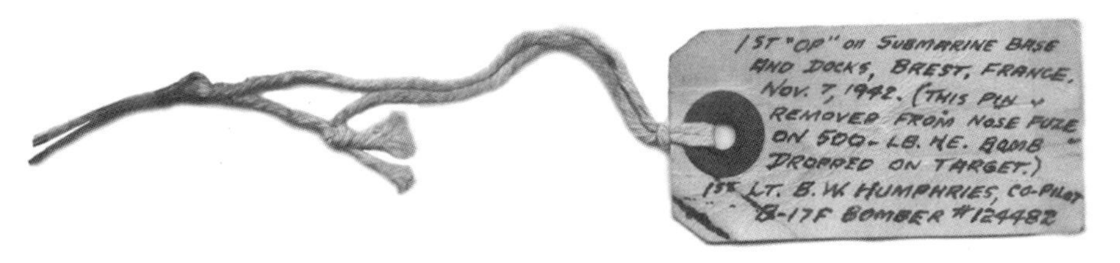

Bomb fuse tag and pin from B-17 #482 "Heavyweight Annihilators." This was from among the very first bombs dropped by the 91st BG(H) on the Germans. (Col. Bert W. Humphries)

Col. Bert Humphries
co-pilot B-17F #41-24482 "Heavyweight Annihilators"
written 7 November 1942:

General Information

The target was the submarine base and docks along the waterfront at Brest, France. Bombing altitude of 19,000 feet with 10 - 500 H.E. bombs per aircraft; twelve B-17 bombers—our group is to make up formation, with two replacements. Target area reported to have heaviest anti-aircraft protection of any point in France; were advised that we would not have any fighter aircraft for protection. Weather conditions were unfavorable with cold front lying along the target areas.

Remarks

We were awakened at 0400 hrs by broadcast on P.A. announcing "briefing" at 0530 hrs. Hurriedly dressed and ate breakfast. At breakfast I was given erroneous info that briefing was postponed to 0730 hrs., which was reasonable as it was raining then. Consequently, returning to barracks and by time I discovered error I was half hour late to briefing and had been replaced by Lt. Cox. Had to do a lot of pleading with Major Z in order to get him to change his mind. Takeoff was scheduled for 10:00, but due to bad weather, did not get off until 10:30 (That half hour on the ground was an undescribable strain—allowing free time for the imagination to run wild.). Settled down, more or less, after takeoff, and everything seemed like just another training flight until we got ten or fifteen miles inside the French coast.

At that point we could see the two bomb groups (301st & 306th) just turning on their bombing run, and some distance to the right, over the target area, there were what seemed to be hundreds of fighters swarming like a hive of bees, undoubtedly engaged in a gigantic dogfight over the target. I can truthfully say that I was scared to death. Looking to the right and left, I could only count eight of our bombers out of the original fourteen to take off (six returned due to some mechanical failure)—our total destruction seemed not only inevitable but immediate! I uttered a prayer and thought longingly of home and family. As we approached the target those specks in the skies, which I thought to be fighters, could now be recognized as "flak" bursts. What a relief.

The bombing run was extremely long, being 90 secs or more—and that minute and a half seemed like hours with the vari-colored red-black-white flak bursts getting closer and heavier all the while; and the tracer bullets and cannon fire of enemy fighters, along with the return fire of our 50-cal. incendiary bullets! The tail gunner of our ship brought down in flames a German FW-190 fighter while we were on bombing run. The effects of our bombs could not be determined due to cloud cover, but reports indicate that the U-boat pens were smashed and warehouses and docks leveled. The B-17s of the 91st dropped 40,000 pounds of bombs on their target. They had met the enemy and returned. Their next mission would be much different.

"Heavyweight Annihilators No.2," the B-17 that replaced the original fortress co-piloted by Bert Humphries. His first B-17 was battle damaged beyond repair on 3 January 1943 after a raid to St. Nazaire, France. ("Memphis Belle" led the 91st Bomb Group that day). Read the unbelievable account of that raid when this crew shot down six enemy fighters and fought their way home after their no. 4 engine had the guts blown out of it. (Frank Donofrio)

91st Bombardment Group Prelude to Mission No. 1
10:15 AM. Windshield wipers were swishing back and forth on the B-17s of the 322nd and 324th Squadrons of the 91st. Bassingbourn was rattling with the thundering roar of fifty-six Wright Cyclone engines warming up and preparing for launch.

In the control tower, General Carl Spaatz was joined by General Ira Eaker as they anxiously looked out at their armada. Moments later, the pilots saw from their bombers a man walk out onto the balcony with one arm high in the air. A signal rocket fired from his very pistol and arced into the sky. The mission was on, and the throttles of B-17 #41-24503 "Pandora's Box" were advanced to their stops as the first B-17 roared down the runway. Co-Pilot of this bomber was Col. Stanley F. Wray, 91st BG C.O. He would lead his men into this, their first taste of combat.

The second bomber to climb into the East Anglian sky was none other than the "Memphis Belle." Lt. Morgan could not know his future at this point. He was destined to become one of the most famous pilots of the century, but today he was just another B-17 driver taking off on his first mission, and he had a B-17 and a crew to worry about.

The "Memphis Belle" and her crew made some history on this first operational mission. The lead bomber "Pandora's Box" aborted the mission because of mechanical difficulty. Most of their guns froze up at high altitude, and they were forced to return to base. Although there is some controversy over whether or not this is true, both the mission logs and the engineering logs state that Army '503 did not complete this mission.

If this is the case, then the "Memphis Belle" led the 324th Squadron, as well as the 91st Bomb Group into their first mission. How could anyone know at this point that the Belle would become the very first bomber of the Eighth Air Force to complete the required twenty-five missions and return to the United states?

On this mission, six of the fourteen B-17s would not complete the raid, turning back to base for various reasons. This left only eight 91st Bomb Group aircraft over the target. Not a very desirable way to fly any mission!

Summary - 91st Bombardment Group
Breakfast was ordered for the combat crews at 0500 hours. Briefing, which began an hour later, was carried through without delay or unusual incident. No important changes were made in the original order. The target to be attacked was the submarine base at Brest, and the 91st was one of three heavy bombardment groups to fly on this mission. Colonel Wray gave the crews an excellent pep talk, and everyone was in high spirits. There had been no delays as had attended the fiasco four days before. At 0945 hours, the fourteen aircraft began to take their positions at the end of the runway. The takeoff was scheduled for 1015 hours. The pilots began to warm up their engines. Finally, at the signal from the control tower, the first aircraft began to roar down the runway on the first operational mission of the 91st Group against Nazi-occupied Europe. At 1034 hours the last of the 14 aircraft became airborne.

T.J. Hansbury (tail gunner B-17 #41-24482 "Heavyweight Annihilators) is credited with getting the first "Jerry" plane by the 91st BG(H). It was the baptismal mission on the U-Boat pens at Brest, France, 7 November 1942. (Bert Humphries)

Looking at this picture, one can almost feel the cool early morning air, hear the revving engines of the ground support trucks come and go, the voices all around, everyone preparing their bombers on a typical mission morning. You are hurrying on your way to your job and stop to light your cigarette. A glance to the right and into one of the four hangars. The light coming from inside the building violates the sheer darkness that lurks from the airfield you were walking towards. "Wonder what put that '17 in the shop? Who was hurt in her? Anybody dead? What's gonna happen today?" A buddy from your crew whizzes up in a jeep, tells you to get in, and you're off to your hardstand. Hopefully your bomber won't be in there tonight. (Joe Harlick)

"Jack the Ripper" **(Frank Donofrio)**

"Quitchurbitchin" **(Frank Donofrio)**

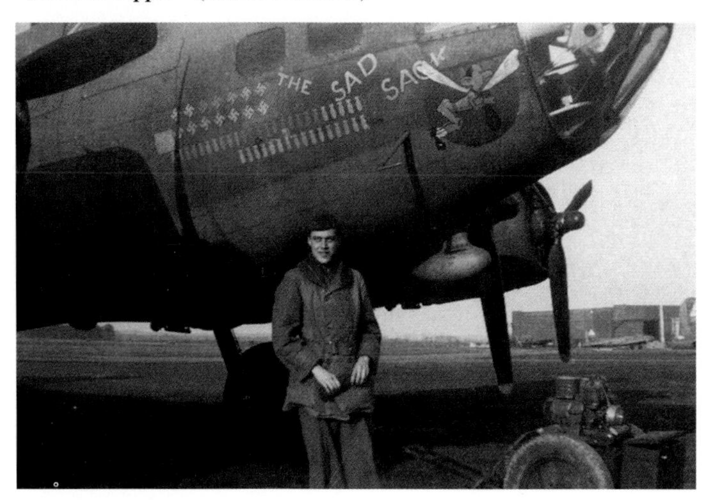

"The Sad Sack" (Frank Donofrio)

"Mizpah - The Bearded Lady" **(Frank Donofrio)**

Take off order for the 91st BG (H) - First Operational Mission 7 Nov. 1942

324th Squadron

(No. 1) B-17F #41-24503 *"Pandoras Box"*	disposition - MIA on 23 Nov. '42 / crashed in channel - 11 killed.
(No. 2) B-17F #41-24485 *"Memphis Belle"*	dsposition - War memorial - Memphis, Tennessee (Survived).
(No. 3) B-17F #41-24490 *"Jack the Ripper"*	disposition - MIA on 22 Feb '44 - crashed Munster- 1killed, 9POW.
(No. 4) B-17F #41-24504 *"The Sad Sack"*	disposition - Returned to the U.S. / War bond Tour after 42 missions
(No. 5) B-17F #41-24505 *"Quitchurbitchin"*	disposition - Severe battle damage. Re-built/Returned U.S./reclaimed 11-22-44
(No. 6) B-17F #41-24515 *"Jersey Bounce"*	disposition - MIA 21 May 1943. Crashed in North Sea. 10 killed.

(No. 7) B-17F #41- available records do not indicate which B-17 flew this reserve position for the 324th squadron.

322nd Squadron

(No. 8) B-17F #41-24499 *"Fury"*	disposition - Crash landed after mid-air @RAF Weston 18 Nov '42.
(No. 9) B-17F #41-24481 *"Hells Angels"*	disposition - MIA 14 May '43-ditched North Sea - 10 killed
(No.10) B-17F #41-24482 *"Heavyweight Annihilators*	disposition - battle damage - crash landed 3 Jan.'43 / salvaged 27 Feb.'43. Gunners claimed 6 enemy aircraft this raid alone!
(No.11) B-17F #41-24479 *"Sad Sack"*	disposition - MIA 18 Nov.'42 midair w/ #499*"Fury"* - 10 killed.
(No.12) B-17F #41-24483 *"Spirit of Alcohol"*	disposition - MIA 19 May '43. 6 POW / 4 killed (inc. British newsman)
(No.13) B-17F #41-24453 *"Mizpah"*	disposition - MIA 17 August '43 Schweinfurt. 5 killed - 5 POW.
(No.14) B-17F #41-24545 *"Motsie"*	disposition - destroyed in first ever ground accident. Cigarette ignited oxygen bottles on 23 March '43. Had flown 16 missions.

"We got the hell shot out of us!"

B-17F #41-24485	"Memphis Belle"
Pilot	Robert K. Morgan
Co-Pilot	James A. Verinis
Bombardier	Vince Evans
Navigator	Charles Leighton
TT / Engineer	Leviticus Dillon
Radio Operator	Robert Hanson
Waist Gunner	Clarence Winchell
Waist Gunner	Harold Loch
Ball Turret	Harvey McNally
Tail Gunner	John P. Quinlan

9 November 1942
Target - Submarine Pens at St. Nazaire, France
Alert no.4 - Mission no.3 for the 91st BG(H)
The second mission for the "Memphis Belle" and her crew would be nothing like the first. While the men figured that they had seen the brunt of the German opposition two days earlier, they all agreed that war could indeed be hell after this one.

Takeoff was at 9:45 AM. Fourteen B-17s of the 91st BG(H) were launched out of all four squadrons. The two previous missions had seen many of the Group's airplanes damaged and in need of repair. The formation made a straight in approach to St. Nazaire under a cloudless sky 500 feet off the water all the way from England. After crossing the French coast, the bombers clawed their way to about 8,500 feet in an effort to surprise the Germans. The Luftwaffe was indeed caught off guard, and none of their fighters were able to get off the ground. The flak gunners, however, were not sleeping. The "Memphis Belle" departed the bomb run with no less than 62 holes punched all over the plane.

Navigator Chuck Leighton reported that it felt like all the flak bursts were detonating inside the Belle with them, but that the five 1,000 pounders in the bomb bay of the "Memphis Belle" were delivered with precision. Co-pilot Jim Verinis saw holes appearing all over the plane and one directly beneath him. On the way out of the target area, the formation attempted evasive maneuvers to throw off the gunners on the ground below. Soon afterwards the B-17s were out over the water and headed for home. Running low on fuel, Leighton plotted a course for the field at Exeter, where the "Memphis Belle" landed at about 5:30 that evening. All of the B-17s suffered some damage, and many men were wounded. The 91st Bomb Group lost three bombers on this mission, and Commanders decided not to try low-altitude bombing on any more raids.

On the way to Exeter it was noted that one engine on the "Memphis Belle" was running rough. An inspection after landing revealed that shrapnel had taken out its oil line. Repairs to the line were made, but the balky engine refused to start, and it looked like the Belle and her crew would stay there for the night—at least. This would not do for Morgan. Both he and Bombardier Vince Evans had dates that night, and they were going to get back to base one way or another. It was strictly prohibited, but Morgan discussed a three-engine takeoff with Verinis, who would have no part in it. It was too dangerous for a battle-damaged B-17. The idea was to try and coax the faulty engine to life by windmilling the propeller in the slip-stream, then land to pick up the rest of the crew. Morgan convinced Flight Engineer Levi Dillon to try it. With the two men on the flight deck, the "Memphis Belle" rolled down the Exeter runway on only three engines in near darkness and fog. Verinis remembered that about halfway down the runway the propeller on the bad engine was windmilling enough to allow Morgan and Dillon to bring it to life, and the Belle became airborne moments later at full power from all four engines. The big bomber again landed, the crew boarded, and they headed home where, with almost no visibility, Morgan landed his bomber intact.

"Memphis Belle" Crew Chief Joe Giambrone's post-mission report read: "Flak holes entire plane, #1 eng won't start, #2 eng won't start, #4 eng won't mesh, #1 eng throws oil, short circuit in electrical sys, rt. brake leaking fluid, replace flares, left landing light out, running light out, check all control cables." (And there were twenty-three missions to go!)

Summary - 91st Bombardment Group
Five Heavy-Bombardment Groups were called to take part in this mission. Three Groups would fly in low and attack from altitudes ranging between 7,500 to 10,000 feet. The two remaining Groups would make their bomb runs immediately afterwards from 20,000 to 21,000 feet. The two high Groups were intended to gain the attention of the enemy's radio detection stations, leading the Germans to believe that this constituted the entire attacking force. The plan was to draw the enemy fighters to the high Groups while the bombers down low would penetrate the target perimeter free from their attacks to make their bomb runs without interference. This plan, however, did not take into account the German ground defenses.

"The first real spanking"

B-17F #41-24485	"Memphis Belle"
Pilot	Robert K. Morgan
Co-Pilot	James A. Verinis
Bombardier	Vince Evans
Navigator	Charles Leighton
TT / Engineer	Leviticus Dillon
Radio Operator	Robert Hanson
Waist Gunner	Clarence Winchell
Waist Gunner	Harold Loch
Ball Turret	Cecil Scott
Tail Gunner	John P. Quinlan

Note: According to available records the chart above displays the men who flew aboard the "Memphis Belle" on this day. Both Leighton and Verinis reported that two enemy fighters were shot down by gunners aboard their bomber, as is stated in their diaries from the period. Leighton wrote that "Quinlan and McNally each got one." This would lead us to believe that one of the regular "Memphis Belle" crew did not participate on this mission and that Harvey McNally (Ball turret on mission number two) flew the raid.

17 November 1942
Target - Submarine Pens at St. Nazaire, France
Alert no.6 - Mission no.5 for the 91st BG(H)
Third Mission for the "Memphis Belle" and crew. The 91st had flown two others while Morgan and his men stood down. The "Memphis Belle" was positioned as the number two ship in the last flight. No enemy action was noted as Morgan and his crew crossed into France. Bombardier Vince Evans would have to sight on a bomber in front of them because the gyroscope on his bombsight failed. This meant that he would open the bomb bay doors and then release the weapons from the "Memphis Belle" the instant the B-17 in front of them did.

Departing the bomb run, the "Memphis Belle" was amidst a very tight formation when they were jumped by German fighter planes, resulting in a twenty minute air battle. Tail gunner John Quinlan was credited with his first kill during the melee. The "Memphis Belle" was rocked by a blast at the left wing that left a huge hole. The bomber also received an enemy .30 bullet in the propeller on the number three engine.

British Spitfires met the formation on the way back to England. Flight Engineer Levi Dillon was reluctant to report that he received a grazing wound during the mission. Later that evening, he was forced to seek aid while visiting a pub in Cambridge, where he was bandaged by a Red Cross worker there named Adele Astaire—the famous Hollywood dancer's sister!

Summary - 91st Bombardment Group

Twenty bombers were assigned and started their takeoffs at 9:15 AM. Six of the B-17s returned to base due to mechanical difficulty, and the remaining fourteen airplanes attacked the enemy at 12:45 in the afternoon from an altitude of 18,000 feet. Some 140 500 pound bombs were delivered by the 91st bombers onto storage sheds holding torpedoes, spare parts, and ammunition. They were hit with good results, leaving the facility ablaze. Flak was moderate to intense and reported as very accurate. Enemy fighter aircraft opposition numbered between 90 to 100 Focke Wulfe 190s. There were also some reported attacks by ME-110s. A majority of enemy attacks were made on different Groups. Six bombers were damaged and two airmen wounded. The 91st gunners claimed five enemy aircraft destroyed, two probable, and another eight damaged.

With all four "fans" turning, "Memphis Belle" (left) is seen just about to take the active runway at Bassingbourn. Always a tense moment for the ground crews, two men watch from their bicycles while their bombers head off into danger. Strategic aerial bombardment had never been attempted before these men took these machines into the maw of death and destruction over the European continent. (Bert Humphries)

"First crack at an inland target"

B-17F #41-24485	"Memphis Belle"
Pilot	Robert K. Morgan
Co-Pilot	James A. Verinis
Bombardier	Vince Evans
Navigator	Charles Leighton
TT / Engineer	Eugene Adkins
Radio Operator	Robert Hanson
Waist Gunner	Clarence Winchell
Waist Gunner	Harold Loch
Ball Turret	Cecil Scott
Tail Gunner	John P. Quinlan

6 December 1942
Target - Locomotive Works at Lille, France
Alert no.15 - Mission no.9 for the 91st BG(H)
What was to be the fourth mission for the Belle was routine on paper, but not for the men of the 91st BG. The "Memphis Belle" had indeed launched on 23 November along with nine other 91st BG planes, but was forced to return to base because of engine difficulty. This action actually may have saved Belle and her crew, as no less than five others did the same. Morgan and his men were not given credit for a completed mission for their short participation on 23 November. One B-17 was lost over the English Channel. This left only four 91st BG B-17s, which proceeded alone to the target at St. Nazaire. These B-17s failed to meet the accompanying Bomb Groups and appeared over the target area by themselves, where they met devestating opposition. Three were lost, including Lt. Duane Jones, who was the very first pilot to take off on the first mission of the 91st in Army '503 "Pandora's Box."

Squadron Commander Major Harold Smelser, who was flying with Jones in "Pandora's Box," would have been justified in scrubbing the mission, but obviously made the decision to continue on. Perhaps he was hoping for a successful rendezvous with the other Group, but no one will never know, as he paid for this decision with his life and that of his men. Smelser's plane was last seen struggling on two engines and smoking. Nevertheless, the losses suffered on the 23 November raid were felt by everyone in the Group. It seemed that all personnel knew at least one of the twenty-five men killed that day out of all of the participating groups. The loss of Smelser affected the young 2nd Lieutenant Bob Morgan in a way he had not expected. From early on in his affiliation with the 91st, Morgan and Smelser had not been the best of friends. Smelser was a rigid military man who demanded military behavior. Morgan never took to that and concentrated on being the best pilot he could be. Smelser never appreciated Morgans "tilted hat and scarf in the wind" mentality, but credited Morgan for being a terrific pilot. Back in the States during training Morgan was not known for extending a whole lot of military courtesies, and was even reprimanded for buzzing things from time to time. One General vowed to make Bob Morgan the oldest 2nd Lieutenant in the Eighth Air Force. After Smelser's death, promotion papers were found in his desk with Bob Morgan's name on them. A couple of weeks later, Morgan was promoted to the rank of Captain, spending just one week as a 1st Lieutenant. Although no longer under the thumb of an old feud with Major Smelser, Morgan could not help but miss a little bit of the daunting Smelser, because within his mind, the man had lost his life honorably, holding onto the strict guidlines he had spent so much effort enforcing.

An all too familiar sight as an ambulance chases a B-17 down a snow-covered runway. When B-17s arrived back at their bases with wounded on board they were given first priority to land. Note that the bomber is taxiing with just engines one and four. (Frank Donofrio)

On 6 December, with the memory of the losses from before, twenty-one bombers began taking off from USAAF station 121 (Bassingbourn). Three would have to abort before the remaining planes hit the Locomotive works and steel mill at Lille, France. Leighton remembered that he was able to get quite a few shots at the attacking enemy fighters that jumped them just after crossing the coast of France. They had been making their now popular head-on attacks. Later British Spitfires met the formation over the target area, which was a rare break for the bomber crews. The egress from the target took the formation directly over Dunkirk, where the famed "little fleet" had so bravely evacuated Allied forces the previous year. Having endured terrific flak that was considered accurate and intense at times, as well as stinging fighter attacks, the bombers flew back towards the safety of England with this mission complete.

Summary - 91st Bombardment Group (Heavy)
Twenty-one bombers from the 91st took off. Eighteen made it to the target area with three aborts. The bombers carried 500 pound H.E. Bombs, which were released over the target at 12:15 PM from an altitude of 20,000 feet. The results were not satisfactory, as only six of approximately 150 bombs fell on the aiming point; most were short and to the right of the target. Five enemy fighters were dispatched and nine B-17s were damaged. Two from another Group were lost due to enemy action. A post mission meeting was held to determine the cause of the poor results but did not result in a determination as to why.

"They circled us like Indians"

B-17F #41-24485	"Memphis Belle"
Pilot	Robert K. Morgan
Co-Pilot	C.W. Freschauf
Bombardier	Vince Evans
Navigator	Charles Leighton
TT / Engineer	Levi Dillon
Radio Operator	Robert Hanson
Waist Gunner	Clarence Winchell
Waist Gunner	Harold Loch
Ball Turret	Cecil Scott
Tail Gunner	John P. Quinlan

Note: According to available records the chart above displays the men who flew aboard the "Memphis Belle" on this day. Jim Verinis ("Memphis Belle" co-pilot) was assigned to command a different bomber this day.

20 December 1942
Target - Airfield at Romilly Sur Seine, France
Alert no.18 - Mission no.11 for the 91st BG(H)
Bob Morgan commented some time later that he remembered this mission as one where they were attacked "on the way in, while they were over, and on the way away from the target." The German fighters followed them, and the attacks went on for an hour and fifty-eight minutes. During one attack, two enemy planes surprised Morgan when they rolled in from the two o'clock position. His first clue to the attack was when Dillon opened fire on them from the upper turret. "We were caught sleeping that time. Had they been able to shoot straight, we would have been shot down. We learned to keep our eyes open on that mission. We later learned that more than 100 fighters were destroyed, a German Officer's mess was hit at lunch time, and one of the worst casualties of the war—a cellar full of Gognac was destroyed."

Cecil Scott down in the ball turret remembered more than 300 fighters attacking in relays, circling then attacking from every direction. He thought that two fighters fell to his guns but was not credited for them.

While Verinis reported that "our ship got two," the official papers show the "Memphis Belle" was credited with only a single fighter downed from Bombardier Vince Evans' gun. Verinis was likely speaking of the B-17 that he was aboard that day, because available records do not indicate that he was aboard the "Memphis Belle" on this mission. Another mystery arose, as there are no other records to indicate Evans receiving credit for shooting down an enemy fighter plane.

Summary - 91st Bombardment Group
Several briefings on this particular target already. As of yet the 91st had not been able to reach it. This created many uneasy feelings among the crews. The mission would require more than two and one half hours over enemy territory. The raid was routine and uneventful until the formation passed Paris, France.

German fighter formations intermittently engaged the bombers, creating some of the fiercest opposition yet seen. Two B-17s from the 91st BG(H) were lost over the continent, and several members were seen to bail out. The gunners of the 91st claimed some twenty-five enemy fighter planes destroyed.

All of the remaining bombers returned to England safely, despite the fact that every one had sustained damage. (That report does not agree with the diaries of the "Memphis Belle" crew. Both Leighton and Scott aboard the Belle said they did not receive a single hole in their plane.) The B-17 piloted by Major Bruce Barton had lost the ailerons and was so badly damaged, including an eight foot hole in the vertical stabilizer, that a forced landing was necessary just inside the coast. The resulting casualties were two sheep and a rabbit. Repairs to his bomber were not practical, and the B-17 was salvaged in the meadow where they put down.

"Generally good" and "certainly better than average" were terms used to describe the results of this raid, with many of the bombs taking effect in the hangar and dispersal areas.

Bomb nose fuse pin and tag which flew over the French target aboard B-17 #41-24545 "Motsie."

Col. Bert Humphries
co-pilot B-17F #41-24482 "Heavyweight Annihilators"
written 20 December 1942:

General Information
The target was the great Nazi aircraft pool base at Romilly Sur Seine, 80 miles south of Paris and 180 miles from the coast of France, to be attacked by four groups (91st, 306th, 303rd, 305th) composed of both Flying Fortresses and Liberators. The 306th Group led the formation and the 91st Group followed, attacking the target at 21,000 feet with 10 500 pound H.E. bombs per aircraft. We were advised to expect heavy flak concentrations all along the route. The latest intellegence reports indicate the recent arrival of approximately 100 fighter aircraft near target area, and since we could expect our friendly escort of fighters to escort us only 50 miles inland, it meant we could be assured of 260 miles of continued enemy fighter attacks.

Remarks
I was awakened a few minutes before 6 AM by the alert officer, and I was dead for sleep due to the late hours of the dance last night. Fortunately, I had not had anything to drink—but there would be plenty that had. After a hurried breakfast I rushed over to the briefing room, where we were given all of the pertinent facts concerning the target, etc. Our crew was slated to fly Lt. Beasley's ship since he and his crew were on "pass." Also, our ship was out of commission and still awaiting to have the guns test fired at high altitude. We taxied out of the dispersal pens at 9:50 and ran into quite a bit of confusion when one of the 323rd ships got stuck in the mud, holding up three other ships behind it. Consequently, only 17 of the proposed 21 aircraft took off, and since our ship was considered a spare, we immediately filled in for the ship that got stuck in the mud.

The climb to our cruising altitude was without incident, except for the fact that we lost number three supercharger due to breakage in exhaust stack, making it difficult to maintain our position in formation. As we crossed the coast of France we could see the long, thin, white vapor trails of fighters far above us. Whether they were friendly or not we were unable to tell, but we were on the alert for the worst, and just a few minutes later, the worst came screaming down on us in a head-on attack!

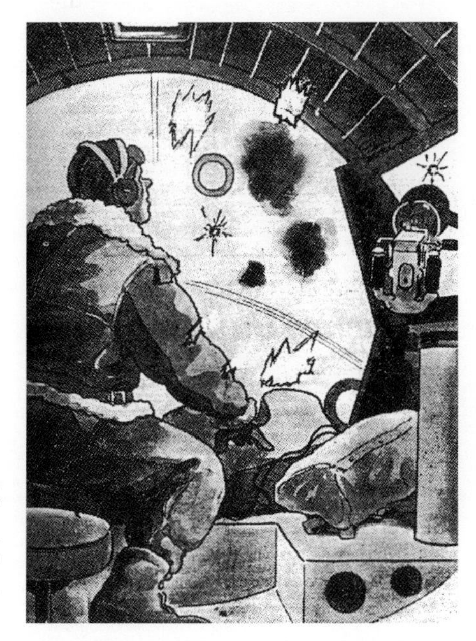

"The damn nose needed remodeling anyway." This cartoon appeared in a British newspaper on January 5, 1943. (Humphries)

At first they came in rather orderly in flights of four to six, and "peeling off" would rake our formation with machine gun and cannon fire. It was after the second or third of these attacks that our rear gunner announced seeing two bombers of our group go down. (Presumably, these were ships piloted by Lt. English and Lt. Corson of the 401st). Then the attacking fighters seemed to come in from two or three directions at once. It was during one such attack that our ship got raked with three .30 cal. machine gun shells finding their place on our nose. One came in above the nose and apparently exploded in the control panel, sending up a sheet of flame between Don and me. Don called over the interphone that he had been hit, since there was blood all over the floor. At the same time I thought it was me who had been hit, for I was suffering from acute pains in the stomach. It turned out that neither of us were scratched; the blood on the floor was found to be red hydraulic fluid, and the pains in my stomach were due to the "bends." The second bullet entered the right side of the ship not more than a few inches from me and penetrated five bulkheads, giving Sgt. Phiepo (waist gunner) a flesh wound in the thigh. Then the bullet veered out the right side of the ship just before reaching the tail gunner. The third bullet did little or no damage.

It was shortly after this attack that the ball turret (Sgt. Budzisz) caught an Me-109E trying to sneak up on us from the rear and shot the fighter down. The intensity of the attacks decreased as we flew further inland. Apparently, the fighters were running out of ammunition and fuel. But we knew, as we saw them break away and head for home, that they were just going down to reload, and would be back up to meet us on the way out! But what an enjoyable "breathing spell" those few minutes were. Our bombing over the target was conducted without interference. The expert job done of camouflaging the landing field was worthy of mention, and I fear if we had been less thoroughly "briefed" the target would have escaped detection.

As it were, very good results were seen on the target. Our tail gunner reported seeing both of the hangars afire! We turned away from the target and headed back for the coast, and had flown but a very few minutes when we were re-engaged by enemy fighters. The preponderance of the attacks were made against the far side of the formation, though we suffered several surprise head-on attacks that almost got us. This trip back was far more nerve racking than the trip in, for I was able to observe at least 75% of the enemy fighters as they climbed for altitude just out of range of our guns. They positioned themselves above us and to the beam, and finally, I could see and hear their screaming dives on the formation. That rolling dive is hard to forget, with the wing guns and nose guns spitting tracer bullets and cannon fire. It appeared as though the fighters were on fire. You would hopefully wait for them to explode from the flames—but on they would come, still spitting tracers and cannon!

Arming the B-17. Often this was done at night, and even in the cold and rain. Ten 500 pounders are slated to be the one way passengers on this fort. (Frank Donofrio)

The super-secret Norden Bombsight was the heart of the bombing platform aboard most of the B-17s. When connected with the C-1 autopilot, the unit performed much better and faster than the competing Sperry S-1B, which was often mounted aboard B-24 Liberators. Many Bombardiers could make their computations in around thirty seconds.
The device would put the plane over the target with the data (including true altitude and airspeed) that the Bombardier entered into the sight mechanism according to differing bomb types. The aircraft would then become a stable bombing platform when the autopilot took over actually flying the plane. (The pilot had to switch control from the flight deck to the Bombardier and was more or less "hands-off" during this time.) Peering through the rubber eyecup atop the unit, the Bombardier would make visual sights on the ground below and ahead of the plane, putting the cross hairs on the aiming point. Minor course corrections could then be made to the flight path with the use of the control knobs, thus keeping the cross hairs firmly on the target. The bombsight would compute the drift and release point. Two indicating needles would be moving toward each other during the bomb run, and when they touched an electrical connection was made and the bombs were dropped.
Accuracy often depended upon a weatherman's skill. The visual device required a clear picture of the conditions below the plane so the Bombardier could see the ground.
Bursting flak could bounce the plane around the sky, so the unit was mounted gyroscopically.

It was on the trip out that I saw, for the first time, an aircraft get shot down—and both of them were bombers from the 306th Group! The first had its tail shot away, and it immediately plummeted to earth like a falling rock. Only one person was seen to bail out, and then the crash! The second ship was seen to gradually slide out of formation with one engine dead and another afire, and enemy fighters swarming over it, waiting for the kill. It was a comforting sight to see eight of the crew parachute to safety. I am still "sweating out" those other two men (my guess is that they were pilot and co-pilot, for they have quite a time getting out.) One interesting and amusing episode during the flight back to the coast was Capt. Wallick's trickery against the Hun. After losing an engine he was forced to drop out of the formation, and his two wing ships, piloted by Lts. Barton and Baird, courageously followed him down. Of course, their small number of 3 attracted the fire of a great number of fighters, and as a result Barton's ship got badly shot up and headed for the ground (he managed to get across the channel and crash-landed in South England—killing two sheep and a rabbit while doing so.)

The Station 121 Control Tower is dwarfed by the immense hangars near the perimeter track at Bassingbourn. Ground vehicles dash about the base carrying crews, bombs, ammunition, and oxygen, among other things. No less than nineteen B-17s are staging around the base in this photo. Also note the P-47 Thunderbolt and the P-51 Mustang.

Meanwhile, Lt. Baird had given his compatriots up for lost, so he pointed his nose down and streaked for home—and made it! Capt. Wallick, on the other hand, having lost a second engine, was losing altitude rapidly and never expected evading capture. So in desperation, he lowered his landing gear (having heard that to be the international sign of surrender) and spiraled down with an FW-190 on his tail. But the tail gunner had not known of Capt. Wallick's armistice, so when the trailing enemy fighter came within range he shot him down! When Wallick reached cloud cover, he raised his landing gear, started up one of his dead engines, and staggered home. It is said that the propeller on his damaged engine flew off the hub (while he was making a landing) and beat him to the field.

The fighters broke off their attacks on us over the coast of France, so we were able to cross the channel unmolested and started our descent. This was the first opportunity to assess the damage to our ship and crew. Since our main hydraulic line was damaged we would have no brakes on landing, and consequently we ordered all the crew to radio room (safest place in case of crash) just in case. We had to circle the field until all the other planes had landed. Then we came in, making a dead-stick landing on a grassy area; it was a very slow, tail first landing, and we scarcely had to touch the emergency brakes to bring the ship to a stop. Taxied over to control tower where an ambulance met us to pick up Sgt. Phiepo. My, what a relief to set foot on earth again, for this was one afternoon I truly never expected to see (Intelligence reports disclosed the loss of 6 bombers versus approximately 50 fighters).

A typical "F" model nose gun installation, with one .30 and one .50 machine gun each. ("Memphis Belle" was fitted with twin caliber fifties in her nose.) These forward firing weapons enabled the Bombardier to handle the crushing frontal attacks from the enemy fighter planes when he was not on the bombsight. Note the straps which were used to help balance the machine gun. Also the tubular steel mounts that were installed to keep the gun's recoil from breaking the plexiglass nose piece. The view from this vantage point on a B-17 was spectacular. Many Bombardiers felt, however, that this position was often paid for with the danger, injuries, and sometimes the very life of the crewman. Note that the bombsight is covered and the .50 gun is not armed with the bullets below it. (Frank Donofrio)

This is how a B-17 with no wheels sometimes had to be moved. Note that only her right side props appear to be damaged. While all four engines may have been running when the bomber landed, it is clear that the damage to the no. three and four engines occurred because only the left main wheel is extended. The trailer supports the plane by connecting to the wing spar attach points.

"Tragedy"

B-17F #41-24485	"Memphis Belle"
Pilot	James A. Verinis
Co-Pilot	J.S. Jackson
Bombardier	Vince Evans
Navigator	Charles Leighton
TT / Engineer	Eugene Adkins
Radio Operator	Robert Hanson
Waist Gunner	Clarence Winchell
Waist Gunner	Harold Loch
Ball Turret	Cecil Scott
Tail Gunner	John P. Quinlan

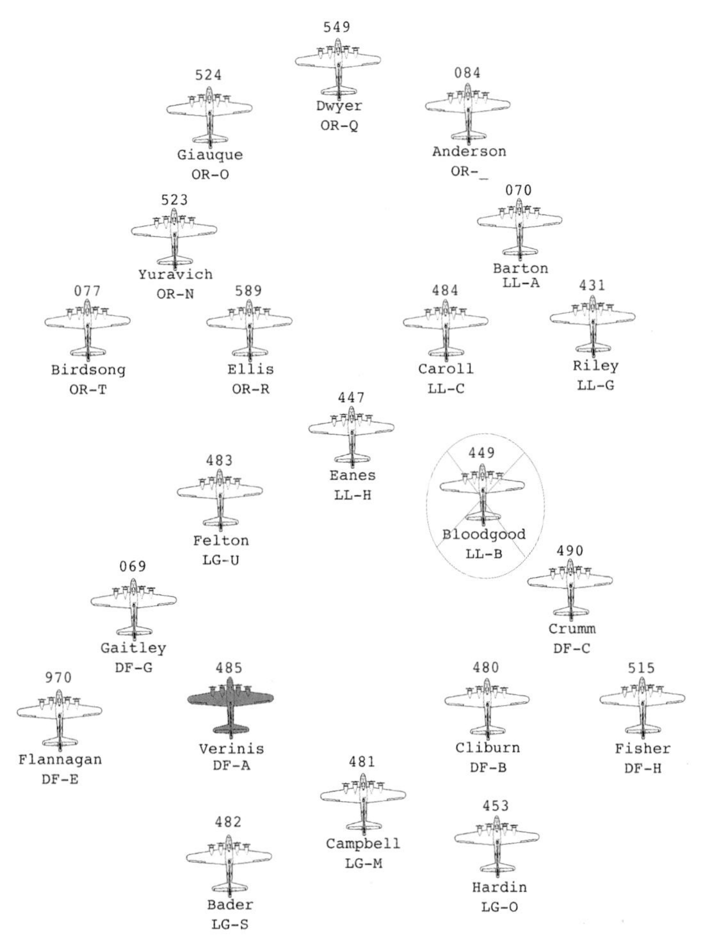

30 December 1942. Target - Submarine pens at Lorient, France
Alert no.19 - Mission no.12 for the 91st BG(H)
Bob Morgan, the usual pilot of the "Memphis Belle," was sick this day and could not take part in this raid. Jim Verinis, who had actually been trained as an Aircraft Commander and was normally the co-pilot of the Belle, was going to be in the left seat for this one. Heavy unexpected winds blew the weary formation off course on the return trip, which was to be made entirely over water to avoid enemy encounters. Many bombers were caught off guard when they realized with sudden horror that they were not flying over the southern tip of England, but had wandered back over the Brest peninsula of France! Many Navigators were actually fooled by the false radio beacons being broadcast by the Germans and began to let down out of the formation. At least three B-17s were lost in the ensuing maelstrom.

Summary - 91st Bombardment Group
Good weather over the target. Snow and flurries in England. Two aborted aircraft. Formations of enemy fighter aircraft began their attacks shortly after reaching enemy-occupied territory. The German fighters held off during the bomb run to allow for huge concentrations of flak barrages. Immediately following the bomb release and just beyond the range of the anti-aircraft cannon, the fighters re-engaged the bomber formation. Each B-17 carried two 2,000 pound bombs, of which many made direct hits on their reinforced concrete roofs but failed to destroy them. Many surrounding buildings, which felt the brunt of the falling tonnage, were destroyed."

Lt. Bloodgood in ship #449 "Short Snorter" went down off the coast of France. All aboard were killed. Major Myers aboard #070 "Invasion 2nd" was mortally wounded. The 91st also had an additional seven airmen injured this day.

Twenty-three formation mission stack diagrams (left) are provided within this book. Within them are all of the 91st Bomb Group B-17s which were scheduled to fly on these raids. It should be remembered that this was just one Group among several taking part in each raid. Multiply the number of the planes in each diagram by three or four and one will have a better idea of what a complete combat wing looked like. The number at the nose of each illustrated plane in the diagram corresponds to the last three digits of that plane's tail number. The name below the tail is the man scheduled to command that plane on that day. The lettering below the pilot's name matches that bomber's "waist code," which is an indication of the squadron he belonged to. A B-17 within a circle indicates an early return. This does not, however, mean that his bombs were not dropped; it simply means that the plane returned before the others. This could be for a variety of reasons, from mechanical difficulty to combat damage. A bomber within a circle and an "X" indicates that the bomber was lost on the raid. It does not mean that all aboard were killed, only that the plane was gone. The fate of the men was recorded in Missing Air Crew reports.

It should be understood by the reader that all of the formation diagrams within this book are authentic. They were compiled from actual handwritten records supplied by 322nd Squadron Operations Officer Bert W. Humphries. Also make note of the fact that as the raids progressed, the formations changed as the bombers moved into varying positions to cover the open spaces in the formation that were left when a bomber was shot down or returned to base early with difficulties.

Col. Bert Humphries
co-pilot B-17F #41-24482 "Heavyweight Annihilators"
written 30 December 1942:

General Information

The target was the important sub base and installations located on the Atlantic coast of France at Lorient. Four Bombardment Groups (306, 91, 303, 305) composed of Flying Fortresses and Liberators were taking part, with the 306th BG leading. Our Group was to be the second, bombing from 21,000 feet with two 2,000 pound bombs per aircraft. This was to be our eighth attack on military objectives in that locale, so we knew from past experience that we could expect dense, heavy, and accurate flak, as well as the best of the Luftwaffe fighters. Weather forecasts were uncertain, so this mission might end up like the last over Lorient, where we went on a "cooks tour" of France and no bombs were dropped due to overcast skies obscuring the target.

Remarks

Was awakened shortly before 6:00 this morning, and since briefing was to occur at 6:45 AM I hurriedly dressed and rushed downstairs for breakfast. As I left the club for the briefing room, I was amazed to find the ground covered with snow—the first time in my life I remembered seeing such a sight—and it was truly beautiful. However, the strong wind blowing made it quite uncomfortable, and I withdrew tortoise-like into my fur-lined flying suit. The briefing was completed in short order, and after last minute preperations Don (pilot) and I secured a truck, picked up Jim and Jack at their room, and went out to the ship. It was snowing quite heavily when we arrived, and since the ground crew was pre-flighting the ship, we remained huddled in the truck, watching the snow fall and wondering if we would be able to get off the ground.

The snow finally subsided, and the next 45 minutes were busily spent assembling guns, etc. Taxiing and take off were without unusual incident, but we had climbed only several hundred feet when we ran into snow flurries that put us on instruments—we almost lost the formation. Our Group was scattered all over the sky and never assembled correctly until almost mid-channel. The other Bomb Groups must have had the same difficulty, for they were pretty well scattered and mixed up, too. (The 306th had established an abortive record—all but one lone ship turned back before reaching the mid-channel point, and it tacked on to the rear of our formation.) We circled the rendevous point at 21,000 feet, looking in every direction for the 306 Group, who were supposed to lead. At 11:05, five minutes after the "zero hour," the 91st assumed the lead and started across the channel.

While no one had been shot down over the target, Verinis reported later that one possible explanation for the bombers' appearance over France was a terrific 100 mph head wind. The return trip took more than five hours, while the trip in took only two. The Navigator for the lead plane in the Group (401st Squadron) was ultimately blamed by many for the error in navigation which led to the slaughter over France while coming home. It was reported that one of the pilots of that bomber (Major Myers) noticed the error in time to veer the formation sharply away from France, but not before his ship was badly hit and he was wounded—fatally. His evasive maneuvers did show the following bombers their error and likely saved many. While it was probably an error in judgement for Myers' navigator, many also felt that another cause for concern was the fact that the crews let down their guard too early. The practice of allowing only the lead bomber's navigator to plot a return course for the formation would not do any longer, and plans would change for all future missions. All navigators would be given complete briefings for both the target ingress and egress. They would stay at their jobs—navigating their own ship—all the way home. There would be no relaxing until after they landed.

The coast of France was obscured by clouds as we passed over it, but they soon dissipated, allowing a full clear picture of the target, still twenty miles away! The bombing run was interrupted by several fighters making head-on attacks at the lead ships, and the flak was beginning to get dense, but we flew straight and level through it, all of us making the smoothest bombing run I have ever seen. All of this time I could see the fighters off in the distance climbing feverishly for altitude. They had positioned themselves for the attack, which came at the very instant we dropped our bombs. The next fifteen or twenty minutes were so filled with activity that it is difficult to single out any particular episode. We managed to keep in close formation with our element leader Capt. Campbell, and by using violent evasive action we weathered attack after attack from enemy fighters without being hit once. Our gunners claimed one enemy FW-190 probably destroyed.

Towards the latter part of the melee I noticed a lone B-17 quite a ways beneath us, with his no. 1 engine on fire and his no. 2 engine feathered, and fighters still attacking him. Don called Capt. Campbell (element leader) on the radio set and asked him to drop down with our three-ship element and render support to the unfortunate bomber beneath us. After several unsuccessful attempts to contact Capt. Campbell, Don and I decided to leave formation and go down by

Another nose fuse pin and its tied tag. This one flew aboard the B-17 which was normally assigned to Don Bader's crew on the "Heavyweight Annihilator." Less than a week after this raid, that B-17 returned from a mission with severe damage which assured that it would never fly again.

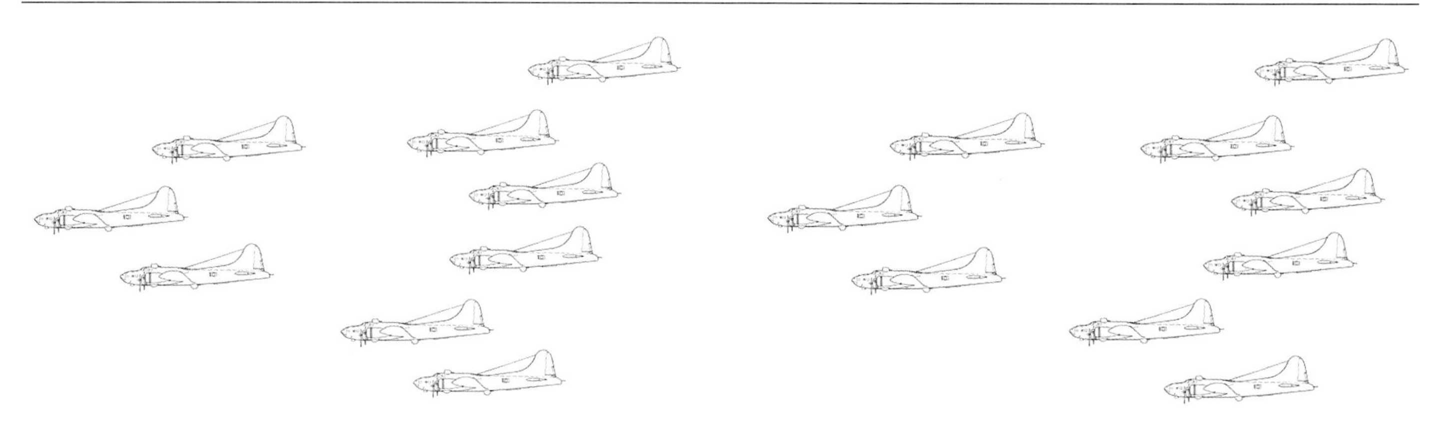

When viewed from the side, the combat box formations were stacked vertically as well as horizontally. This provided a good deal of mutual fire support from all of the planes in the wing. As they were echeloned out across the span of the wing's formation the aircraft were segregated into low and high squadrons at marginally different altitudes. By the end of the war, many formations measured thousands of feet across, hundreds of feet high, and many miles long.

ourselves. We had just positioned our ship on the other wing in time to receive the attacks of two enemy fighters, which we successfully drove away. We recognized the lone B-17 as belonging to the 306 (the same bomber mentioned earlier). After it had put out the fire and gotten both of the engines running again, and since no fighters were to be seen, our ship joined the formation then nearest to us. But close examination revealed the markings on the ships in that formation not to be ours, so we tried another and another, and finally the fourth attempt proved to be the remnants of our group. There were remnants, or bunches, of ships scattered all over the sky, so that it was difficult to determine who belonged to what!

That is the tendancy after a hard fight, to relax by spreading out the formation—and worse yet, to head for home as individual ships. But this was one day where such a practice would prove extremely dangerous, for we were flying through overcast clouds, and a freak wind change made our navigation quite unreliable. (We were returning from the target by water route rather than re-crossing France.) Nevertheless, it seemed that few of the crews realized the seriousness of the situation. I know ours didn't. The pilot (Don) left his seat and went down to the nose of the ship to stretch and have a bite to eat. The Bombardier came up and sat in the pilot's seat and proceeded to fly the airplane under my instruction. It was such a relief to let go of the wheel, remove my feet from the rudder pedals, and take a long wanted stretch and smoke!

Soon we flew over some small islands and land was just ahead when all of a sudden a red ball of flame shot past the nose and exploded uncomfortably near! I immediately diagnosed it as "light flak" and called up the Navigator to make note of the event on his log, and warned him to get busy with the signal lamps and answer any further challenges from the ground so we would not be fired on again! I thought to myself that it was bad enough to be fired at by the enemy, let alone to be fired upon when we returned to England. Hardly a minute had elapsed when we had flak bursting all around us. I immediately snatched the controls from Jim, started violent evasive turns, and sped up to catch the formation just ahead. It was then that I glanced over the right wing and saw a tremendous dog fight. I could make out a burning B-17 (later believed to be the same B-17 that we had rescued over an hour ago—see footnote) making its death spin into the sea, and another B-17 was frantically fighting for its life with five or six attacking from every side. (It seems like this was a long and drawn out incident the way it is told, but really the time required was only a few seconds.)

Then it suddenly dawned on me that wind change had blown us off course, and away we flew. Although we were never attacked by fighters, we came pretty close to some flak bursts. Needless to say, the trip home from there on was in dead earnest, and I dare say that no one relaxed until we set foot on the ground.

Note:
On 12 January Humphries read in a newspaper that the B-17 that they rescued did indeed return to England, albeit a crash landing, and the men were evidently saved. They were from the 306th Group, and their pilot was Capt. Clyde B. Walker from Tulsa, OK. The B-17 was named "Boom Town."

"Slapping flak"

B-17F #41-24485	"Memphis Belle"
Pilot	Robert K. Morgan
Co-Pilot	C.E. Putnam
Bombardier	Vince Evans
Navigator	Charles Leighton
TT / Engineer	Eugene Adkins
Radio Operator	Robert Hanson
Waist Gunner	Clarence Winchell
Waist Gunner	Harold Loch
Ball Turret	Cecil Scott
Tail Gunner	John P. Quinlan

3 January 1943
Target - Submarine Pens at St. Nazaire, France
Alert no.20 - Mission no.13 for the 91st BG(H)
The events of the previous week, when the 91st's planes were caught off guard over the Brest peninsula, served the "Memphis Belle" and her crew—not only by teaching the all important lesson that a mission is not over until the wheels are safely on the ground, but also because of the spectacular job of Leighton navigating the Belle on that return trip. Evans had also displayed great skill when his bombs fell very accurately on the target. Because of these efforts, Leighton would be given the all-important task of not only leading the 91st Bomb Group, but the entire formation on some later raids as navigator. All other bombers in the 91st assembly would release their weapons on the cue of the "Memphis Belle," and Evans would receive the Air Medal for his accuracy on this mission. There was a cold harsh feeling of reality among the men before this mission even launched. As soon as Col. Wray advised those taking part that they were headed back to St. Nazaire for the third time to hit the sub pens again, an almost stoic feeling swept through the ranks. The Nazis had displayed great determination in repairing the installations after previous raids, and the German Navy had hardly slowed down their output of the horrible Wolf Pack submarines that were sending American ships to the bottom of the Atlantic at acute levels. Many of the men took deep breaths; they were growing tired of going back into that hell, and vowed amongst themselves that they would hit this damned target hard and level it for good.

During his post-mission debriefing, Evans remembered that when they turned to begin the run on the target, an unanticipated very strong head wind met them and reduced their ground speed to only eighty-five or ninety miles per hour! This would create a much longer bomb run than the normal average of fifty seconds or so, while giving German gunners on the ground a much greater chance of killing B-17s with their flak guns. The winds caused the bombers to go wide on the turn into the target, and Wing Commander Major Putnam (riding in the right seat of the "Memphis Belle") coaxed Evans to "settle down on it."

This was remembered to be a terribly long two and a half minute bomb run, of which the bombers on the approach could not deviate at all. Flashes of anti-aircraft fire were visible in the eyepiece of Evans' bombsight. This could not be allowed to be a distraction. The explosions intensified as the bombers clawed on through the bursts. Even a moment's hesitation from Evans could mean missing the objective, and consequently a return trip to St. Nazaire. Evans could not take his eyes from the bombsight, as he now controlled the flight path of the thirty ton bomber. Fighters flew through his view of the ground below, and from the pre-mission briefing Evans could make out familiar points from photos as he watched for the very building that he would target on. Holes were appearing all over the "Memphis Belle," and still Evans did not flinch. Even Morgan got on the interphone and asked him "When the hell are you going to drop the bombs?" Evans calmly told him to "Take it easy."

One of the more dramatic events of the war took place on this day and within this combat wing aboard B-17F #41-24620 *"Snap, Crackle, Pop."* Assigned to the 360th Bomb Squadron in the 303rd Bomb Group, she was flown by Lt. Arthur Adams to St. Nazaire to hit the sub pens there. (The 303rd BG(H) led the Wing this day. "Memphis Belle" in the 91st BG(H) was in the third Group over the target behind the 303rd.) On the departure from the target the ship was bracketed by intense and accurate flak, which killed the bombardier and seriously wounded the Navigator. A sweeping attack from two FW-190s finished the crippled bomber in a single pass, taking almost an entire wing from the plane. Only three of her crew could escape the B-17, which was falling to earth in a flat spin. The Navigator and the tail gunner parachuted into the water just off the French coast and were quickly captured. S/Sgt Alan Magee (ball turret gunner), who was already wounded in the face, had managed to get out of his turret and into the bomber, only to find himself falling free in the air with no parachute after the bomber broke apart. He lost consciousness as he fell more than 20,000 feet through the roof of a train station in St. Nazaire. His captors found him severly injured, but alive! (Frank Donofrio)

The sweating Bombardier put the cross hairs where he knew the target to be. The moment those "passengers" fell from the "Memphis Belle," he knew they were right on. They hit squarely. Strike photos later revealed that Evans' bombs actually impacted less than ten feet from the center of the target area!

Navigator Chuck Leighton remembered that as the formation departed the bomb run and headed back out over the Atlantic, enemy fighters engaged them almost immediately, making their now notorious head-on attacks. Major Putnam (co-pilot for this raid) was flying the "Memphis Belle" at that point, as was recalled in the diaries of the crew when they made note that his evasive maneuvers with the Belle allowed them to come through this part without a scratch.

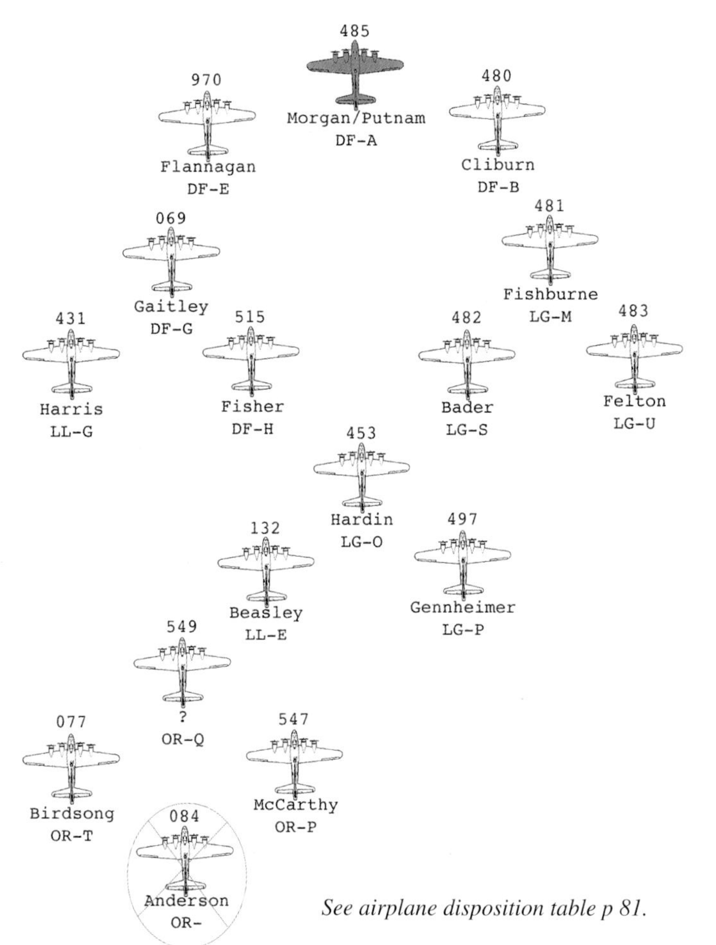

See airplane disposition table p 81.

B-17 #084 "Panhandle Dogey," flown by Lt. William M. Anderson of the 323rd Sq., was directly hit by a round of flak that detonated inside the bomber while they were over the target. Only Navigator Lt. Roten survived.

B-17 #481 "Hell's Angels" was piloted on this raid by 322 Squadron Commander Major Fishburne, with Capt. Campbell in the right seat. Make note of how #481 dropped out of the formation to assist Lt. Don Bader in #482 when the latter bomber was nearly shot out of the sky and limping back to England.

A similar event occurred when #069 "Our Gang" was hit and #515 "Jersey Bounce" left the flight to render assistance. However, #515 would not fair as well as #481 for this act of valor. It received terrific damage, but made it back with #069!

The "Memphis Belle" landed in Southwest England at St. Eval because of a flat tire. The crew stayed there overnight while the bomber was repaired. Another bomber from the 91st apparently put down there as well. It did not fare as well as Morgan's B-17, however. Lt. Fisher, who commanded B-17 #41-24515 "Jersey Bounce," was flying a plane that was badly hit. A 20 mm shell had detonated inside his airplane, creating severe damage, and Fisher's navigator was hit as well. The "Jersey Bounce" would be repaired and then flown by Morgan and his crew a month later to hit targets at Emden, Germany. This was necessary because, unkown to the crew of the "Memphis Belle," their bomber would require serious repair after an upcoming mission to Lorient. But for now they were safe.

Summary - 91st Bombardment Group

The first mission of 1943 and the seventh mission for "Memphis Belle." The torpedo sheds and submarine pens on the Atlantic coast of France at St. Nazaire were to be hit. In particular, one small but important building in the complex. The 91st sent sixteen B-17s. Three of them aborted. The remaining bombers scored several direct hits on the complex and the aiming point. Until this time, the 91st had not done a better job of delivering weapons to a target. Reports indicate that the 91st (which led the entire formation) performed much better than the accompanying Groups. Two men were killed on this raid, and ten wounded.

A graphic way to represent the incredible losses suffered by the 91st BG alone would be to reveal the destiny of each of the bombers that took part on this January 3, 1943, mission. Of the sixteen bombers that were flying that day, only two lived through the next nine months. They were #480 & #485. The chart (above, opposite) shows their fates.

In all, the B-17s listed in this table represent significant losses for the Group. Through a nine month period, sixty-seven airmen were killed as these bombers were methodically destroyed. Notable bombers here are the following: #480, which was utilized by William Wyler as a camera ship during the filming of his documentary "Memphis Belle"; #970, which was taken over by Jim Verinis, who was "Memphis Belle" co-pilot for the first missions; and #482, which was the plane assigned to Bert W. Humphries, who contributed his diaries for research towards this book. Also, the "Memphis Belle" crew flew ships #480 and #515 while the Belle was laid up for repairs on later missions.

Col. Bert Humphries
co-pilot B-17F #41-24482 "Heavyweight Annihilators"
written 3 January 1943:

General Information

The target for this mission was the installations of the important U-boat base at St. Nazaire, France. The same four Bombardment Groups (303, 305, 91, 306) as in the last few raids were participating, with the 303rd Group leading. Our Group was to be the third over the target, flying at an altitude of 22,000 feet with five 1,000 pound bombs per ship. Our past experience over St. Nazaire had

1-3-43	#084	*"Panhandle Dogey"*	crashed continent. 9 KIA 1 POW
1-3-43	#482	*"Heavyweight Annihilators"*	badly damaged - Salvaged
2-16-43	#431	*"The Saint"*	Crashed on take off. Salvaged
3-4-43	#549	*"Stupen taket"*	Crashed Munster, Germany 8 KIA 2 POW
5-1-43	#547	*"Vertigo"*	Enemy action, Ditched Channel 5 KIA 5 POW
5-14-43	#481	*"Hell's Angels"*	Ditched North Sea 10 KIA
5-19-43	#483	*"Spirit of Alcohol"*	Crashed Kiel fiord 4 KIA (incl. British newsman) 6 POW
5-21-43	#515	*"Jersey Bounce"*	Ditched North Sea 10 KIA
6-9-43	#970	*"Connecticut Yankee"*	Crash landed crew returned - Salvaged
6-22-43	#132	*"Royal Flush"*	Enemy action crashed North Sea. 9 POW 2 escaped 1 killed
8-12-43	#077	*"Delta Rebel no.2"*	Crashed 4 KIA 6 POW
8-17-43	#069	*"Our Gang"*	Crashed Schweinfurt 10 POW
8-17-43	#453	*"Mizpah"*	Crashed Schweinfurt 5 KIA 5 POW
9-6-43	#497	*"Frisco Jenny"*	Ditched Channel 10 Returned
10-19-44	#480	*"The Bad Penny"*	Fatigued VIP transport Salvaged
Today	#485	*"Memphis Belle"*	Survived

been grievously memorable, so we were expecting the worst—both in flak and fighters. Weather forecasts were very good, with the exception of very high winds (100 mph) opposing us on our bomb run—but if the flak isn't too heavy the resultant low speed on the run should make for some accurate bombing.

Remarks

I spent a very restless night, being awakened by wild dreams and nightmares. In one I dreamed that I met Lt. Frazier and Capt. McCormick in some strange and distant land. This particular dream really bothered me, for both of these men were shot down over St. Nazaire on the November 23 raid. So I awakened this morning with the premonition that disaster was close on my heels.

The briefing was scheduled for 6:00 AM, which was too early to allow us to have breakfast, so I was far from being in a pleasant mood. Since we had seen the target pictures so often and knew the fighter and anti-aircraft defenses almost by heart, the "briefing" lived up to its name and was very short. We returned to the club and had breakfast, then secured transportation and went out to the ship to get it in readiness for take off. While the ground crew was preflighting the ship, I strolled over to a nearby fallen log and sat down as the first streaks of dawn broke through and silhouetted our bomber. It was an awe-inspiring sight, and I murmered a prayer for the safe-keeping of the crew on this mission.

The top turret guns gave us considerable trouble, and when we took off neither of them were operating. But Sgt. Hall (replacement engineer for this mission) worked feverishly on them, and by the time we had reached mid-channel, he had both of them working, and all other guns were working sweetly (except radio gun) so we continued on. There was scarcely a cloud in the sky, and the visibility was unlimited—it seemed as though you could see clear across France. It was by far the prettiest day we have yet had, and the flight inland was not marred by one burst of flak or one enemy fighter! It must be the proverbial "calm before the storm" thought I, to myself. This was the first chance that I had been able to study the landscapes beneath me. It looked so green and pleasant, with the little villages nestled in the valleys of the rolling hills; all seemed so quiet and peaceful beneath.

The bombing run approached St. Nazaire from the east and was much too long (14 minutes), allowing the flak batteries to make very accurate adjustments on us. We flew straight and level for the last several minutes (that seemed like ages) and should have gotten some excellent hits on the target. Just as the last bomb left the ship, we received a direct hit of "heavy" flak in our right wing, tearing a huge hole in it. Other near misses were bouncing us around like a cork on a stormy sea. Finally the formation turned off the target to the left and headed for the water; but just as we were crossing the coast, we flew right through a dense barrage of mixed-colored heavy flak, and one of the shells had our number on it. It made a direct hit on the number four engine and blew the guts out of it. It seemed as if the whole right wing burst into flames. Don and I were busy beavers in the cockpit, feathering the props, trying to extinguish the flames, evading the attacking fighters, and trying to keep formation. (Just then someone was talking over the interphone, and I learned that the flak had torn numerous holes in the plexiglass nose and both Jim and Jack were injured.)

What a condition we were in: No. 4 engine blazing with fire and sending out a stream of smoke that could be seen for miles, inviting every fighter in the sky for the kill. Steadily we were falling further and further behind and dropping lower and lower beneath the formation, which was our only hope.

Just about that time, I saw a B-17 off in the distance, burning like a torch and tumbling downward. Just before hitting the water, it exploded into what seemed a hundred pieces. You can imagine the fear that gripped me looking out my side and seeing the number four engine blazing more fiercly than ever, with the flames enveloping the entire wing and gasoline tanks just beneath the skin surface of the wing. Our end was very near, only I was not exactly sure of how the picture would close. I was certain, however, that there could be only two alternatives: either we would explode into a million pieces when the flames ate their way into the gasoline tank; or the fighters would save us from the torture of burning by riddling us with their cannon and machine gun fire. It seemed as though the judge could not decide our fate without further cross examination and testing, so to speak, for the fighters started diving in on us from every direction, while Don and I were trying every trick we knew to evade them. How long these withering attacks kept coming, I don't know, but it seemed like hours. The first breathing spell I had, I gave a quick glance to number four engine and the fire was out!

What a feeling of elation I experienced! That was our first ray of hope for escape and our lives! Some say it was "Vulcan," the good gremlin (who eats fire), that came to our rescue. But I have my own convictions—the same convictions I held while sitting on that log early this morning while watching the crew pre-flight the airplane. Needless to say, this new stroke of luck gave new life and courage to the whole crew, and time after time we fought off the attacks of the fighters by returning their withering gunfire with tracers of our own that often found their mark, sending another Focke-Wulfe 190 spinning to the ground in flames. But still the fighters came screaming in, not one at a time, but two and three at a time and coming from all directions. Did someone say you could not maneuver a B-17? Don't let 'em kid you. On several occasions we had the ship in a seventy degree bank that left the attacking fighters firing at thin air. We were trying every evasive action we could think of, and they must have been good, for we had fought off approximately twenty or twenty-five fighter attacks without receiving a single damaging shot to either our ship or crew.

Both Don and I were getting so tired, and I had severe cramps in both of my legs from kicking the rudder so continuously and violently. However, during all of our maneuvering we made it a point to keep track of our formation, now quite a distance ahead and above us. Suddenly, there came an excited warning over the interphone "Fighter approaching from the twelve o'clock position—below." I immediately nosed the ship down, but it was too late, for he had us in his sights. I could hear the crashing glass, the breaking metal, and could see a spurt of flame in the cockpit, and then it was over! A light yellow smoke having an acrid sickening smell hung over the cockpit. I saw Don reach downward to his trouser leg. He had been hit! The wounds were not serious, but still it impaired his flying. It would be only a matter of minutes now before the fighters would have to turn back, for we were quite some distance out to sea, but we needed help if we were to last those final few minutes.

I started calling the lead ship (Major Putnam in the "Memphis Belle") by radio—once, twice, three times and no reply. I tried again, pleading for him to drop down some and reduce his airspeed so we could catch up with the formation. Finally, I received an answer. It was Capt. Campbell's voice (calling from Fishburne's B-17 #481 "Hell's Angels")! He had heard our call and was now making a large circle and dropping down to cover our tail, and just in time to shoot down a fighter coming in to attack us. Whew! We were safe from the enemy; now if we could only reach England. It was questionable at times, for number four engine was dead and windmilling, and number three engine was coughing and sputtering! We finally caught up with the formation and limped home—sometimes on three engines, and in short breathtaking intervals—sometimes on two engines. Since our flight was a long one and our gas was running low, we had to land at St. Eval Aerodrome, on the southern tip of England. The hydraulic system was shot out on our ship which meant no brakes, so we were last to land. It was a difficult landing but successfully accomplished, with nothing more than a mild ground loop at the end of the runway due to lack of brakes.

As the ship came to a stop and the wheels settled in the soft ground, I can remember now the tired sigh of relief Don and I exchanged. An ambulance came driving up and transported Don and Jim to the hospital before I got energy enough to climb out of my seat and make my exit through the front hatchway. Then I walked slowly around the ship and surveyed the many battle scars and damage. I was a bit sad as I drove away in the truck and looked back at our ship "Heavyweight Annihilator," for I had fears that she would never fly again. Whether she does or not, I will never forget that she lived up to her name against overwhelming odds, and she "slugged it out" to a victorious finish and brought her crew safely home.

The P-47 Thunderbolt was a welcome sight to any bomber flying into combat. Often called "little friend," the sturdy attack plane was necessary to keep the German fighters at bay. This plane has been described as probably the best attack aircraft ever built. -Frank Donofrio

Passing Comments for the January 3, 1943, Mission to St. Nazaire, France

There is much to learn from the formation diagram on page 80, for it is exactly the positions that the bombers flew in at least part of this particular mission. Jim Verinis (co-pilot) for the first few raids on "Memphis Belle" witnessed one unidentified B-17 vaporized by a direct hit in its bomb bay. Left Waist Gunner Bill Winchell saw it, too. Both of these men later commented about the dwindling condition of B-17 #069 "Our Gang" flown by Captain Ed Gaitley. That it had fallen out of the formation with No. 2 engine burning furiously, and that Lt. Fisher in B-17 #515 "Jersey Bounce" went down to help him fight off the enemy. (Pilots were not allowed to render assistance in this manner. Commanders believed that it jeopardized the entire formation when one of the bombers was not in its place in the assembly.) The "Jersey Bounce" consequently paid for this act of valor by being badly shot up. Both ships, however, made it back.

Also worthy of note is that the "Memphis Belle" actually flew with an extra gunner this day. Evans (Formation Bombardier) could not be distracted by having to man his guns there in the extreme nose, so a Sgt. Cornwell would stand behind him and operate his guns while Evans tended to his bombsight.

The events of this raid were not to be forgotten. A total of seven bombers and their crews were lost—one of them from the 91st. (#084 "Panhandle Dogey" flown by Lt. Anderson of the 323rd sqdn.)

38 enemy fighters destroyed. Of these...

24 fell at the guns of 91st BG B-17s. Of these...

17 fell at the guns of the 322 sqdn (91st BG). Of these...

6 fighters were dispatched by B-17 #482 "Heavyweight Annihilators"

Thirty-eight enemy fighters to seven bombers. To a statistician counting aircraft, this appears acceptable. To those who count people, it is not. Seventy American airmen lost, wounded, or held captive compared to only thirty-eight enemy airmen. The rumor that had been going around the 91st that they would be home in the U.S. by March 1st must have been welcome to these weary men.

As Lt. Bert W. Humphries feared, his B-17 "Heavyweight Annihilators" would never take to the skies again. She would stay at St. Eval, where her undamaged parts would slowly be stripped away to be transplanted into other B-17s that needed them. It would be exactly one month later when replacement B-17s began arriving at Bassingbourn and he was able to "pick" a bomber to replace #482. The January 3rd raid to St. Nazaire was to be Humphries' very last combat mission. His pilot and navigator were wounded and recovering in a hospital near Taunton, England. Humphries had taken some duties as Assistant Squadron Operations Officer and was hoping to become assigned in this role permanently. It was not until the end of February that a replacement co-pilot arrived at Bassingbourn to eventually be assigned as co-pilot for "Heavyweight Annihilators no.2" (B-17F #42-5712). Coincidentally, this B-17 would suffer nearly the same fate as its predecessor when it was badly damaged during the August 17th raid to Schweinfurt and crash landed near Manston, Germany. The crew was not so fortunate. Two were killed, three became prisoners, and three crewmen were returned.

Forts destroy 38 F.W.s

GERMAN FIGHTERS SWOOP IN 'SUICIDE' ATTACKS

By MONTAGUE LACEY

A BOMBER STATION, Monday.—United States Flying Fortresses shot down 38 German fighters yesterday over St. Nazaire, U-boat base on the French Atlantic coast.

Above: Noted artist of the 91st Bomb Group, Cpl. Tony Starcer. (Frank Donofrio) Right: British newspaper clipping, January '43.

"13 bombers on the 13th"

B-17F #41-24485	"Memphis Belle"
Pilot	Robert K. Morgan
Co-Pilot	Col. S.J. Wray
Bombardier	Vince Evans
Navigator	Charles Leighton
TT / Engineer	Eugene Adkins
Radio Operator	Robert Hanson
Waist Gunner	Clarence Winchell
Waist Gunner	Harold Loch
Ball Turret	Cecil Scott
Tail Gunner	John P. Quinlan

13 January 1943
Target - Railroad Marshalling Yards at Lille, France
Alert no.22 - Mission no.14 for the 91st BG(H)
As many as seventy-three bombers made up the Bombing Wing of this, the eighth mission for the "Memphis Belle" and crew. Four Bomb Groups were participating, with thirteen B-17s from Bassingbourn taking off at 12:35 in the afternoon. The only B-17 from the 91st that was damaged during the mission belonged to Lt. Felton. During the morning briefing, he made mention of the fact that thirteen bombers were flying and that the date was the thirteenth! Felton's co-pilot was Lt. Harold Kious, who was Bert Humphries' roommate! Humphries also remembered this incident in his wartime diary.

"Memphis Belle" navigator Chuck Leighton remarked that they were the formation lead again, that the raid was relatively simple, and that the 91st Commanding Officer Col. Wray was flying the Belle as co-pilot. Leighton remarked in his diary that he had witnessed two bombers go down (possibly one struck in a mid-air collision with an enemy fighter), and that he felt that they were about the only original crew still intact, and this was just their eighth mission! Mainly, Morgan's crew was still together. However, Jim Verinis (co-pilot) had been assigned to a different bomber ("Connecticut Yankee" #970), and Levi Dillon (TT/Eng) was re-assigned to another Group.

The "Memphis Belle" received many flak and bullet holes, but nothing that would amount to serious damage. Much of the enemy attention was drawn to other Groups in the formation. It was thought that the German pilots knew the reputation of the 91st's tight formations and preferred to attack different Groups instead.

"Memphis Belle" crewmen reported short enemy fighter attacks immediately following the release of the bombs. Also, light and inaccurate flak over the target was encountered. Felton's B-17 returned to Bassingbourn completely riddled and with three wounded crewmen. While the formation reported three B-17s missing, the 91st had only this one damaged. Felton had been flying in the notorious "tail-end Charlie" position in the formation, which was a favorite for the German Luftwaffe fighter pilots.

#42-2970 "Connecticut Yankee." This B-17 was assigned to Jim Verinis, who had completed flight training as a First Pilot. Assigned to the crew of Lt. Morgan, Verinis flew the first several missions of the "Memphis Belle" as co-pilot. Because of his proficiency as a pilot, it was inevitable that a Fortress would be assigned to him. Jim Verinis actually became the very first Eighth Air Force Airman to achieve the twenty-five mission mark aboard this B-17 on 13 May 1943. The bomber was then assigned to 2/Lt. Pelgram after Verinis came home to the U.S. with the "Memphis Belle" in June of that year. The Yankee was crash landed in a field that September near Winchelsea, England, damaged beyond repair. (Jim Verinis) Right: Bomb tag which flew aboard the "Memphis Belle" during the raid of 13 January 1943. Note that it was signed by Col. Wray, who was co-piloting the plane that day. The tag has been in the memoirs of "Memphis Belle" Radio Operator Bob Hanson since WWII.

"Sheer Drama"

B-17F #41-24485	"Memphis Belle"
Pilot	Robert K. Morgan
Co-Pilot	Lt. Col. Lawrence
Bombardier	Vince Evans
Navigator	Charles Leighton
TT / Engineer	Eugene Adkins
Radio Operator	Robert Hanson
Waist Gunner	Clarence Winchell
Waist Gunner	Harold Loch
Ball Turret	Cecil Scott
Tail Gunner	John P. Quinlan

The 91st was launched on the 27 January raid into Germany four days after this mission. The "Memphis Belle" pilot and crew took off, but they were flying the B-17 "The Great Speckled Bird." Records indicate problems with this ship forced them to return to base shortly before 11:00 AM. This mission did not count as "completed" for them.

23 January 1943
Target - Submarine Pens at Lorient, France
Alert no.27 - Mission no.15 for the 91st BG(H)
Thirteen B-17s took off from Bassingbourn at 10:35 AM in very poor weather conditions. All of them were loaded with a combination of 500 and 1,000 pound High Explosive bombs. This was the ninth mission for the "Memphis Belle" and crew, and it was not going to be pretty. Only eight bombers of the 91st reached the target for bombing. The other five had returned to base shortly before lunch. They were unable to maintain formation during the initial climb to the rally point and became separated in the mile thick overcast. The Belle was assigned as a lead bomber on this raid and did not turn back. An extra gunner was also aboard "Memphis Belle" so Vince Evans could focus his efforts as lead Bombardier for the other bombers again on that day.

Of those that pressed on, six tacked on to a Group that had successfully marshalled and was on its way towards Lorient, France. The remaining two found a different Group and flew with them to hit targets in Brest, France. This all resulted in a very bad situation for the men of the 91st, if not the entire formation!

Five B-17s were shot down (believed to be from the 303rd Group). "Sad Sack," a B-17 of the 324th squadron (#504) in the 91st BG(H) flown by Lt. Fisher, was badly shot up and had to make a forced landing in Southern England. One of his men died and two were wounded. Fisher received glass splinters in his eyes when his windshield disinegrated. For at least one twenty-two minute period the B-17s of the 91st suffered terrible attacks from enemy fighters that were probably drawn to them because of their small formation.

During the attacks, the German pilots guided their fighters right in on the noses of the B-17s. One of these broke through the American defensive firepower and headed—guns blazing—directly for the "Memphis Belle." Morgan emembered that the usual procedure was to put the thirty ton bomber into a dive to avoid severe damage. Because of the formation below, that tactic would not work this time. In only a few short seconds Morgan thought about evasive flying, then heaved on the control wheel of the Belle and raised her nose up and into the attacking FW-190. Quinlan, back in the tail of the "Memphis Belle," had to just sit back there and take it all. The enemy gunfire missed the front end of the Belle and began to tear into the underside of the tail. In sheer seconds metal was punctured, ripping, and tearing. Pieces were falling into the slip stream, and the "Memphis Belle" caught fire.

Quinlan was letting everyone on board know what he was going through, yelling that the entire tail was "shot off," and that it was leaving the airplane and it was burning! Morgan attempted to call his tailgunner on the interphone. No response. Almost a minute had gone by when Quinlan finally reported that the fire had gone out and they might be all right.

When the formation turned from the bomb run, Evans jumped up to one of the .50 machine guns mounted above his bomb sight and began firing at approaching fighters. The extra gunner with them there in the nose compartment was busy on the other machine gun. Navigator Chuck Leighton got in there to assist with the extra .30 machine gun, but it had to mounted before he could add to the forward defenses of the "Memphis Belle." The place to mount the weapon in the plexiglass nose cone of the B-17 was covered by a rubber plug that had frozen solidly in its place. When Leighton tried to force the plug out, he knocked it pretty hard and a large area broke away around the plug. Freezing 200 mile-per-hour winds howled through the nose of the Belle. To make matters worse, the elevators back on the tail had jammed and there was a two-hour flight back to base.

In the post mission report to his Crew Chief, Morgan listed a few defects in the "Memphis Belle." He wrote: "Vertical stabilizer, Rt. Horizontal stabilizer, #2 supercharger out, co-pilot slide panel split." It is clear from the formal forms used to report what repairs were needed after missions that the drama was always left out, leaving the reader to wonder just how bad a mission had to be in order to create damage such as this to a B-17 Flying Fortress.

"Into Germany"

B-17F #41-24515	"Jersey Bounce"
Pilot	Robert K. Morgan
Co-Pilot	Col. S. J. Wray
Bombardier	Vince Evans
Navigator	Charles Leighton
TT / Engineer	Eugene Adkins
Radio Operator	Robert Hanson
Waist Gunner	Clarence Winchell
Waist Gunner	Harold Loch
Ball Turret	Cecil Scott
Tail Gunner	John P. Quinlan

B-17 #544 "Pennsylvania Polka" was blown out of the formation with the no. 2 engine smoking. It was then jumped by two Fw-190s which shot out engines 3 and 4. The bomber spun in as the tail tore away. All aboard were killed. Bomber #589 "Texas Bronco" also received much damage just after reaching the target area. The gear dropped and the B-17 fell behind. A Bf-110 attacked the Bronco, shooting it down. The German plane also went down during its attacks. Two aboard this bomber were killed, while seven became Prisoners of War.
B-17 #545 "Motsie" returned early due to mechanical difficulties.

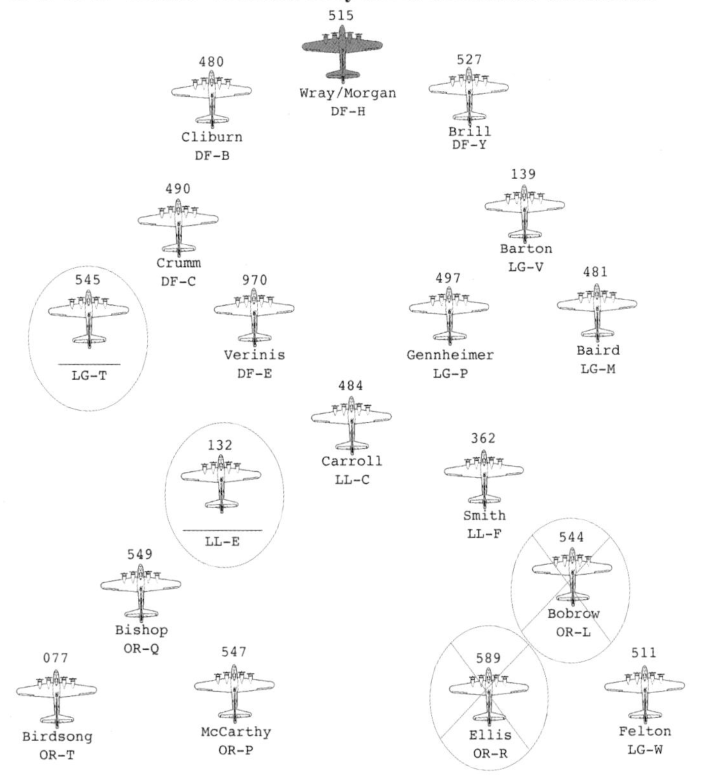

4 February 1943
Target - Submarine Pens at Emden, Germany
Alert no.35 - Mission no.17 for the 91st BG(H)
Primary Target: Hamm, Germany - Railroad Marshalling Yards
Secondary Target: Osnabruck, Germany
Alternatives: Emden, Germany - Submarine Pens
Any populated German town.
Enemy defenses: Fighters - (single engine) 55
(twin engine) 125
Flak Batteries - Heavy and Accurate
Time Schedule: Sta. 0805; Taxi 0835; T.O. 085O; Leave 0912; zero hour 1015; Target 1141; etr 1345
Bomb Load: 10 X 500 pound High Explosive
Bombing Altitude: 91st BG(H) at 19,000 ft & 306th BG(H)
Other Bomb Groups: 303rd & 305th in 102nd C/W
Order of Combat Wings: 101st; 102nd

This was the tenth mission for Morgan and his crew and their first trip into Germany. On this raid, the "Memphis Belle" would stay parked at Bassingbourn while her crew would fly into combat aboard a different B-17, the "Jersey Bounce." The Belle was still licking her wounds from the 23 January mission. In the right seat for his second combat trip with Morgan was none other than the 91st BG Commander Col. Stanley Wray.

Bob Morgan's crew in the "Jersey Bounce" was again leading the formation. This would be a difficult trip into the target, for it included a 150 mile trip inland towards the uppermost corner of the industrialized Ruhr Valley. Fighter opposition was expected to be unusually heavy, not only because of the obvious reasons, but because the Americans were now making strikes directly onto German soil. The Germans would fight much harder to prevent the bombers from succeeding at their mission.

The weather was ideal over East Anglia, England, and engine start and take off was uneventful. The Allied bombers formed up successfully, and at zero hour all sixty airplanes in the wing turned towards Germany. As the bombers approached Holland, the crews could see that the continent was obscured under heavy clouds. This would prevent them from hitting the primary target at Hamm. Similarly, the secondary target at Osnabruck was under the protection of cloud cover. The formation would not be able to hit these areas, and from the right seat of the "Memphis Belle" Col. Wray made the decision to "hit the nearest target of any size." They pressed on over Germany to hit the Submarine pens at Emden. As the bombers turned towards the third target area, the crews began to wonder if they would ever turn back towards England. The Luftwaffe threw everything they could at the bombers, and sweeping attacks from Me-110s, 210s, Ju-88s, as well as both Me-109s and Fw-190s struck hard at the Americans.

Winchell remembered that one fighter came in very fast from the left of the formation over Lt. Cliburn, flying "Bad Penny" (#480). That bomber was positioned on the left wing of the "Jersey Bounce." The enemy fighter somehow managed to penetrate the defenses of all the gunners, coming right over "Bad Penny" and then right be-

#41-24490 "Jack the Ripper" was among the very first B-17s to arrive in England with the 91st BG(H). She was also the very last bomber of the Group to be lost in combat. On February 22, 1944, the B-17 crashed at Munster, Germany. One crewman was killed, while nine others became POWs. See the formation diagram (previous page) to locate the position of this bomber during this particular raid. (Joe Harlick)

low the "Jersey Bounce"! (See formation diagram previous page). This particular mission would become known as the "Cooks Tour" because of the time spent over German ground. Even over Emden, the clouds were found to be nearly solid. The Bombardiers did see several breaks which would be enough to allow them to do their jobs. The formation released their weapons over Emden, causing considerable damage to the harbor and surrounding areas. However, the after-mission strike photos indicated overall poor results from this mission. Many felt that it would have been just as good to bring the bombs back to England.

Two of the bombers were lost from the 91st element, and a total of five from the Bombing Wing never returned. More than half of the aircraft returned with damage ranging from light to substantial. The 91st BG(H) mission summary regarded this run as "most unsatisfactory."

Aboard the "Jersey Bounce," Gene Adkins up in the top turret was busy working on his two .50 machine guns. Because of the tight space, he removed his gloves for about two minutes in order to get the covers off the guns. At more than fifty degrees below zero it was not long before he began to feel the effects of serious frostbite. The rough part was that this happened at the beginning of the mission and he would have to ride out the trip—and stay at his station—with frozen hands. Between fighter attacks Adkins was going back to the radio room and sticking his hands between his legs to keep them warm. When they landed, he found out just how bad his hands were. He was shipped off to a hospital where doctors worked to save them. Although they were successful, Adkins was off the crew, and his replacement was the man who had been manning the right waist gun as Second Engineer on the "Memphis Belle"—Harold Loch.

#42-29837 "Lady Luck" was the B-17 that filled in and replaced the Belle after the Memphis Belle was ordered home. This bomber returned to the U.S. in July 1945.

"Mission recalled - Return to Base"

B-17F #41-24485	"Memphis Belle"
Pilot	Robert K. Morgan
Co-Pilot	H.W. Aycock
Bombardier	Vince Evans
Navigator	Charles Leighton
TT / Engineer	Harold Loch
Radio Operator	Robert Hanson
Waist Gunner	Clarence Winchell
Waist Gunner	E. Scott Miller
Ball Turret	Cecil Scott
Tail Gunner	John P. Quinlan

14 February 1943
Target - Railroad Yards at Hamm, Germany
Alert no.40 - Mission no.18 for the 91st BG(H)
Raid #11 for the "Memphis Belle" and crew—but not much of a raid after all. The 322nd Squadron Operations Officer Bert Humphries was awakened late by the night-duty orderly and had to rush to schedule, then awaken the combat crews in time for the 0515 briefing. (Several "Valentines" were found in bed with some combat crew members in the Number Two Officer's Quarters!)

For the fourth time the 91st was scheduled to smack the Rail Yards at Hamm, Germany, and each B-17 was loaded with five 1,000 pounders. The bombing altitude was to be from 23,000 feet by some twenty bombers from Bassingbourn. The launch was uneventful, and the 91st marshalled in trail behind bombers of the 303rd, 305th, and 306th to make up the bombing wing.

Building cloud layers caused concern for the men in the formation as they flew high over the North Sea. After crossing into Holland, the crews could no longer see the ground and the mission was aborted. Because the bombers had penetrated more than twenty miles inland, this would count as a completed mission. The B-17s endured some anti-aircraft blasts over enemy territory, but this resulted in no damage or losses. No bombs were dropped, and all the airplanes returned to base with them still secured in the bomb bay shackles.

On the "Memphis Belle" Harold Loch was now riding in the top turret as Flight Engineer. He was now the official replacement for Gene Adkins, who had endured terrible frostbite ten days earlier. There was also a new waist gunner aboard named Emerson Scott Miller. He would fly on the Belle in this position through the end of its missions but would be unable to return to the U.S. with the bomber later in June with the rest of the crew. He would not have his twenty-five missions in and had to finish aboard different B-17s.

Bill Winchell in the left waist position remembered that the mission was more like a cross-country training sortie back in the states. But at least they chalked up another raid. Hopefully the next one would be as easy.

"Mags" off on two and three, the "Memphis Belle" taxis into position during her 1943 War bond and good will tour of the United States. This photo was snapped during her stop at Wright Field, Ohio. Engines one and four are providing turning power for the thirty ton B-17. It was common Flying Fortress procedure to taxi with the outboards, as differential power settings to these engines made turning much easier on the brakes and tires. Note the B-23 and B-25 behind the "Memphis Belle." (Frank Donofrio)

"We've lost number two"

B-17F #41-24485	"Memphis Belle"
Pilot	Robert K. Morgan
Co-Pilot	H.W. Aycock
Bombardier	Vince Evans
Navigator	Charles Leighton
TT / Engineer	Harold Loch
Radio Operator	Robert Hanson
Waist Gunner	Clarence Winchell
Waist Gunner	E. Scott Miller
Ball Turret	Cecil Scott
Tail Gunner	John P. Quinlan

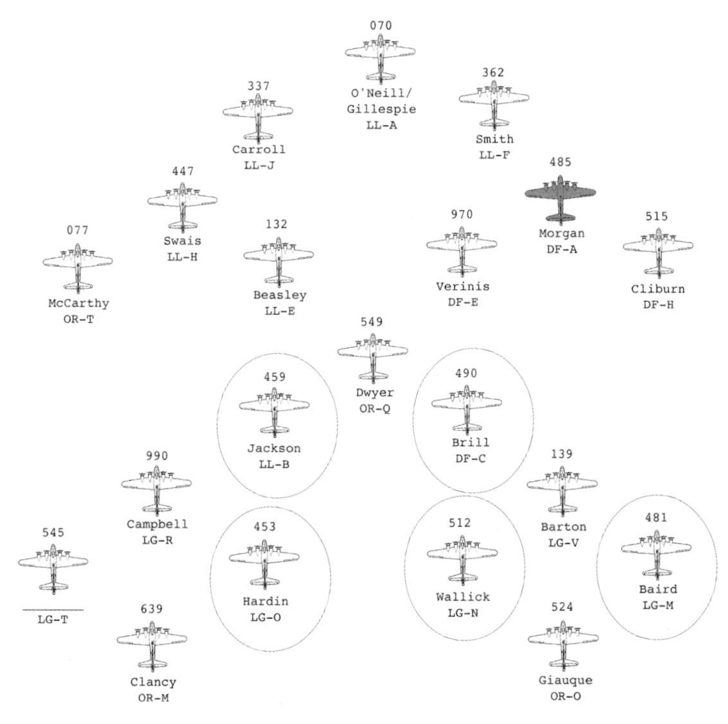

Of note here is lead ship #070 "Invasion 2nd" piloted by Capt. Oscar O'Neill, the bomber which first received the attention of Hollywood film director William Wyler when he arrived at Bassingbourn to produce his now-famous 1943 documentary. Wyler would have to switch the focus of his cameras to the "Memphis Belle" when later, on the 17th of April during a raid to Bremen, this bomber went "missing in action" (its 18th mission and O'Neill's 24th mission) and crashed in Germany. O'Neill and his crew became Prisoners of War. Capt. O'Neill is the father of Hollywood actress Jennifer O'Neill.

16 February 1943
Target - Docks and Navy Installations at St. Nazaire, France
Alert no.41 - Mission no.19 for the 91st BG(H)

Battle Journal Input

Target: (1) St. Nazaire, France - Docks and Navy Installations
(2) St. Nazaire, France - Turning Basin and Locks
(3) Lorient, France - Harbor Installations

Enemy Defenses:
Fighters (Single engine) fifty
(twin engine) fifty
Flak Batteries - Heavy and Accurate
Smoke Screen at Lorient

Friendly Fighter Support: Seven Spitfire Squadrons
Time Schedule: Sta 0750; taxi 0815; takeoff 0830; leave 0855; zero hour 1000; Target 1219; estimated return 1504.
Bomb Load: 5 X 1000 High Explosive
Bombing Altitude: 91st BG(H) 24,000 feet
Other Bomb Groups: 306th, 305th, 303rd

#459	"Hellsapoppin"	Returned Early
#490	"Jack the Ripper"	Returned Early
#453	"Mizpah"	Returned Early
#512	"Rose O'Day"	Did not take off
#481	"Hell's Angels"	Did not take off

Mission #12 for the "Memphis Belle" and crew, and the fourth time they would visit the German Navy at St. Nazaire. (This was the sixth time the 91st had visited this particular target.) The crews were surprised that they were ordered to hit French targets after assurances from Col. Wray that it would be German raids from now on. Rumors that the best Luftwaffe squadrons were relocated to the Russian front were put down when some of Göring's best aces met the bombers over the target area in sweeping frontal attacks. The flak corridors were reported as heavy, intense, and accurate.

The crews reported enemy fighter attacks all the way back across France. And "Memphis Belle" left waist gunner Bill Winchell reported in his diary that the Belle "Got two bad hits, one in the right wing, and another put our number two engine out. Got it feathered and came home on the remaining three. Number three leaked oil badly, but we made it home. Crew all okay, but the old girl will be laid up for a while."

Post mission reports indicated that some of the bombs of the 91st Bomb Group exploded with good results within fifty feet of the intended targets. Six B-17s from the wing were lost—none from the 91st. Sadly, Lt. Brill, flying B-17 #490 "Jack the Ripper" (324th Squadron), received very heavy damage and a direct hit from a 20mm that killed his radio operator S/Sgt. Middleton. (Note: in the diagram for this mission his bomber is listed as returning early. This bomber would be lost on a later mission—*see* table p. 90.)

In another attempt to describe the enormous losses suffered by the aircrews of the Eighth Air Force in this stage of the Second World War, one can look at the twenty B-17s that flew with the "Memphis Belle" on the Emden mission of 16 February.

Of these bombers only the "Memphis Belle" survived the war (#485 in the formation diagram). Sixteen of them would not see the end of 1943, and the remaining three did not make it through August 1944.

Imagine, if you will, the two hundred men inside these airplanes on this mission. If they had been assembled and briefed before this February 16th mission that eighty-eight of them within seven months would be dead, and that another sixty-four would become Prisoners of War, how many would have quit right there?

One hundred fifty-two men out of two hundred from these bomber crews would not return to Bassingbourn. This represents more than seventy-five percent, and these numbers are not hypothetical—this actually happened!

#362	"Short Snorter II"	Feb 26, 1943	crashed N. Sea, 10 Killed
#549	"Stupentaket"	March 4, 1943	8 killed 2 Prisoners
#512	"Rose O'Day"	March 4, 1943	ditched-Channel, 7 killed / 3 POWs
#545	"Motsie"	March 23, 1943	ground crew cigarette fire, destroyed
#070	"Invasion 2nd"	April 17th, 1943	crashed - Germany, 10 POWs
#337	"Short Snorter III"	April 17, 1943	8 killed 2 Prisoners
#481	"Hell's Angels"	May 14,1943	ditched N. Sea, 10 killed
#459	"Hellsapoppin"	May 17, 1943	crashed - Germany, 5 killed / 5 POWs
#515	"Jersey Bounce"	May 21, 1943	crashed N. Sea, 10 killed
#132	"Royal Flush"	June 22, 1943	crashed N. Sea, 1 killed/9POWs/1Evade
#077	"Delta Rebel No.2"	August 12, 1943	4 killed / 6 POWs
#453	"Mizpah"	August 17, 1943	Schweinfurt, 5 killed / 5 POWs
#524	"Eagles Wrath"	August 17, 1943	Schweinfurt, 3 killed/5 POWs/2 Evade
#990	"Dame Satan"	August 17, 1943	Schweinfurt, 2 killed/4 POWs/4 Evade
#139	"Chief Sly II"	August 17, 1943	Schweinfurt, 6 killed / 4 POWs
#970	"Connecticut Yankee"	Sep. 6, 1943	crashed England, crew okay
#490	"Jack the Ripper"	Feb 22, 1944	crashed - Germany, 1 killed / 9 POWs
#447	"Kickapoo"	Feb 26, 1944	crashed N. Sea, 10 killed
#639	"The Careful Virgin"	August 4, 1944	Sacrificed as part of the Aphrodite Project. Filled with highly explosive nitrostarch and remotely flown into target.

"Memphis Belle" Radio Operator Bob Hanson shows his lucky rabbit foot. The lucky charm flew all twenty-five raids with him. Bob had befriended a British Spitfire pilot who had used it on his fifty-two missions before giving it to Hanson. The rabbit foot flew no less than seventy-seven missions over Nazi occupied Europe!

"Memphis Belle crew flies #13"

B-17F #41-24515	"Jersey Bounce"
Pilot	Robert K. Morgan
Co-Pilot	H.W. Aycock
Bombardier	Vince Evans
Navigator	Charles Leighton
TT / Engineer	Harold Loch
Radio Operator	Robert Hanson
Waist Gunner	Clarence Winchell
Waist Gunner	E. Scott Miller
Ball Turret	Cecil Scott
Tail Gunner	John P. Quinlan

26 February 1943
Target - Shipping in Harbor at Wilhelmshaven, Germany
Alert no.45 - Mission no.20 for the 91st BG(H)

Battle Journal Input
Target: (1) Bremen, Germany - Focke Wulfe Aircraft Plant
(2) Wilhelmshaven, Germany - Shipping in Harbor (Reports state that German Navy Cruiser *Admiral Von Scheer* is harbouring here)
(3) Emden, Germany - or any other populated town
Time Schedule: Sta 0730; taxi 0755; takeoff 0815; leave 0838; ero hour 0930; Target 1103; estimated return 1400.
Bomb Load: 10 X 500 pound High Explosive
Bombing Altitude: 91st BG(H) 23,000 feet
Other Bomb Groups: 306th, 305th, 303rd

#970	"Connecticut Yankee"	did not take off
#225	"Stormy Weather"	did not take off
#447	"Kickapoo"	Missing in Action
#362	"Short Snorter II"	Missing in Action

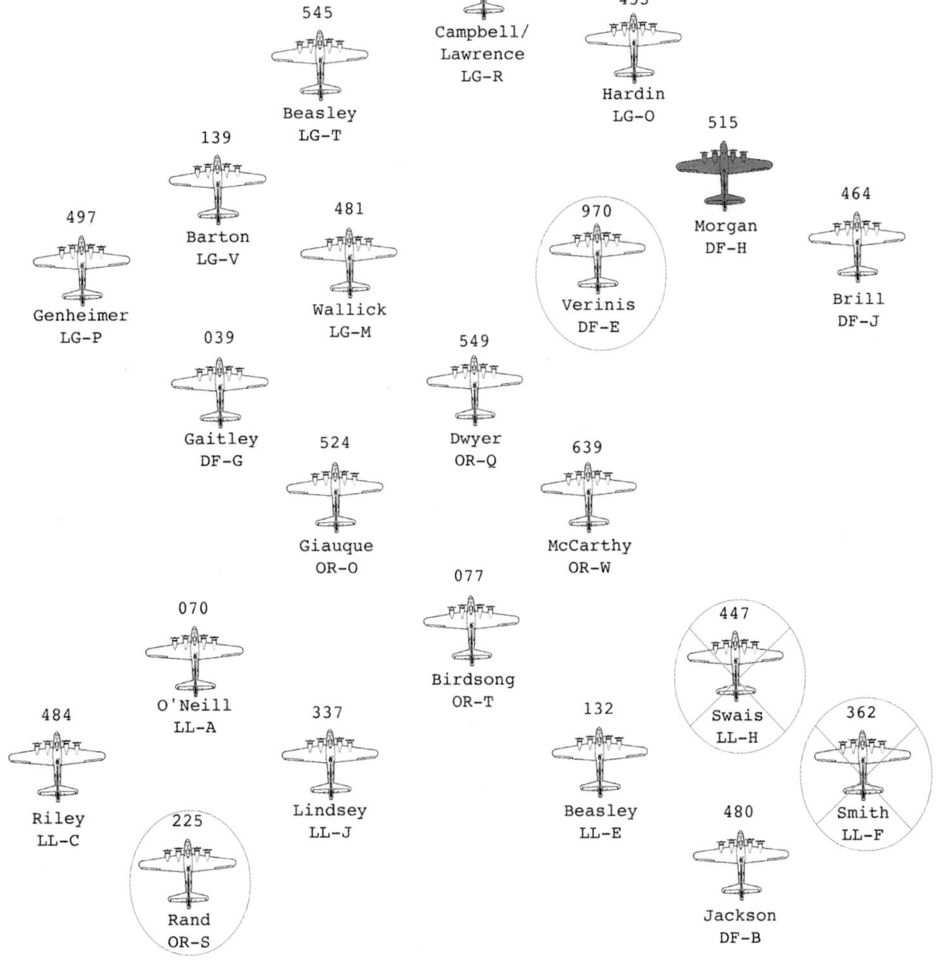

Note on this raid that the "Memphis Belle" crew is again aboard the "Jersey Bounce"—the Belle was still undergoing repair work. Also, Jim Verinis in #970 is positioned in the same spot in the formation just on Morgan's left wing as in the 16 February mission ten days earlier. He did not take off because of mechanical trouble. Lt. Smith in #362 "Short Snorter II" and his crew were last seen headed back to England near the Frisian Islands. A sea crash was assumed, and their fate was never known. The number 4 engine was smoking on "Kickapoo" (#447) when it was last seen. There was also no word as to what happened to Johnny Swais or his crew aboard that bomber.

A fairly thick haze covered Station 121 as the bombers launched to hit the primary target at Bremen, Germany. The mission was on schedule, but the assembly was a little difficult because of the visibility. The Navigator in the lead bomber failed to check the upper level winds, and this resulted in the entire formation appearing over the Frisian Islands, where they endured terrific flak barrages.

The enemy aircraft factory to be hit was protected from the American planes under a complete overcast, so the formation turned and headed for the briefed secondary—the Naval harbor at Wilhelmshaven. The harbor was partially covered by clouds, but a protective smoke screen was also making sighting the bomb drop even more difficult. All of the bombs were dropped on the shipping basin, but strike photos later showed that the results were very poor.

The crews also reported that the Germans were making another attempt at bombing the American formation from above. This was as inaccurate as the previous attempt and did not cause too much concern. Some reported that the Luftwaffe seemed to be concentrating their attacks on the B-24s below them. In B-17 #481 "Hell's Angels" Capt. Wallick had to drop down out of the formation. Flak had blown a large hole in the plexiglass nosepiece and half of his oxygen system was shot out, along with severed control cables and more.

No less than five B-17s and two B-24s were lost from the total U.S. bombing force. Two of these B-17s were of the 401st Squadron (91st BG). Nineteen combat crewmen suffered frostbite, putting a strain on the already depleted flying rosters. Forty-one men of the 91st were lost, equivalent to the crews of four heavy bombers.

B-17F #42-2970 "Connecticut Yankee." This bomber was flown by Jim Verinis (Memphis Belle Co-Pilot for the first missions.) Jim was also the very first Eighth Air Force crewman to achieve the coveted twenty-five mission mark! This feat was accomplished on 13 May 1943 in the bomber you see above. After Verinis returned to the "Zone of the Interior" (the United States) with the "Memphis Belle" crew, #970 was assigned to Lt. Pelgram, who was at the controls on 6 September when a crash landing was made in the U.K. The B-17 was damaged beyond repair and was salvaged soon thereafter. Note that the weary bomber has thirty-six missions and eleven enemy fighter kills credited. "Connecticut Yankee" is seen near the center of the formation diagrammed on the preceding page. -Jim Verinis

"Second day in a row"

B-17F #41-24515	"Jersey Bounce"
Pilot	Robert K. Morgan
Co-Pilot	Jackson
Bombardier	Cornwall
Navigator	Ehrenberg
TT / Engineer	Robbins
Radio Operator	Current
Waist Gunner	Kirkpatrick
Waist Gunner	Pope
Ball Turret	Cole
Tail Gunner	Nastal

27 February 1943
Target - River harbor at Brest, France
Alert no.46 - Mission no.21 for the 91st BG(H)

Battle Journal Input
Target: (1) Brest, France - Port Militaire
Time Schedule: Sta 1127; taxi 1152; takeoff 1212; leave 1232; zero hour 1400; Target 1429; estimated return 1630.
Bomb Load: 5 X 1,000 pound High Explosive
Bombing Altitude: 91st BG(H) 24,000 feet
Other Bomb Groups: 306th, 305th, 303rd

As a result of missing a prior raid on the "Memphis Belle," Pilot Bob Morgan would be able to catch up with the rest of his crew by flying a mission on the "Jersey Bounce." Morgan was certainly aboard the "Bounce" this day, but it is unknown if he was the pilot or co-pilot, as the available records are conflicting. This was indeed, though, his thirteenth raid. "Memphis Belle" Navigator Chuck Leighton also flew on this raid (on yet another B-17), while the rest of the regular "Memphis Belle" crew stayed at Bassingbourn. The "Belle" was still being repaired from damage received on 16 February.

Morgan, in B-17 #515, was positioned just on the left wing of the 91st BG lead ship. Some said that poor lead ship navigation on this raid again led to landfall some fifty miles from the intended point. This naturally altered the "initial point," and the crews hurried to make adjustments for an unbriefed bomb run. The target was bombed through heavy broken clouds and no ships were lost.

Many of the crews considered this to be a "milk run," likely due to excellent friendly fighter escorts by both RAF and Allied fighter squadrons. Three friendly fighters, however, were shot down.

Post mission reports indicated overall poor results, with only several bombs successfully damaging military targets. Sadly, some weapons fell into a residential area, likely doing considerable harm to French civilians and housing.

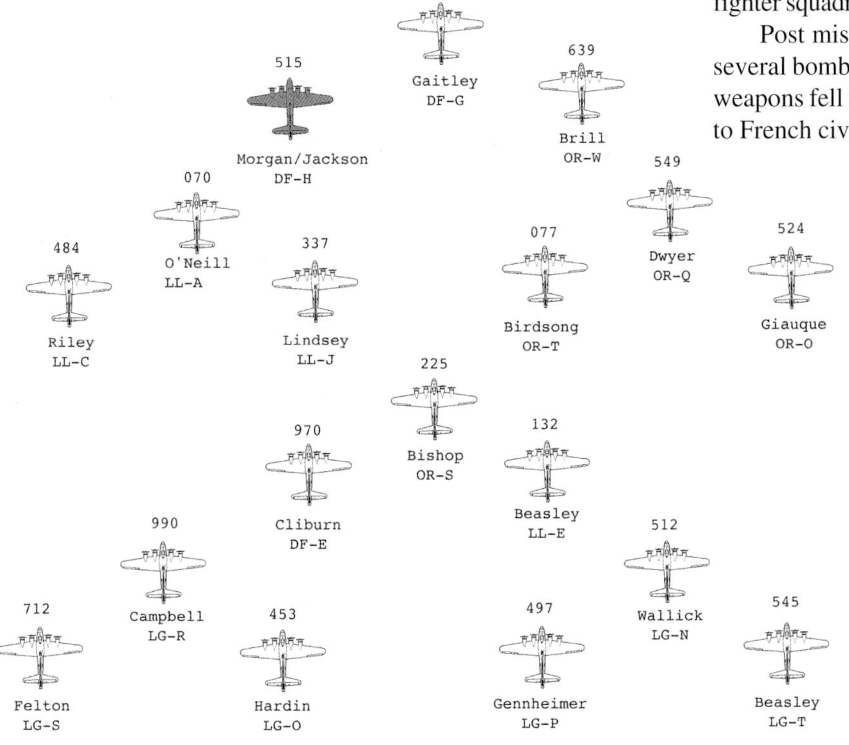

Note: The crew that flew with Morgan this day was made up of mostly regular members of the "Jersey Bounce" (#515) crew. The tail gunner here (Casimer A. "Tony" Nastal) would eventually be assigned to return to the U.S. with the crew of "Memphis Belle" for the famed War bond tour.

"What have we learned?"

(Col. Bert W. Humphries)

Summarized by Col. Bert Humphries
Operations Officer 322 Squadron / 91st BG(H)

While the Eighth Air Force strove to perfect the art of daylight precision bombing, the value of these raids was still questioned in both British and American circles. During January 1943 Allied leaders conferred on war plans at Casablanca, where the British again expressed their belief that the U.S. heavy bombers would be better employed in aiding the RAF on night assaults.

General Arnold had General Eaker flown down from London to Casablanca in order to present the case for continued daylight bombing raids. Apparently Eaker presented his case quite effectively, for one of the results of the conference was a policy directive calling for a day and night strategic bomber offensive in which the RAF and the USAAF would combine. (Not until late spring 1943 were detailed plans for such an offensive completed and put into operation. Meanwhile, the Eighth Air Force's bombing campaign continued to be regarded as on probation by Allied planners.)

The basic question after the VIII Bomber Command's twenty-one missions (through February 1943) remained the same; could the Eighth's bombers make effective attacks without prohibitive losses?

So far the B-17s and B-24s had shown a lower loss than the RAF on night raids. But it was on the night bomber that the Germans had been concentrating their defensive efforts, due to their belief that large bombers were too vulnerable for daylight operations. They had been caught off guard when the U.S. bombing formations appeared. Nevertheless, the Luftwaffe had been quick to rally, for whereas 3.7% of the attacking daylight bombers had fallen to enemy fighters in November 1942, the figure had risen to 8.8% in December.

Bombing results appeared to vary from good to very bad. In any case, the effect of the early daylight bombing missions upon the German war effort could hardly be more than a matter of extreme irritation, since the amount of damage created was readily repaired owing to the small force of bombers. Even so, the Eighth's commanders looked for better bombing results as their tactics improved, for it had been as much a case of finding the best method of attack as actually making the attack.

In seeking to prove the worth of U.S. bombing doctrine, the Eighth was beset with many other problems, the greatest of which was undoubtedly weather conditions. The essential good visibility for precision bombing was seldom available in the winter months. Cloud cover had to constantly be reckoned with, for a cloudless or near-cloudless sky occurred on the average of about once every twenty days during the winter of 1942-1943. This was the primary factor conditioning the slow pace of operations. Even on a day that started with a clear blue sky, such was the variable nature of the weather that a complete overcast could develop within hours. During January-February 1943 operations, twenty-six combat missions were planned but only eight were completed. Often the bombers would be airborne when deteriorating weather over the continent caused the abandonment of a mission. This became a frustrating and morale dampening occurrence for the aircrews, yet was to be a regular feature of daylight bombing operations.

Mounting a mission was a hectic and laborious task that commenced hours before the bombers took off and entailed meticulous staff planning. As soon as the "combat mission alert" was received at a bomber base, engineering sections would ready their available aircraft, checking and testing most of the night. The ordnance crews who handled the high explosive bombs would begin loading while it was still dark, as ammunition was put aboard and turrets were given a final go-over by the armorers. Radios were checked by technicians, and the still closely-guarded Norden bomb sights were installed in the nose compartment. Oxygen bottles were replenished, cameras for strike photos were loaded and set, fuel tanks were topped off, and a variety of other essential tasks were also performed.

Meanwhile, squadron operations personnel were involved in determining the make-up of combat crews and their assignment to in commission aircraft. Then those combat crews would have to be awakened and cooks would prepare breakfast that would fit into a schedule that would allow a "briefing" period by S-2 intelligence officers, and yet allow time for crews to collect and prepare their flying equipment.

Transportation had to be arranged to carry the crews to the "briefing," and from there to the supply section for the issuance of parachutes, flak vests, helmets, "Mae West" flotation devices, escape packets of continental currency, maps, compass, signal mirror, etc. The supply section transport then carried the crews to their respective aircraft in the dispersal areas. A number of specialist tasks had to be squeezed in, as well—the Chaplains had their moment for a brief religious service immediately following the "briefing" session; Navigators would have to study routes and schedules; Bombardiers studied the relationship of the "initial point" to the target from the latest available photographs; pilots reviewed the radio frequencies and beacons to be employed in gathering the formation by bomb groups into combat wings at the designated "zero hour" for departure; and gunners checked the unloaded weapons for ease of operations while the Radio Operator set up the required radio

"Memphis Belle" in the air over England. Look for another photo taken during this same flight near the end of this book. Note: someone is sticking their head out of the radio room hatch! - Frank Donofrio

frequencies for the mission and made the required check-in to the Control Tower.

As the minutes ticked by and the crucial moment of takeoff approached, nearly every man on the bomber base played his part in preparing and executing the mission, knowing that at any time—from the field order coming over the teletype machine to the Group's formation approaching the enemy's coast—that the mission could be canceled, or "scrubbed," to use the idiom of the day.

To the crews under pre-combat tension and stress, a mission "scrub" was no relief, for it meant that they would have to face the same procedure again before they could cross off another mission from the twenty-five each was usually expected to fly to complete a "combat tour."

The restored "Memphis Belle" cockpit as it appears today, down to the photo of Margaret Polk in the compass card holder. From this flight deck men witnessed the vain struggle of many B-17s as they tried to claw into the air before often falling in flames to the European earth. Of the surviving Flying Fortresses, "Memphis Belle" is the only B-17 that flew her particular tour of duty that remains today. As mentioned before, the 91st BG had 428 B-17s assigned to it. Before the *Ragged Irregulars* returned to the United States, 197 were lost. This image is available as a full color poster through the courtesy of Chuck Pearson, Knoxville, TN.

To the ground crews it meant that guns, bombs, bomb sights, cameras, ammunition, and other equipment had to be removed, and topped-off fuel would have to be drawn off from each involved bomber. And possibly a few hours after this stand-down activity had been completed, the whole process might begin again. A "scrubbed" mission was the most morale-sapping problem of the Eighth's first winter.

The English winter also hindered operations in another way. Dampness affected the functioning of the aircraft, for moisture and water that found its way into the components of the aircraft while parked in the dispersal areas would turn to ice at 15,000 feet. At combat altitudes gun-actuating components, gun turret and bomb door mechanisms, as well as trim tabs failed to operate due to lubricants freezing.

Also, superchargers often could not be operated because of congealing oil in the regulator lines. New oils with anti-freeze elements eventually solved most of these problems. The original U.S. oils proved unsatisfactory, and British products filled in pending the supply of new types from America. It was necessary for guns to be completely covered with oil to ensure their opertation at high altitudes. The sub-zero conditions at extreme altitudes were also responsible for landing lights burning out, and the glass covers cracking through sudden changes of temperature.

The temperature gauge installed in the nose compartment of the B-17 had -50° F as its lower limit, and the majority of time spent at bombing altitudes the gauge was "pegged" at that mark, so the real temperature was seldom known.

Air crews suffered much during this first winter. Oxygen masks had a nasty habit of becoming frozen closed with ice, and if a malfunction was not noticed by a crew mate, death would follow quickly. The electrically heated shoes and mittens were particularly plagued with problems, resulting in hundreds of cases of frostbitten feet and hands.

Another factor influencing VIII Bomber Command's rate of operations was the demands of "junior," as the Twelfth Air Force was dubbed. Apart from training units and supplying personnel, the Eighth had parted with seventy-five percent of its stocks of aircraft spares and servicing equipment. The cupboard was indeed bare, and there was little hope of replenishment in the near future, for the North African campaign continued to have priority on the limited shipments from the U.S.

The situation would have been even worse had it not been for the British assistance in providing transport, servicing equipment, tools, bombs, dinghies, and flying clothes, not to mention items for base and personal use. The transport included several thousand bicycles, which acted as personal transport for Colonels down to Privates. Much of this equipment was counted a poor substitute, yet some was retained in preference long after defencies were made up from the U.S.

Above all, the few aircraft and crews reaching the Eighth in the closing months of 1942 were inadequate to replace operational losses. In consequence, combat crews saw their units visibly shrinking with every mission, a situation not conducive to good morale.

All twenty-one missions so far undertaken by VIII Bomber Command were to targets in enemy-occupied territories. There had long been a desire amongst all ranks of the Eighth Air Force to attack targets in the enemy's homeland. Apart from wishing to be on equal terms with the RAF, whose bombers raided Germany almost nightly, the bombing of enemy targets in France and the low countries had often resulted in casualties among the local civilian population, and the German propaganda agencies were quick to exploit these occurrences.

When a B-17 was no longer flyable, the crew had to know how to get out. Diagrams like this were included on posters and in training manuals, as well as placed all over the bases. Combat situations sometimes made new openings that the crew could use to affect their escape from a plane, as flak could blow new "doors" into the body of the bomber. Getting out of a spinning and falling B-17 was often impossible for some aircrew. Wearing bulky flight gear associated with high "G" loadings could pin a man to the floor or wall of the plane, making it impossible for him to get to any opening that might be available for him to jump through. (Frank Donofrio)

Inevitably bombs went astray, and it was understandable that air crews preferred this to happen over Germany. Additionally, the first offensive action against the enemy homeland would bring badly-needed publicity to the Eighth Air Force's efforts. Now, only time would tell whether or not the Eighth was ready for the "tough" targets in the homeland of Germany.

4 March 1943 (A notorious day for the 91st BG)

The men of USAAF Station 121 would reel from one of the most punishing days they had yet known. On March 4, tragedy struck the B-17s from Bassingbourn when they hit the Railroad Marshalling Yards at Hamm, Germany. (Robert Morgan and Jim Verinis both flew this mission, even though "Memphis Belle" was still under repair. Morgan was in "Jersey Bounce," and Verinis was in "Connecticut Yankee." Morgan was forced to return to base due to mechanical difficulty with his B-17.) Of the twenty bombers launched that morning, three returned early, four were shot down, and one was damaged beyond repair. (Even though "Memphis Belle" was not on this mission it is still included in this examination, because three members of her regular crew took part aboard other B-17s.

Returned Early	**Missing in Action 4 March 1943**
#453 "Mizpah" LG-O	#512 "Rose O'Day" LG-N (7 killed/3 POWs)
#077 "Delta Rebel No.2" OR-T	#464 "Excaliber" DF-J (3 killed/7 rescued)
#515 "Jersey Bounce" DF-M	#370 *not named* DF-K (9 killed/1 POW)
#549 "Stupentaket" OR-Q	(8 killed / 2 POWs)

Severly Damaged - #225 "Stormy Weather"

One of the more unusual incidents of the Air War happened on this day when Capt. Birdsong and his crew took off for the raid aboard "Delta Rebel No.2," but finished the mission aboard "Stormy Weather" later that afternoon. Birdsong would end up with the Distinguished Flying Cross for his part in the mission. During the takeoff Birdsong experienced trouble with "Delta Rebel No.2" and headed back to Bassingbourn. The Captain was able to land his B-17 just moments after the last bomber from the 91st BG launched, and in less than twenty minutes Birdsong and his crew had transferred all of their gear and were again airborne in the B-17 "Stormy Weather." They managed to catch up to the 91st formation before reaching the Combat Wing assembly point. The 305th and 303rd Bomb Groups were in the lead, and the 306th and the 91st were in trail. Again the bombers were on the all too familiar route over the Norfolk coast before making a turn to the northeast out over the North Sea. It was hoped that enemy detection would lead the Germans to believe that their coastal ports were in danger. Some fifty miles from the English coast the formation turned southeast to head across Holland towards the target. A thick haze obscured visibility, and the pilots were struggling to see even a thousand yards from their position above the overcast. With little sign of improving, the lead Combat Wing made the decision to turn back. Their southerly course for England brought them above clearing conditions, and the decision was made to hit the tertiary target at Rotterdam. The 306th elected to return to base with their bombs, leaving the sixteen B-17s of the 91st alone to face the defenses surrounding Hamm, Germany.

In the lead ship for the 91st BG was Major Paul Fishburne, and he found his small band of Fortresses winging their way in clear skies towards the fierce Luftwaffe fighters that would soon pounce them. He had every right to abort the mission, but decided to press on with the attack. Moderate flak caused little trouble for the bombers, and after bomb release a good turn had them headed for home. Now it was time for the fighters. Lt. Henderson in B-17 #370 was last seen in the area of the target on fire and spinning to earth. Nine aboard his bomber were killed, and only one lived to become a POW. Lt. Brill in B-17 #464 "Excalibur" received the brunt of a frontal attack from a Bf-110 that showered the B-17 with machine gun fire, knocking out his number one and three engines. Number four was running rough when one and three succumbed and stopped, and then number four ran away, leaving Brill with just one engine to try to get home on.

"Excalibur" slowed to just over one hundred miles per hour, and the bomber fell far behind the formation as it descended under little control. Brill ditched "Excalibur" in the chilling waters of the North Sea, but his men had trouble inflating the life rafts. Brill, his co-pilot, and the ball turret gunner were swept away beyond rescue, but not before they managed to get both rafts inflated and the injured crewmen into them. Lt. Brill and his co-pilot, Lt. Lowry, both received posthumous Distinguished Service Crosses for their sacrifices.

As the remaining bombers made the Dutch border all hell broke loose, as an estimated 175 enemy fighters swarmed over the six-

George Birdsong started this mission flying "Delta Rebel No.2," and ended the raid flying "Stormy Weather!" (Frank Donofrio)

Major Paul Fishburne. (Frank Donofrio)

teen bombers. B-17 #524 "Eagles Wrath" was damaged with injured on board, and #990 "Dame Satan" was getting by on only a single properly operating engine! In B-17 #225 "Stormy Weather," Capt. Birdsong had his hands full. An exploding 20mm shell ripped through the nose compartment, starting a fire in the oxygen system. His number three engine was out of commission and its propeller would not feather. Birdsong reported later that the fighters were so intense for him that they actually followed him into their own flak, and violent evasive flying was necessary to keep him from being rammed on two occasions.

"Stormy Weather" was again rocked by a direct cannon barrage (this time from an FW-190), and the co-pilot's side of the cockpit was opened in an instant. The co-pilot was hit. leaving a huge gash in his head. In a fit of pain, "Stormy Weather's" right seater was thrashing about so violently that he actually slipped out of his seat and jammed the rudder pedals and flight controls. Birdsong was hit and blinded on his left side. Up in the nose, the navigator too had a bad head wound.

"Stormy Weather" began to fall behind the formation. Her engineer was helped by the radio operator in moving the injured and despondent co-pilot back to the relative comfort of the radio room. Up front, Birdsong saw two more stragglers and tacked on to them. They were B-17s #512 "Rose O'Day" and #549 "Stupentaket." "Stormy Weather" took the center position, and the three Flying Fortresses headed for England. Now it seemed that the three would protect each other on the flight back. This feeling would not last long at all as an FW-190 tore through the group out of the three o'clock position, and a burst from him found its mark on "Stupentaket." B-17 #549 exploded, killing McCarthy and all of his crew, except the bombardier and ball turret gunner.

Felton was next. His bomber (#512) was on the receiving end of a Bf-110 attack that lethally wounded the airplane, and "Rose O'Day" began a sickening spin to earth. Only a single waist gunner, the radio operator, and co-pilot Harold Kious survived. Birdsong was now alone again and determined to get home. He put his wounded bomber into a dive for the deck. It was all "Stormy Weather" could take, and as the battered Fortress leveled off on the water, enemy fighters continued to attack, with bullets and cannon fire still finding their mark on the already perforated B-17. "Stormy Weather's" gunners kept the Germans at bay until the ammunition was gone. A lone Bf-110 came alongside "Stormy Weather," its pilot inspecting the limping Fort; he shook his head in disbelief, saluted, and peeled off.

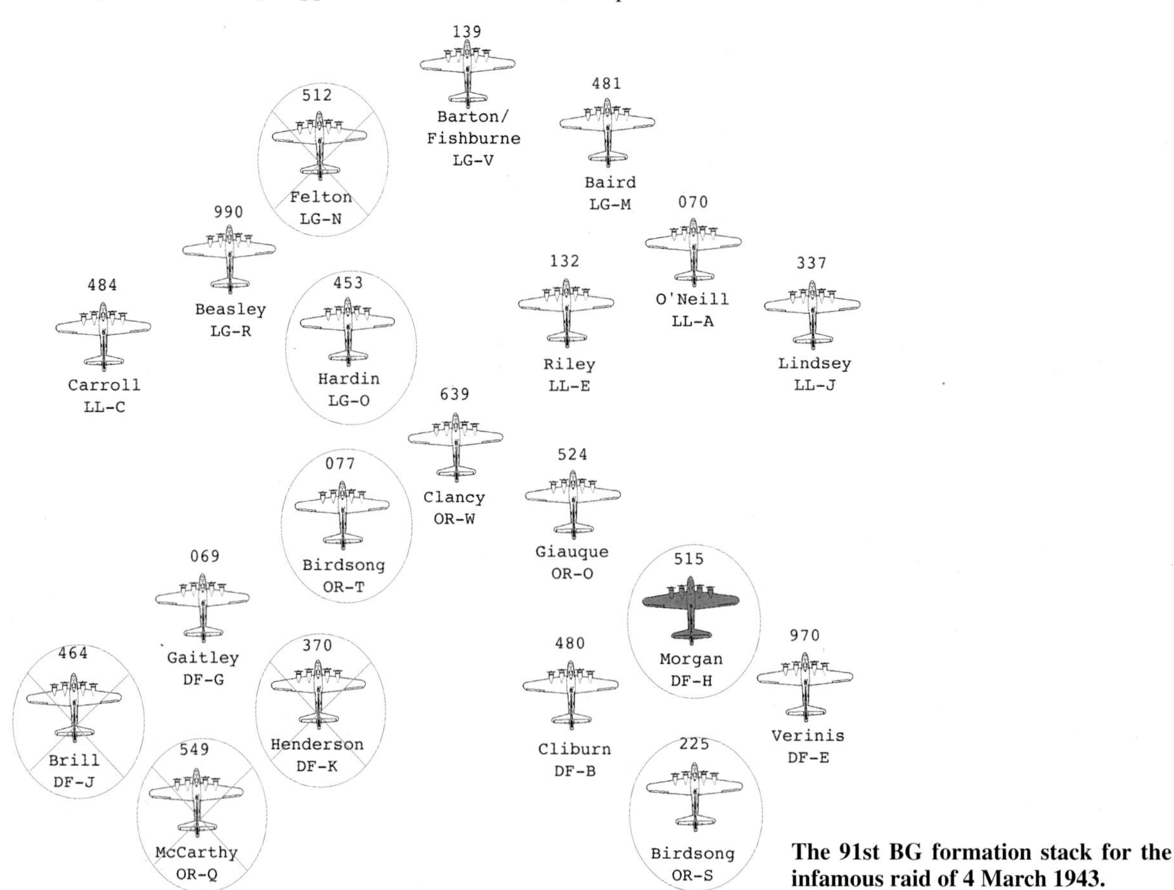

The 91st BG formation stack for the infamous raid of 4 March 1943.

An exploding bomber created a sight which could never be forgotten. This B-17 from the 483rd Bomb Group went down, taking its entire crew. Some from the 91st remembered that it was a scene very similar to this when one of their own, #549 "Stupentakit," was lost on 4 March 1943. (Frank Donofrio)

Her bombs spilling away, this 465th Group B-24 endures a violent death from a direct flak burst over Blechhammer, Germany. All aboard perished. (Frank Donofrio)

It was not over yet, The navigator was unable to plot a course for England, and the bombardier informed Birdsong that the bombs were still on board! The bombs were made "safe" by reinstalling the pins (how the bombardier even found them through the melee of the preceding hour or so is incredible). On their arrival over England, Birdsong informed his men that they could elect to stay with the bomber for a landing attempt or parachute, and they all stayed. With no brakes, Birdsong lined up for the Bassingbourn runway but was unable to get the B-17 to settle down to the concrete in time to roll out for a stop. He overshot and cobbed the throttles to his remaining good engines, but they did not provide sufficient power to get the plane safely in the air. "Stormy Weather" tore through bushes and barbed wire, jumped a roadway, and ground looped in a field just off the station—all this with the sensitive load of RDX bombs still aboard! Everyone wondered if Birdsong should have just stayed at Bassingbourn when he returned earlier that morning unable to fly the mission on "Delta Rebel no.2"!

The story of the "Memphis Belle" could easily have ended with this mission. Had Morgan continued on in #515 it is not beyond belief that he could have been killed or taken prisoner. Jim Verinis remembered feeling that no one would make it back of the sixteen bombers that appeared over Hamm that day—the running hour and forty-five minute fight severely damaged just about every B-17 up there. However, his bomber #970 actually received little damage!

The 91st BG gunners claimed thirteen enemy fighters destroyed, with three "probables" and four damaged. The cost was high, however, with four bombers lost, thirty-four missing or dead, five seriously wounded, and a fifth B-17 seriously damaged. #225 "Stormy Weather" was eventually repaired and renamed "V-Packette," only to be shot down on the notorious mission to bomb Schweinfurt on 17 August 1943.

"My Gun was a Brownie!"

Many of the books and writings that include photos of aerial combat today have the photo labs of the Eighth Air Force to thank for their hard work and dedication. They did not know it then, but they were creating image archives for many generations who would go on to cherish what WWII air combat was really like. So important was their work that mention of them is prudent whenever the air war is remembered. One of these guys was named Joe Harlick, and included here are his very words:

> The photo unit photographed each flight crew member in civilian clothes. He would carry these for fake passports, improving his chances for escape if he should be shot down behind enemy lines. Our primary objective was for bomb strike photographs, but there were a thousand other pictures made with hand-held cameras of the base activities, such as: battle damage after a mission; flight crew group shots; nose art on the planes; and visiting dignitaries, which included Royalty, high-ranking Generals, movie celebrities, and USO performers. Many times we would be given a photograph of the next day's target to copy, and we would have to have prints made for handouts at the briefing.
>
> All of the film processing was in black and white, with the exception of one roll of experimental color film sent to us by Eastman Kodak Company. With a shortage of 4X5 cut film, I designed a film splitter from some rollers and razor blades. We saved the ends of 9 1/2 inch wide film and split it into four inch widths and five inch lengths (in the dark, of course). This also worked great for my personal 120 size roll film camera. I carried my own camera everywhere, and that's why I have so many candid photos around, on, and off the base.

91st BG(H) Bassingbourn, England. Photographic Specialist Corporal Joe Harlick mounts a K-17 strike camera into its position beneath the radio room floor of a B-17. Joe was responsible for saving thousands of precious WWII negatives at his home in Ocean Shores, Washington.

We also had several loan type hand-held cameras that crew members could check out for combat missions, if they were interested in photography. The K-20 was a hand-held 4X5 roll film camera, and we also had a 16mm Bell & Howell Filmo camera and a Kodak 16mm camera. We lost 86 strike and hand-held cameras in combat before the D-Day landings.

As for working hours, we didn't punch a clock. If there was work to do, we stayed with it until completion. Many times I worked for thirty hours straight, got a few hours sleep, and then our planes would return from a mission and we would be back at it. The strike cameras would receive new loads of film after they were checked mechanically, usually about midnight. It would be 1:00 AM before we would get the information on the target for the next mission. Headquarters would give us the flight plan and serial numbers of the bombers designated to carry strike cameras. I would designate which camera would fly in each plane. The B-17s were well dispersed in case of a German air raid to prevent major losses in case of strafing or bombing. We would have to find the designated B-17 in the dark, and usually in fog, over a seven mile perimeter on the base. Many times we would get lost just trying to cross the runway. Because of blackout rules we had only a two-inch slit of light from our headlights on our 4X4 truck. You would have to be within a few yards of the plane before it was possible to see the numbers on the tail. At 3:00 AM the base was a beehive of activity. The planes were being loaded with bombs, high-octane gasoline, oxygen, cameras, ammunition, life-saving gear, etc., all at the same time. There were also last minute repairs on the engines, radios, navigation equipment, and cleaning windows. The Norden bombsight was the last item to be loaded in the lead planes, just before taking off.

The floor of the radio room was the only flat piece of floor in the B-17, and was also the warmest. Therefore all wounded personnel, if possible, were brought to this area. Since our strike cameras were located in a pit below this floor, I had many unforgettable bloody scenes of this area forever implanted in my memory.

Since members of the photo unit were not compelled to fly, nor received flight pay, a method had to be found to turn on the strike cameras as the bombs were released and then off after the bomb run. This turned out to be a challenge. For that reason many of the photo personnel flew missions as volunteers. Our first photo officer, Lt. Oakley, did not return from one of these missions.

Since photographers were not there to turn on strike cameras, a switch was installed for the radio/gunner to turn on and off. This did not work well, though. If he was busy manning his gun or his radio, the camera was forgotten, resulting in no photographs. We took this problem seriously, as we were instructed that if we failed to get photographs of the target area on the bomb run the Group would not receive credit for the mission.

Additional camera problems were encountered. The outside temperature could sometimes drop to as much as sixty degrees below zero. The film would get brittle and break or jam, or the lens would frost over with ice, resulting in no photographs. To conquer some of these problems we used yankee ingenuity. The bombsights had 24-volt heater muffs that were plugged into the power supply, so why not camera muffs? A muff was designed for each model camera, slipped over the magazine, and plugged into the 24-volt power supply. Protective filters were attached to the lenses with heater wires between the lenses. This pretty well took care of the freezing problems, but we still could not rely on the personnel on board to turn the strike camera on and off. After studying the wiring diagram in the B-17, I discovered a delay relay in the bomb circuit. By hooking the camera into this agastat it worked well on the ground, but altitude and temperatures affected the pneumatic delay here, also. It was still not 100% reliable. Finally, I came up with a foolproof switch. I called it my "mousetrap" switch. It turned the strike cameras on positively when the first bomb fell from the bomb bay. For inventing this switch, I was awarded a Certificate of Merit from the Eighth Air Force (no raise in pay, though). Then each camera had a micro switch installed in the film magazine that would turn it off when the film ran out. The larger cameras had an intervolameter that could be preset for a certain number of exposures and then shut off. In order to save film, we would re-roll in the darkroom and cut the roll of film into halves or thirds, depending on the length of film. This solved about 95% of our strike camera problems.

Joe Harlick's work has appeared in many magazines and books all over the world. Many of his images can be found in "The Ragged Irregulars of Bassingbourn" by Marion Havelaar. Harlick is now retired after a distinguished career at Boeing Aircraft.

Right: Sequential Strike Camera photos demonstrate the drama of the death of a B-17 (#42-31333 LG-W "Wee Willie" 323 SQ/ 91st BG) on 8 April 1945. The plane is seen in the top frame with her left wing folding over the fuselage. Only 3 seconds later that wing has cleared the right one. Then another 3 seconds elapsed until the entire bomber exploded. She had 128 missions behind her and was the first "G" model in the 91st Bomb Group. Some of her crew survived the crash of this fortress!

"Outta gas, can we make it back on three?"

B-17F #41-24485	"Memphis Belle"
Pilot	Robert K. Morgan
Co-Pilot	H.W. Aycock
Bombardier	Vince Evans
Navigator	Charles Leighton
TT / Engineer	Harold Loch
Radio Operator	Robert Hanson
Waist Gunner	Clarence Winchell
Waist Gunner	E. Scott Miller
Ball Turret	Cecil Scott
Tail Gunner	John P. Quinlan

6 March 1943
Target - Submarine Pens at Lorient, France
Alert no.49 - Mission no.23 for the 91st BG(H)

Battle Journal Input
Target: (1) Lorient, France - U Boat Installations
(2) Brest, France - U Boat Installations
(3) Brest, France - Port Militaire
Enemy Defenses:
Fighters (Single engine) 60
(twin engine) 30
Flak Batteries - Heavy and Accurate
Friendly Fighter Support: None
Time Schedule: Sta 0923; taxi 0943; takeoff 0958; leave 1030; zero hour 1330; Target 1450.
Bomb Load: 5 X 1,000 pound High Explosive
Bombing Altitude: 91st BG(H) 23,000 feet
Other Bomb Groups: 306th, 305th, 303rd

Bob Morgan was back in "Memphis Belle." The 91st was able to launch fourteen bombers on this raid that was to end with a planned landing at a diversionary field at Davidstowe in Cornwall for fuel. It was expected that all bombers would be running on vapors by the time they reached England, and the planners wanted to play it safe.

The mission went as planned with only #459 "Hellsapoppin" returning early because of mechanical difficulty. Post mission strike photos indicated fair bombing results, with hard strikes inundating

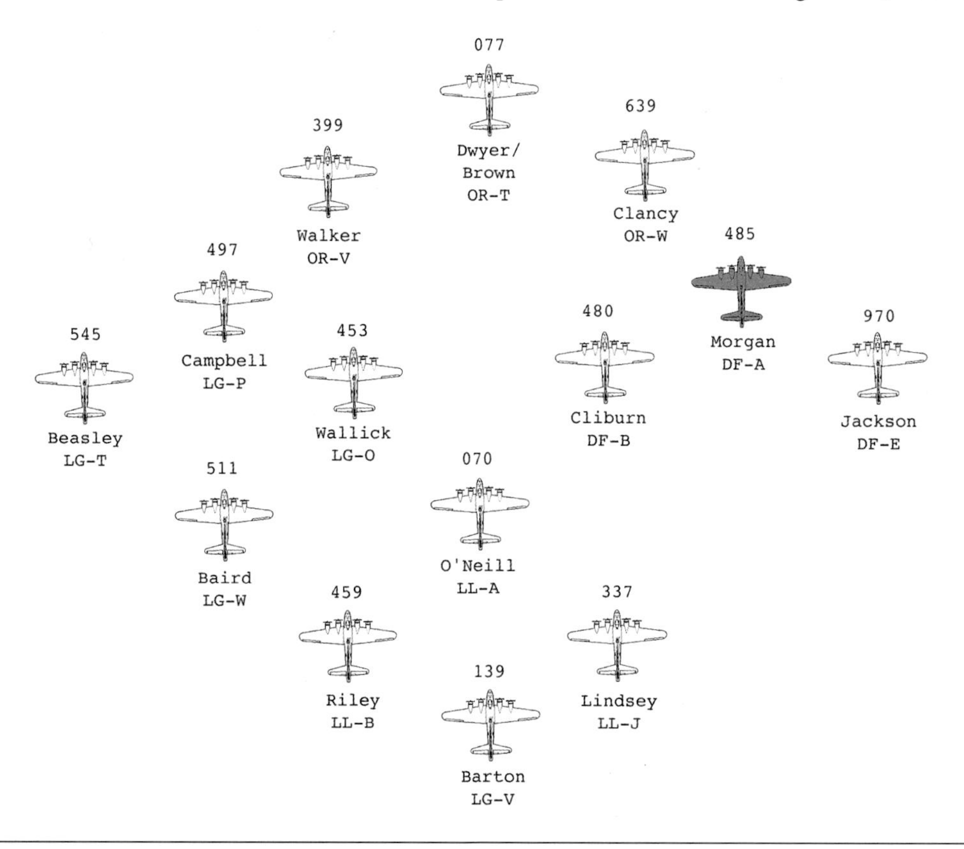

the warehouses and docks. The 306th in the lead caught the brunt of German defenses, with the loss of one bomber to flak and then enemy fighters finishing it off. A B-24 outfit that made a diversionary attack at Brest suffered the loss of two ships. The 91st escaped both casualties and loss of airplanes.

The highlight of the day was a planned party at the 91st Officer's Club that night (another famed party event, for which the "Boys" were becoming well known). With the bombers landing in Cornwall the party would be ill attended, but that did not distract Morgan's thoughts while they headed back towards England. He evidently harbored thoughts of just how he could nurse the "Memphis Belle" on almost no fuel all the way back to Bassingbourn.

Two of Morgan's crew remembered that there was a bad engine on the "Memphis Belle," and also that they were the only crew to make it back to Bassingbourn. It is thought that Morgan must have been remembering that famed three-engined takeoff for a date and possibly figured "Hell, if I can get airborne on three, I can certainly fly on three!"

The only problem was that Major Aycock, the Squadron Commander for the 324th, was riding in the right seat of the "Belle" that day, and Morgan's desire to get back to Bassingbourn on three engines would have been against regulations. Since Bill Winchell back on the left waist gun wrote that one engine was not running, then it is likely that Morgan convinced Aycock to allow him to shut it down for some unknown reason. The real reason for returning to Bassingborun on three engines will likely never be known, but it is known that the "Memphis Belle" crew were the only ones to both take part in the mission and make the party that night!

With her number three engine smoking, this crippled fort makes a run for home. Like a vapor trail, the smoke pointed to a damaged airplane, which quickly gained the attention of enemy fighter pilots. The wolves then jumped the straggler in the hopes that it would be easy prey. (Frank Donofrio)

Another mission - 8 March 1943

"Memphis Belle" was scheduled to take part in this raid, but available records indicate an early return to base because of mechanical troubles. Another note: discrepencies exist between these available records in who actually flew the "Memphis Belle" that day. Without a doubt Col. Wray was her co-pilot, but the 91st BG report states that Capt. Gaitley was the pilot, while the records written directly from the "formation" board show that Capt. Robert Morgan was her skipper. Nonetheless it is inconsequential, as "Belle" #485 returned to base without completing the raid. Along with "Memphis Belle" B-17 #712 "Heavyweight Annihilators no.2" returned early. The first reports of this bomber's demise indicated that they had been shot down. This is because when the formation last saw this plane, two props were feathered and it had turned from the "initial point" with two enemy fighters in chase. What the other bombers could not know was that Lt. Gennheimer was successful in nursing the bomber back to a relatively safe landing at Exeter in Southern England. B-17 #077 "Delta Rebel no.2" is recorded as not taking off.

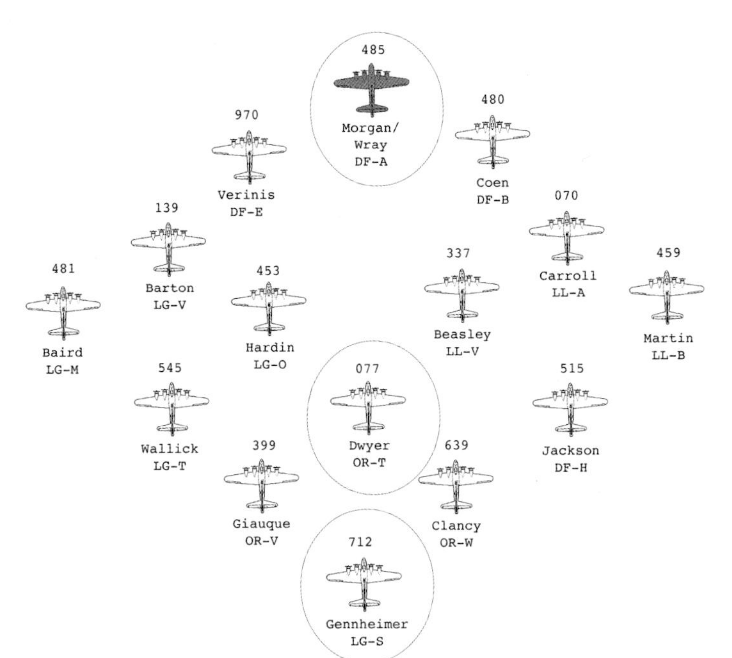

"It's about time"

B-17F #41-24485	"Memphis Belle"
Pilot	Robert K. Morgan
Co-Pilot	C.W. Freschauf
Bombardier	Vince Evans
Navigator	Charles Leighton
TT / Engineer	Harold Loch
Radio Operator	Robert Hanson
Waist Gunner	Clarence Winchell
Waist Gunner	E. Scott Miller
Ball Turret	Cecil Scott
Tail Gunner	John P. Quinlan

12 March 1943
Target - Railway Yards at Rouen, France
Alert no.53 - Mission no.25 for the 91st BG(H)

Battle Journal Input
Target: (1) Rouen, France - Railroad Marshalling Yards
Enemy Defenses:
Fighters (Single engine) 170
(twin engine) 10-20
Flak Batteries - Heavy at target
Friendly Fighter Support: 2 sqdns Spitfires @close
2 sqdns Spitfires @target
6 sqdns Spitfires @withdraw
Time Schedule: Sta 0910; taxi 0930; takeoff 0950; leave 1013; zero hour 1230; Estimated time of return; 1415.
Bomb Load: 5 X 1,000 pound High Explosive
Bombing Altitude: 91st BG(H) 21,000 feet
Other Bomb Groups: 306th, 305th, 303rd

For once it seemed like everything was going to go off as planned. This was the thirteenth mission for the "Memphis Belle," but the sixteenth for her crew. Captain Morgan had been her skipper for all her completed missions except the sixth, when Jim Verinis commanded her back on December 30th over Lorient. This would be a textbook raid, where all eighteen bombers of the 91st would reach the target without one turning back. One where all eighteen bombers of the 91st would return to base without serious damage and with all of their aircrews.

This success was due in large part to the terrific friendly fighter support from British Spitfires. The addition of long range fuel tanks enabled the RAF to launch their fighters in an order where the bombers would receive protection on the way in, over, and away from the target. The bombardiers finally were able to concentrate on their duties without distraction, and they virtually wiped the target out!

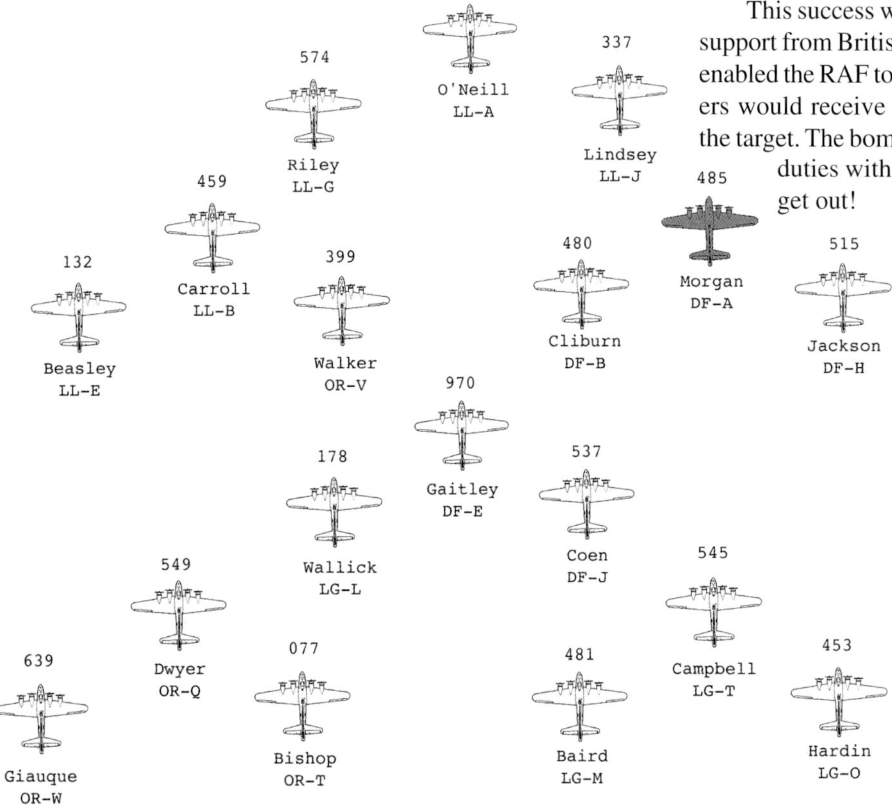

"Who's in the lead anyway?"

B-17F #41-24485	"Memphis Belle"
Pilot	Robert K. Morgan
Co-Pilot	C.W. Freschauf
Bombardier	Vince Evans
Navigator	Charles Leighton
TT / Engineer	Harold Loch
Radio Operator	Robert Hanson
Waist Gunner	Clarence Winchell
Waist Gunner	E. Scott Miller
Ball Turret	Cecil Scott
Tail Gunner	John P. Quinlan

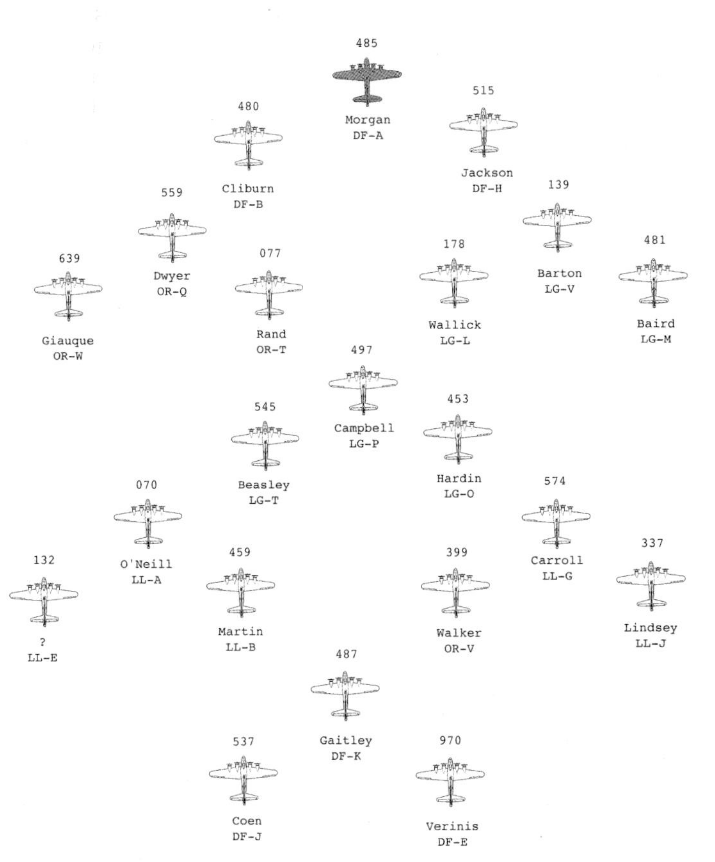

13 March 1943
Target - Rail Yards at Amiens, France
Alert no.54 - Mission no.26 for the 91st BG(H)

Battle Journal Input
Target: (1) Amiens, France - Railroad Marshalling Yards
(2) Abbeville, France - Railroad Marshalling Yards
Enemy Defenses:
Fighters (Single engine) 160
(twin engine) 55
Flak Batteries - Heavy and Accurate
Friendly Fighter Support: 6 sqdns Spitfires @close
2 sqdns Spitfires @withdraw
Time Schedule: Sta 1030; taxi 1050; takeoff 1110; leave 1135; zero hour 1430; Estimated time of return; 1601.
Bomb Load: 6 X 1,000 pound High Explosive
Bombing Altitude: 91st BG(H) 24,000 feet
Other Bomb Groups: 306th, 305th, 303rd

Twenty-one bombers were scheduled to bomb this day. #399 "Man O' War" was unable to take off, and #453 "Mizpah" returned early. The men were facing a "Snafu" (to use a phrase of the day), for this would not be a good mission to be a part of. The entire effort seemed terribly futile to the men of the 91st, and several factors made this feeling common among nearly everyone involved.

Not only did Bob Morgan in "Memphis Belle" lead the 91st Bomb Group, but he was scheduled to lead all four of the participating Groups. This was not, however, the result. The 91st launched without incident after an hour postponement to allow for the weather to clear. Twice the 91st circled Bassingbourn before the other three Bomb Groups came into view and began forming up into the two Combat Wings. The men on the ground were treated to what must have seemed an invincible armada, as some eighty Flying Fortresses with eight hundred men flew over their field and headed towards the enemy.

The 91st's efforts to obtain the lead position for the wing were thwarted time and again by the B-17s of the 306th, who evidently thought they were to either lead the whole Wing or fly at the assigned altitude for the 91st. At least three sources indicate that they were only successful after more or less bullying their way into the lead position.

Landfall was made far from the briefed point (likely due to the confusion on the way in), and the bombers found themselves in flak barrages as soon as they made the French coast. Then, upon reaching the IP, the 91st was again forced to leave the bomber stream by the 306th, who were by this time causing so much havoc that some evasive maneuvers were necessary to prevent midair collisions. Only a single squadron of the 91st was able to release their bombs on the enemy before they were forced out of the Wing.

It was very likely that Morgan was now fed up with how the mission was going and decided to turn the remaining 91st BG planes towards the secondary target at Abbeville. He said later in his post mission debriefing that they hit the airfield there even though they

A lone Bf-110 makes a pass at these two 91st BG forts. As many as twenty .50 machine guns could be pointed at him at any time during his attack on just these two B-17s. (Frank Donofrio)

were not briefed for it. Some records indicate that he may have surrendered the lead to another Group that was also headed for the secondary—the rail yards there. Again closing cloud cover made targeting difficult, and the bombers in front of the "Memphis Belle" were flying an erratic zig-zag course, apparently unable to find their checkpoint. The run on this target was aborted because Morgan feared dropping their bombs on yet more B-17s beneath them.

Although friendly fighters were providing top cover for the bombers, Morgan later recalled that he knew they would soon be out of fuel and headed for England. With the threat of German fighters in the area a quick decision to hit a target of opportunity was made when the airfield there was spotted. It was hit with good results, and then Morgan turned the 91st back toward England.

It was not until later that he would learn as the U.S. bombers approached the French coast on the way out RAF radar stations detected more than one hundred-twenty German fighters headed for an interception. Had they made even one more run on a target they would have met the Luftwaffe over the continent, having to slug it out severely outnumbered.

Post-mission recon showed overall poor results, with three 91st BG squadrons being the only ones to hit the airfield at Abbeville. Moreover, because of the initial melee on the way into the target nearly every Group bombed a different target. Many bombs consequently fell into civilian inhabited areas, probably killing innocent Frenchmen. One account even states that "Everyone agrees that the mission was shameful and a senseless rat race from start to finish."

It seemed that the only bright spot was the ten squadrons of British Spitfire escorts that were assigned to protect the American bomber force.

"Bravery and Courage"

18 March 1943
1st Lt. Jack W. Mathis
359th Bombardment Squadron
303rd Bombardment Group

The Medal of Honor

Jack Mathis of San Angelo, Texas, never flew on the "Memphis Belle." He was a bombardier aboard a different B-17 Flying Fortress from a different Bomb Group. The bomber that he was assigned to was not as famous as the subject of this book. It was called "The Duchess," #41-24561. As a matter of fact, little has been written about #561 or the only man ever to lose his life during the fifty-nine missions this bomber flew.

But such an act of heroism, bravery, and courage occurred on this airplane that mention of it in this book seems prudent. It also seems too little, too late. The bomber crews that heard the story when it happened knew that they would never, and could never, forget the story.

1st Lt. Mathis was hunched over his bombsight and making his adjustments with just under a minute to the bomb release. His B-17 "The Duchess" was positioned in the lead for the 359th Squadron on the way to hit the sub pens at Vegesack. The other B-17s in trail would release their bombs the moment Mathis toggled his. In a sheer instant, a flak burst exploded near the right side of the nose, blowing a large hole in it and throwing Jack's body to the rear of the nose compartment. Knowing his life would soon end, and disregarding the excruciating pain from deep gashes in his abdomen and side, he somehow overcame the pain from his nearly severed right arm, along with his other wounds, and crawled back to his bombsight. As his life ebbed away he made some final adjustments to his sight and toggled the weapons. The last word he spoke was "Bombs," and with the navigator watching Mathis succumb added the crucial word "away." This notified the pilot that "The Duchess" had delivered her payload and could turn for home. The other B-17s, waiting for their cue, followed, and the target was hit with accuracy. This selfless act prevented more bombers from having to return to this target, thus saving untold lives.

No one can be sure just how many other stories like this can be told. The same dignity, honor, and courage was present on all American bombers in WWII. It is certain, however, that not everyone possesses the kind of dedication to duty and sacrifice as Lt. Jack Mathis and the others like him. His uncommon valor would win him the nation's highest honor. He was the first Eighth Air Force Airman to receive the Medal of Honor—posthumously.

To add to this tragic story, Jack's brother Mark, who was at the time flying missions in medium bombers, managed to arrange a transfer into the same squadron. Mark not only flew in the "Duchess," but even used the same bombsight as Jack. Some said that he was trying to finish Jack's missions. Mark never made it, as he went "missing in action" during his fourth trip to the shipyards at Kiel, Germany.

"Another rough one"

B-17F #41-24485	"Memphis Belle"
Pilot	Robert K. Morgan
Co-Pilot	C.W. Freschauf
Bombardier	Vince Evans
Navigator	Charles Leighton
TT / Engineer	Harold Loch
Radio Operator	Robert Hanson
Waist Gunner	Clarence Winchell
Waist Gunner	E. Scott Miller
Ball Turret	Cecil Scott
Tail Gunner	John P. Quinlan

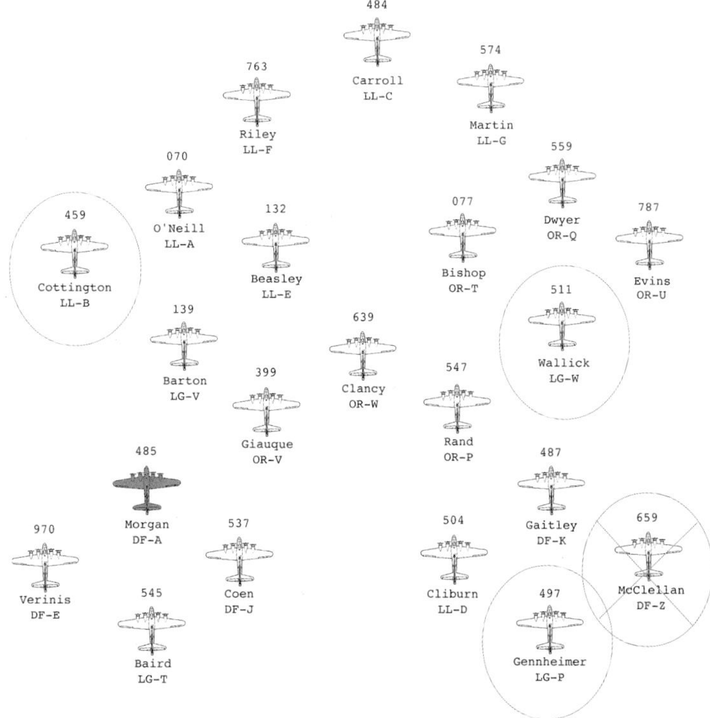

22 March 1943
Target - Pocket Battleship at Wilhelmshaven, Germany
Alert no.60 - Mission no.28 for the 91st BG(H)

Battle Journal Input

Target: (1) Wilhelmshaven, Germany - *Admiral Von Scheer* Pocket Battleship in Dry Dock
(2) Emden, Germany - Center of Town
(3) Any industrial town

Enemy Defenses:
Fighters (Single engine) 35
(twin engine) 130
Flak Batteries - 68 X 4 & 1 X 12

Time Schedule: Sta 1115; taxi 1135; takeoff 1150; leave 1216; zero hour 1300; Estimated time of return; 1720.
Bomb Load: 5 X 1,000 pound High Explosive
Bombing Altitude: 91st BG(H) 25,000 feet
Other Bomb Groups: 306th, 305th, 303rd

#459	"Hellsapoppin"	Returned Early
#511	"Wheel 'N Deal"	Returned Early
#497	"Mizpah II"	Returned Early
#545	"Motsie"	Did not take off
#659	"Liberty Bell"	MIA 11 killed

In this formation B-17 #659 "Liberty Bell" is shown as having been flown by Capt. Hascall McClellan. To this day, "Memphis Belle" pilot Bob Morgan remembers McClellan as a close friend. Of some significance is a fact that went unnoticed for almost sixty years. The Liberty Bell's demise was recorded on motion picture film from the right waist position of the "Memphis Belle" by William Wyler, and the sequence was included in the final edit of the 1943 documentary "Memphis Belle." Morgan remembered the bomber as the plane that is seen tumbling to earth while the crew of the "Memphis Belle" urges the "Liberty Bell" crew to get out of their plane. Eleven men died on #659 that day. Most of the crew was made up of guys from the 92nd Bomb Group that were training with the 91st's crews.

Just about every mission created occurrences that everyone involved remembered. This would be no exception. Typical early returns plagued three of the 91st's bombers. "Motsie" (#545) did not even take off. In B-17 #511 "Wheel 'N Deal" Capt. Wallick was forced to turn back just as the assembly reached the German coast. The propeller shaft on his number four engine broke, and he

B-17F #42-5132 "Royal Flush" (see formation above) of the 401st Bomb Squadron / 91st BG(H) LL-E. This bomber went MIA on June 22, 1943—a mission that saw the loss of fifteen B-17s of a total of almost two hundred participating. "Royal Flush" crashed in the North Sea. One crewman was killed, and nine were taken prisoner. Two of those, however, made escapes. She is seen here at the moment her wheels leave the Bassingbourn runway with her four powerful Wrights pulling her skyward. -Joe Harlick

B-17F #41-24504 "The Sad Sack." This bomber had a long and distinguished career, and like the" Memphis Belle" was returned to the U.S. on March 15, 1944, for a War bond and morale tour. "The Sad Sack" flew quite a few missions in the same formations as " Memphis Belle," often carrying wounded and dead back to England during her forty-two raids on the enemy. (Joe Harlick)

was unable to feather that prop. Violent vibrations shook the B-17 all the way back to England, where the propeller finally wrenched free.

Capt. Cliburn in #504 "The Sad Sack" had two of his engines put out of commission. Available records indicate that he finished the raid. With one engine burning and one not turning, "The Sad Sack" slowed and began to lose altitude. The formation slowed too, and dropped to provide cover for Cliburn. "Memphis Belle" gunner Bill Winchell reported that they were quickly jumped by no less than five Me-110s and two Ju-88s. Morgan in "Memphis Belle" entered a steep dive that left the Germans firing at thin air. Cliburn was at this time just skimming the waters of the North Sea when two Me-110s spotted him and dove in to finish him off. The protective guns of the Group blasted one, sending that German into the sea. The other decided that this was too much and peeled off for home. Cliburn was gaining a reputation for flying damaged bombers, but most of his missions were flown aboard "The Bad Penny"—a ship utilized by Hollywood Director William Wyler to record his famed documentary "Memphis Belle." Over in #970 "Connecticut Yankee" Jim Verinis was dealing with a failed electrical system. Four of his crew endured temperatures more than forty below zero for an hour and a half. They were all frostbitten.

The pocket battleship that was the intended target was well protected by more than fifty fighters, which climbed to kill the attacking B-17s. Just before the "Memphis Belle" released it bombs, one Ju-88 turned and attacked. Again, the "Belle" was nosed over and thrown around the sky in a fashion that would make any fighter pilot envious. Enemy fighters attacked the bomber stream all the way into the target area, then flew away some distance out of their own flak barrages. When the bombers left the target, they returned angry and determined. Nonetheless, the bombardiers were reported as having placed a very concentrated pattern into the harbor area, with several weapons detonating very close to the bow of the Admiral Von Scheer and likely doing great damage. Other bombs fell into harbor installations and docks, creating what many felt was a successful mission.

B-17 #42-29659 "Liberty Bell" was destroyed in the air by an unusual means; the Luftwaffe was dropping air bombs on them from above. This practice had not been very successful in the past, but this day an Me-109 found its mark, and "Liberty Bell" was forced to ditch in the North Sea. Reports indicate that they fell under attack from various types of fighters, and the B-17 was last seen just over one hundred miles from the English coast being fired on by an Fw-190. Capt. Hascall McClellan and all aboard were killed. Back in the states, McClellan's wife had just given birth to their baby girl. William Wyler had been on this very mission in the "Memphis Belle." In the Belle, the interphone came to life as her gunners reported a B-17 going down. Wyler captured McClellan's plane on film as it rolled over out of control and descended toward the surface of the North Sea. This horrific scene was included in the final edit of the "Memphis Belle" documentary, and is a memory that still affects Bob Morgan. Hascall McClellan was a close friend.

Yet another sad note: Coen in B-17 #537 (an unnamed bomber) would not live to see the next week. He and his entire crew would be killed on the right wing of "Memphis Belle" on the 28 March mission to Rouen, France. This next mission would be no picnic. Some members of the 91st felt they had the worst behind them. With every mission, they became more and more unsure of that.

"Belle Tailgunner Hit"

B-17F #41-24485	"Memphis Belle"
Pilot	Robert K. Morgan
Co-Pilot	J.M. Smith
Bombardier	Vince Evans
Navigator	Charles Leighton
TT / Engineer	Harold Loch
Radio Operator	Robert Hanson
Waist Gunner	Clarence Winchell
Waist Gunner	S.J. Spagnolo
Ball Turret	Cecil Scott
Tail Gunner	John P. Quinlan

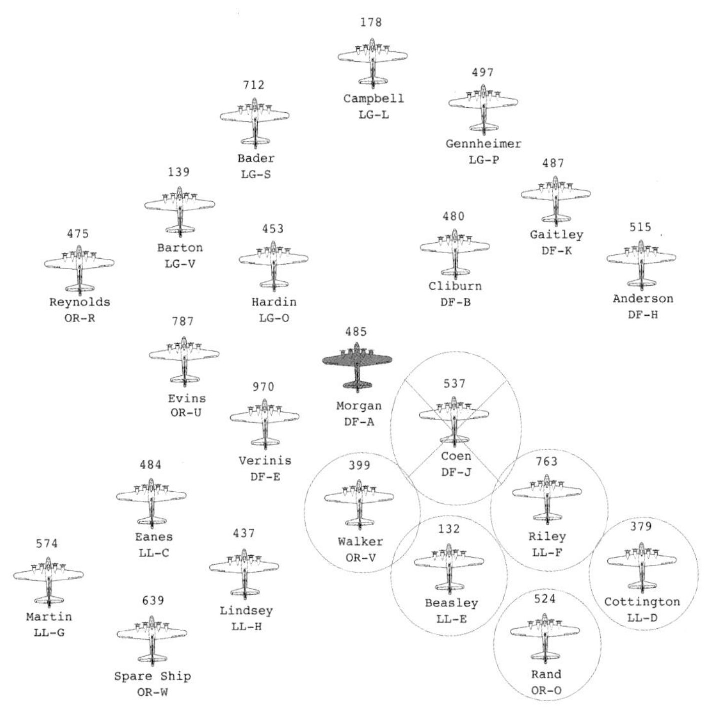

One look at the formation diagram here and it is easy to see why the end of a formation would often be referred to as "Purple Heart Corner." The 91st took a beating on this run—five participating B-17s had to return early because of battle damage on the way into the target. They were:
#475 "Stric Nine" **#379 "Yankee Eagle"**
#399 "Man 'O War" **#763 "Bomb Boogie"**
#132 "Royal Flush"
In #524 "Eagles Wrath" Lt. Rand was experiencing enough trouble to force him to abort the mission. He called the tower to inform them, whereupon #639 launched as the spare.

28 March 1943
Target - Rail Center at Rouen, France
Alert no.63 - Mission no.29 for the 91st BG(H)

Battle Journal Input
Target: (1) Rouen, France -Railroad Yards and Engine Sheds
Enemy Defenses:
Fighters (Single engine) Many
(twin engine) Many
Flak Batteries - Heavy and Accurate
Friendly Fighter Support: 6 sqdns RAF Spitfires @close
1 sqdns RAF Spitfires @target
4 sqdns RAF Spitfires @withdrawal
Time Schedule: Sta 1020; taxi 1040; takeoff 1055; leave 1119; zero hour 1300; Estimated time of return; 1415.
Bomb Load: 6 X 1,000 pound High Explosive
Bombing Altitude: 91st BG(H) 25,000 feet
Other Bomb Groups: 306th, 305th, 303rd

With the constant need for replacements on other bombers, it is safe to say that the mission count for both the "Memphis Belle" as well as her crew was becoming confusing, and it was this way on all of the bombers in all of the Groups. According to the official 91st BG records, this was the sixteenth mission for the "Memphis Belle" and the eighteenth for most of her crew.

The "Belle" was going to need to lick her wounds from the brutal attacks to the formation. Two gunners on Morgan's crew would succumb to lack of oxygen and pass out. Tailgunner Johnny Quinlan would be injured, and the tail of the "Memphis Belle" would again be badly damaged. (Just look at how close he was to two of the bombers that were damaged and shot down on this raid.) By studying the formation for this raid you can see just how close the story of this heroic bomber and crew came to ending.

The mission was launched a full thirty minutes early, with twenty-one B-17s headed towards France. A crystal clear sky was soon thickening, and the continent was beneath a 5/10 coverage. The launch time change created a change in the rally time for the friendly fighter escorts, and when the bombers reached the briefed interception point, the Spitfires were nowhere to be seen. As a result, the enemy took an unseen opportunity and blasted into the bomber stream. Most of the American gunners believed them at first to be the Spitfires they had been waiting on.

These first attack runs through the formation created a terrible melee and inflicted great damage to five B-17s, forcing them to abort and return to base. "Belle" waist gunner Bill Winchell later said that he could not remember the Germans behaving in such an aggressive manner by attacking the formation as they flew down along the French coast.

One new Luftwaffe tactic was to bear down on the rear of a bomber and take a quick shot as the fighter slipped below the formation. From there the German pilot could stand his fighter on its tail and rake the underbelly of a B-17 with cannon fire before falling away. An attack just like this nearly downed the "Memphis Belle"

#25379 "Yankee Eagle," shown here with forty-five missions and six downed enemy fighters credited. This "Memphis Belle" sister made it back to the U.S., and like the "Belle" was re-designated as a TB-17F. (Frank Do-nofrio)

#41-24524 "The Eagles Wrath" flew many raids with "Memphis Belle," but was lost over Schweinfurt August 17, 1943. Three KIA, five POWs, two evaded. (Frank Donofrio)

that day, and it killed Lt. Coen and his crew in B-17 #537, which was positioned just on the right wing of the "Belle." Coen was last seen turning back towards the French coast so that his crew could parachute to land instead of the sea. Several reports indicate that five were seen to bail out of the stricken B-17. Two other "official" reports indicate that all aboard perished. An interesting note: "Memphis Belle" tailgunner Johnny Quinlan avenged the death of Coen and his crew by pouring a long burst into the fighter that attacked Coen. Smoke poured from the enemy plane just before it exploded.

Moments later, 20mm rounds began exploding all around the Memphis Belle's tail. Quinlan's guns were rendered useless, and

#25763 "Bomb Boogie" MIA on September 6, 1943, due to enemy action (fighters) near Stuttgart. The "Boogie" crashed near Laon, France. Four of her crew evaded, while six became German prisoners. (Frank Donofrio)

he was hit in the face and leg—injuries for which he would later receive the Purple Heart. He was the only member of the "Belle" crew to receive the award, and he did not even want it! At a point in the mission the two waist gunners aboard "Memphis Belle" were deprived of oxygen and passed out. Had it not been for the alertness of the crew and damned good leadership by Robert Morgan, Winchell and Spagnolo would have died in the "Memphis Belle." Morgan had not heard any status report from the gunners in some time, and he alerted the radio operator, Bob Hanson, who, when he looked to the back out of the radio room, saw both men passed out on the floor of the "Belle." He immediately reconnected the gunners to portable oxygen bottles and revived them.

Despite the terrific beating taken by the bombers this day, the target was smashed by good bombing techniques through a small break in the cloud cover. Winchell was emotional regarding the loss of a roommate (Jimmy Bechtel) who was aboard Coen's B-17. It seemed to Winchell that Bechtel had seen enough in his short four mission tour of duty. Bechtel had just returned to flying status after having to be fished out of the sea following his third mission. Fished out of the sea on his third raid, and shot down on his fourth. Winchell saw the five parachutes floating down from the descending B-17 and, hoping that his friend Bechtel was in one of those chutes, promised him "Good luck Beck, see you back in the states when it's over. The next Nazi bastard I get is for you, fella."

"Snafus and Gremlins"

B-17F #41-24485	"Memphis Belle"
Pilot	H.W. Aycock
Co-Pilot	J.M. Smith
Bombardier	Vince Evans
Navigator	Charles Leighton
TT / Engineer	Harold Loch
Radio Operator	Robert Hanson
Waist Gunner	Clarence Winchell
Waist Gunner	S.J. Spagnolo
Ball Turret	Cecil Scott
Tail Gunner	John P. Quinlan

An interesting note: 1st Lt. Harold Kious, co-pilot for #512 "Rose O'Day"—shot down March 4, 1943—was heard today on a German short wave radio transmission. "He is a POW in Stalag Luft III and is sound. Evidently his pilot, 1st Lt. Ralph Felton, is dead. It is learned from other sources that at least two others from the 91st are imprisoned at Stalag Luft VIIIB."

31 March 1943
Target - Shipyards at Rotterdam, Holland
Alert no.64 - Mission no.30 for the 91st BG(H)

Battle Journal Input
Target: (1) Rotterdam - Railroad Engine Sheds and Assembly Halls
Enemy Defenses:
Fighters (Single engine) 150
(twin engine) 135
Flak Batteries - 3X6; 4X4; 5X4 at coast
Friendly Fighter Support: none
Time Schedule: Sta 0900; taxi 0920; takeoff 0930; leave 0958; zero hour 1200; Estimated time of return; 1404.
Bomb Load: 6 X 1,000 pound High Explosive
Bombing Altitude: 91st BG(H) 22,000 feet
Other Bomb Groups: 306th, 305th, 303rd

Twenty airplanes of the 91st BG took off on what was to be a routine mission. "Memphis Belle" was the lead ship for the Group, but Robert Morgan was not on board. In the pilot's seat for this, the 17th mission for the plane, was the 324th squadron commander Major Aycock. The regular crew did fly this raid, bringing their totals (for the most part) up to nineteen raids completed—only six to go. It is uknown why Bob Morgan did not make this mission.

Visibility created the largest problem of the day. First, the fog and haze was so thick that forming up was quite difficult. Two B-17s from the 303rd BG collided during their climbout, killing fourteen of the twenty men aboard. The entire bomb load from one of the planes detonated, and it was said that the blast could be felt for a radius of thirty miles! As a result of the difficult climb and consequent grouping of the Combat Wing, the armada could not assemble until after they reached 7,000 feet. A good deal of fuel was spent in this process before they were ready to head for the target.

It was becoming a common practice for the Wings to fly diversionary patterns to throw off the Germans, hopefully confusing them as to where the Americans might be headed. This only brought the fighters out over the ocean, where they began their attacks in earnest. Another problem for the diversionary flights that day was as the B-17s were occupied with this task of flying various heading changes, turning the formation through a series of triangles, the enemy fighters were busy attacking them. The target, which they could clearly see from their vantage point over the ocean, was quickly becoming obscured by a swiftly moving weather front.

By the time they turned for the target area their objective was safely hidden beneath a layer of clouds, and the bombers of the 91st brought their weapons home with them (with the exception of two planes that salvoed their loads into the English Channel).

Bomb Groups flying in trail of the 91st eventually found holes in the clouds and destroyed the target. Capt. Gaitley in B-17 #487 "Ritzy Blitz" was thankful for the clouds. He had an engine shot out and enemy fighters on him. As he straggled behind the formation he was able to tuck into the clouds, thus finding a safe haven. He eventually made it back to Bassingbourn.

One final note about the typical English weather. So socked in was the approach back to base that the Group was forced to request by radio guidance by emergency homing beacon. This was only done in desperate conditions. The Group somehow managed to approach England via the Thames estuary, which was a definite "no-fly" zone. Gunners on the ground were instructed to shoot at any airplane in this vicinity. Fortunately, word got to these alert gunners just before they opened up on the unwary 91st.

The "Memphis Belle" again left Bassingbourn for another mission on 4 April 1943 with her skipper Bob Morgan at the controls. The 91st put up twenty-six B-17s for this raid to Paris and the Renault Motor and armament works. Six airplanes, including "Memphis Belle," would have to return early because of mechanical troubles. This did not count as a completed mission and must have added to tensions among the crew. As they approached the golden twenty-fifth mission their nerves were taking quite a beating, and their thoughts were often of the many friends and "brothers" that had already died. No one said it, but they all were wondering now if they would actually make it.

Just why did so many of the bombers have to turn back? Some wondered if the crews were getting shaky. Reports indicate, however, that the real reason for all the mechanical troubles was that Command had insisted on a "Maximum Effort" AAA (all available aircraft) mission. The already overworked ground crews were probably forced into placing airplanes on the flight-ready line up that were not really sound and ready to fly.

The target was hit with amazing accuracy, with all bombs falling within the target area. A few did manage to fall within the city limits, causing concern to some back at Bomber Command. It was remembered, though, that the French had been warned many times to stay away from areas that could be considered a potential target for the Mighty Eighth Air Force. No one wanted collateral casualties, but the War had to be brought to a close. Bombing would not stop, but would rather intensify until the Nazis were brought to their knees.

"The Bad Penny"

B-17F #41-24480	"Bad Penny"
Pilot	Robert Morgan
Co-Pilot	J.H. Miller
Bombardier	Vince Evans
Navigator	Charles Leighton
TT / Engineer	Harold Loch
Radio Operator	Robert Hanson
Waist Gunner	Bill Winchell
Waist Gunner	Scott Miller
Ball Turret	Cecil Scott
Tail Gunner	John P. Quinlan

5 April 1943
Target - Aero engine repair works at Antwerp, Belgium
Alert no.66 - Mission no.32 for the 91st BG(H)

Battle Journal Input
Target: (1) Antwerp, Belgium - Aero engine repair works
(2) Antwerp, Belgium - Ford Motor Company
Enemy Defenses:
Fighters (Single engine) 175
(twin engine) 90
Flak Batteries - 60 heavy guns at Ostend
- not much at target
Friendly Fighter Support: 11 squadrons staggered
Time Schedule: Sta 1210; taxi 1230; takeoff 1250; leave 1319; zero hour 1500;
Estimated time of return; 1634.
Bomb Load: 6 X 1,000 pound High Explosive
Bombing Altitude: 91st BG(H) 24,000 feet
Other Bomb Groups: 306th, 305th, 303rd

Did Not Take Off	Return Early
#069 "Our Gang"	#712 "Heavyweight Annihilators No.2"
	#475 "Stric Nine"
	#497 "Mizpah II"

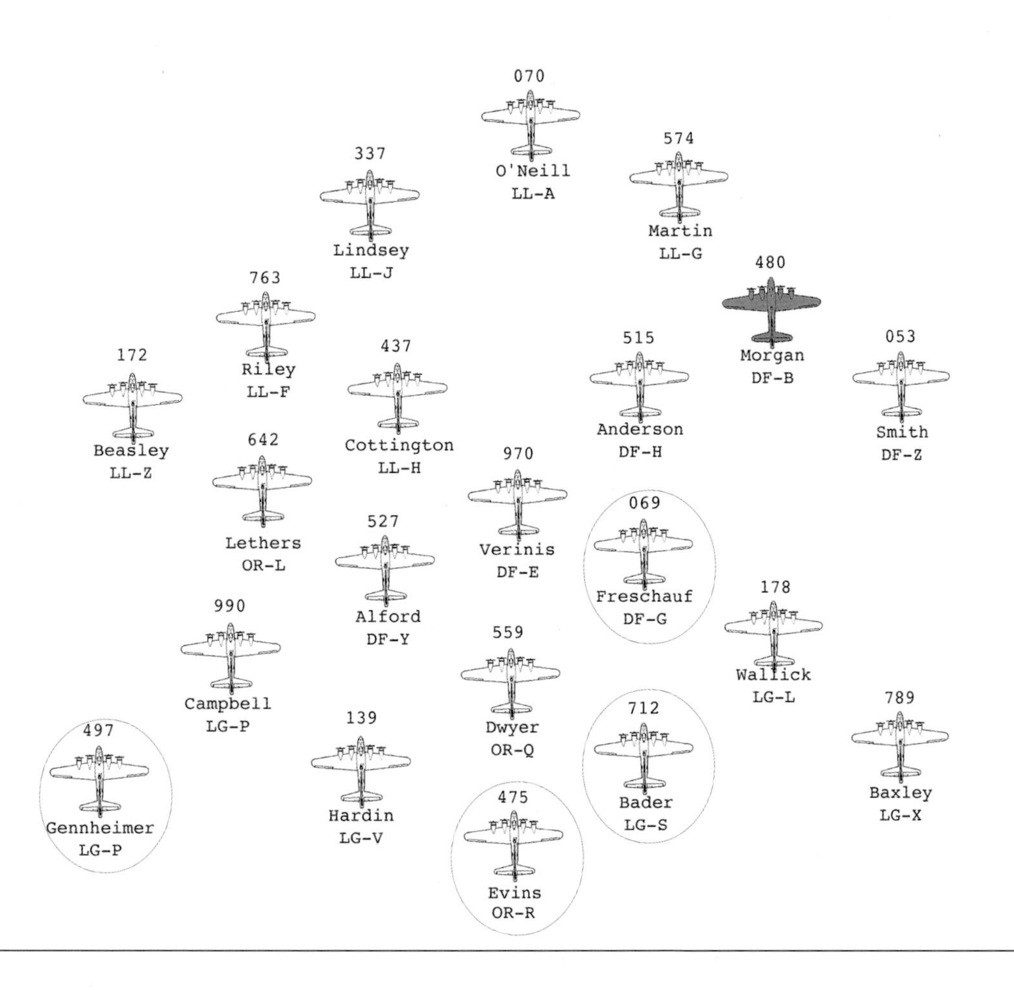

Two "triangles As" of the 91st BG in textbook formation. Flying tight groupings enabled the gunners aboard the bombers to concentrate their fire on enemy fighters, and prohibited the German planes from flying through the bombers. "Get your wingtip in here with me so I can sit on it" was a popular request during formation flight. Unfortunately, this sometimes resulted in bent metal when the big fortresses ran into each other. Sometimes overly excited gunners would accidentally send a few fifty caliber rounds into a B-17 next to them. As important as tight flying was, the inherent dangers were present from take off to landing. The Mighty Eighth Air Force was thought to be the very best at tight formations, as they had more practice than anyone else! (Joe Harlick)

The twentieth mission for a majority of the Memphis Belle's crew—but they would not be flying their beloved "Belle" today. Still under repair after having to turn back from the previous day's mission, she remained at Bassingbourn while crew chief Joe Giambrone made repairs. B-17 #480 "The Bad Penny" would carry the boys over Belgium this day.

Twenty bombers launched without incident. The pilot of "Our Gang" was not able to take off, and "Stric Nine" quickly fired up and moved in to the take off position as a spare ship. The target was of special importance to the bomber crews, because the facility to be bombed was capable of turning out around twenty enemy fighters every week. It was hoped that a good strike here would reduce the opposition they were facing every time they went on a mission.

There were clear skies over Bassingbourn, and the recon flight reported that the target was clear as well. The 306th BG was leading the Combat Wing, and through some mix-up they actually hit the secondary target on the northern edge of Antwerp. The lead bombardier for the 91st recognized their error and toggled his bombs on the proper point. The 91st's bomb pattern effectively covered the target, and later strike photos revealed very accurate bombing.

German fighters were present in force, and seemed to focus on the B-17s of the 306th in the lead. Four of their planes were shot down, while the 91st lost none. The attacking Germans were hitting the lead Group and then flying either above or below the 91st in order to hit the two Groups at the rear of the Combat Wing. The 91st gunners claimed six fighters shot down, three "probables," and one "damaged." Bill Winchell at the left waist window of "Memphis Belle" reported that it seemed like there were millions of fighters over Antwerp. The Luftwaffe fighters followed the B-17s out of the target area, continuing to hit them in swarms in a running fight that lasted until British Spitfires appeared and fought them off.

On his last mission (28 March), Bill Winchell vowed to avenge the loss of his roommate, gunner Jimmy Bechtel, and he did so on this mission. While leaving the target, one Focke Wulfe flew into the range of Winchell's gun, who wasted no time in pouring fifty or more rounds into it. A spectacular stall broke the wing off the fighter, and it fell in a tight spin.

Back at Bassingbourn, "Memphis Belle" bombardier Vince Evans achieved the rank of Captain and also became Group Bombardier. Navigator Charles Leighton was also promoted to the rank of Captain. As the air war droned on, comments regarding the tail numbers of the planes arose. It seemed that when the 91st became operational back in early November, the bombers on the field had similar tail numbers. Now, it was becoming a mixed group of planes, as replacements for lost bombers were arriving.

"Burnout in the climb"

B-17F #41-24485	"Memphis Belle"
Pilot	Robert Morgan
Co-Pilot	J.H. Miller
Bombardier	Vince Evans
Navigator	Charles Leighton
TT / Engineer	Harold Loch
Radio Operator	Robert Hanson
Waist Gunner	Bill Winchell
Waist Gunner	Scott Miller
Ball Turret	Cecil Scott
Tail Gunner	John P. Quinlan

16 April 1943
Target - Submarine Pens at Lorient, France
Alert no.68 - Mission no.33 for the 91st BG(H)

Battle Journal Input
Target: (1) Lorient, France - Powerplant Installation
(2) Brest, France - Commercial Basin (Flak Towers)
Enemy Defenses:
Fighters (Single engine) 125
(twin engine) 75
Flak Batteries - 42 heavy guns
Friendly Fighter Support: none
Time Schedule: Sta 1035; taxi 1055; takeoff 1110; leave 1136; zero hour 1300; Estimated time of return; 1735.
Bomb Load: 5 X 1,000 pound High Explosive
Bombing Altitude: 91st BG(H) 27,000 feet
Other Bomb Groups: 306th, 305th, 303rd, plus B-24s on a diversionary raid.

Did Not Take Off	Return Early
#057 "Blonde Bomber"	#485 "Memphis Belle"
	#515 "Jersey Bounce"
	#970 "Connecticut Yankee"
	#657 (not named)
	#178 "The Old Stand By"
	#511 "Wheel 'N Deal"
	#724 "Thunderbird"
	#712 "Heavyweight Annihilators no.2"

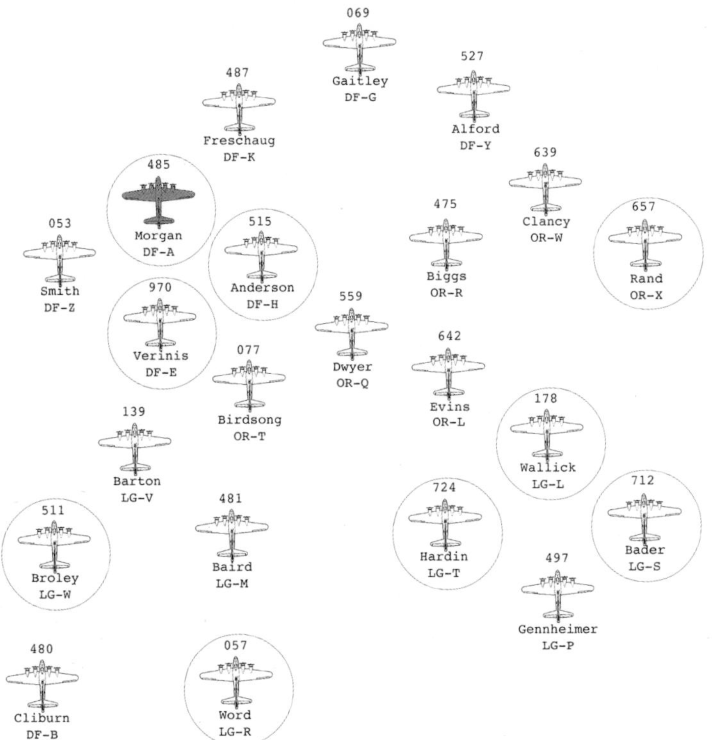

Someone forgot the five "P" principle on this mission (Proper Planning Prevents Poor Performance). The flight plan called for an immediate climb to 23,000 feet just after the Combat Wing rendezvous at a speed of 170 mph. The 91st was also scheduled to bomb from an altitude of 27,000 feet over the target. Some wondered if the turbo-superchargers would survive this extreme altitude, since this was the highest assignment yet for the 91st. It was known among flight crews that these units often failed at these heights—not to mention the severe cold the crews and their guns would have to endure.

During the climb, straining engines began to overheat and superchargers ran away and disinegrated. The "Memphis Belle" made it to a point just inside the French coast when Morgan was forced to return. Jim Verinis suffered two Supercharger failures on "Connecticut Yankee." One went through the top of a wing, and he lost a tire from flying supercharger fragments when another just gave way. Verinis reported that one enemy fighter chased him but gave up, as his B-17 was nosed over and building tremendous speed going out of France. No less than eight 91st BG planes returned to Bassingbourn with engine and supercharger problems. The other Bomb Groups also went through the same troubles.

With only thirteen B-17s from the 91st in a very loose formation left to bomb the target, the fears that the Germans could easily attack them were relieved when the inevitable German defenses failed to appear. However, they showed up later in earnest, but only after the bombers were able to tighten up the Wing.

Ground haze prevented the bombardiers from tracking the target with precision, and various post-mission reports indicated results ranging from poor to satisfactory.

The prescribed return to England was a direct flight at an altitude of only 500 feet off the water over the Bay of Biscay. Many of the bombers had used too much fuel during the dramatic climb on the way in and were now running very low. Several had to make unexpected landings at Davidstowe in Southern England. All B-17s were returned to Bassingbourn before nightfall.

On another note, back in England the first daylight air raid alert sounded just before the bombers began to return to Bassingbourn. The German planes were very high and were probably taking photos to find which Groups were involved in the raids that day. The Germans dropped no bombs on the airfield—this time!

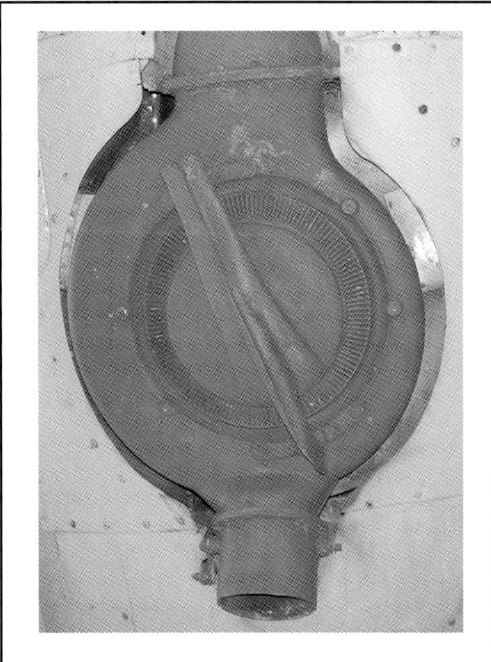

Flight operations at extreme altitudes reduced the efficiency of the radial engines. The Army Air Force solved the problem with the turbo supercharger. They were fitted to all heavy bombers, as well as the P-38 Lightning and the P-47 Thunderbolt. The four units suspended behind and below each engine provided sea-level air pressure to the engines, making them operate as if they were much lower when the airplane was operating in the rarified air at altitude. Intercooler air intakes were built into the leading edges of the wings where outside air was drawn into ducting, compressed to the proper pressure, and then fed into the engine's carburetor. Careful monitoring of the units was necessary throughout the flight. At first this was partially done by manually controlling the intercoolers. Because of this, pilots were unable to efficiently offer their attention to other cockpit duties. The manually controlled superchargers took too much tasking. This was ultimately solved with the addition of electronic supercharger control units that greatly reduced the intricate operations. Great care had to be taken for proper operation of the units, as an overspeeding bucket wheel could explode and disinegrate, causing terrible damage to the airplane. The bucket wheel would glow red and sometimes white from the heat generated by the unit. Adding to this heat were the very high speeds at which the bucket wheel would spin. The Westinghouse and General Electric turbo superchargers added a great deal to the outcome of the War. High-altitude bomber operations would have been impossible without them. To keep the technology from falling into enemy hands, the crews were instructed on how to destroy the superchargers if their plane was forced down behind enemy lines. This was futile, because the enemy eventually ended up with enough crashed B-17s to assemble several operating examples of their own. These planes were used for clandestine operations, as well as training Luftwaffe pilots on attack techniques.

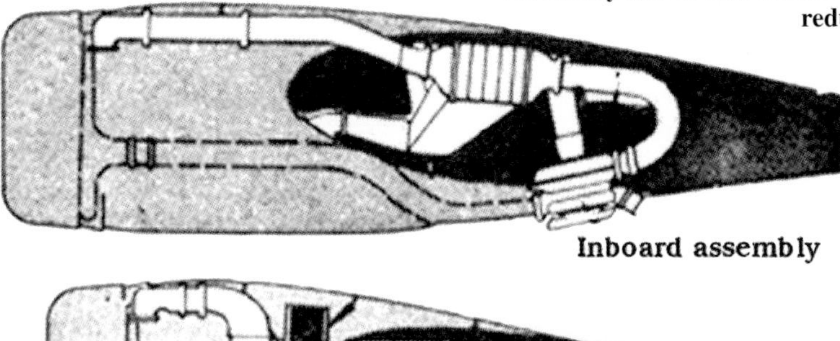

Inboard assembly

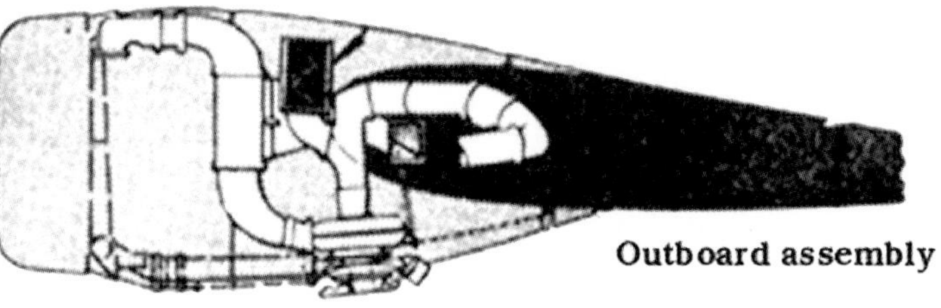

Outboard assembly

"One hell of a beating"

B-17F #41-24485	"Memphis Belle"
Pilot	Robert Morgan
Co-Pilot	J.H. Miller
Bombardier	P. T. Palmer
Navigator	Charles Leighton
TT / Engineer	Harold Loch
Radio Operator	Robert Hanson
Waist Gunner	P. J. Shook
Waist Gunner	Scott Miller
Ball Turret	H. W. Warner
Tail Gunner	John P. Quinlan

17 April 1943
Target - Airplane Factory at Bremen, Germany
Alert no.69 - Mission no.34 for the 91st BG(H)

Battle Journal Input
Target: (1) Bremen, Germany - Focke Wulfe plant
Enemy Defenses:
Fighters (Single engine) 30
(twin engine) 100
Flak Batteries - 76 heavy guns
Friendly Fighter Support: none
Time Schedule: Sta 0925; taxi 0940; takeoff 0955;
zero hour 1100; Estimated time of return; 1735.
Bomb Load: 5 X 1,000 pound High Explosive
Bombing Altitude: 91st BG(H) 25,000 feet
Other Bomb Groups: 306th, 104th (bastard), 305th, 303rd, plus 44th BG B-24 Liberators.

Return Early
#057 "Blonde Bomber"
#763 "Bomb Boogie" MIA
#070 "Invasion 2nd"
#459 "Hellsapoppin"
#172 "Thunderbird"
#574 "The Sky Wolf II"
#391 "Rain of Terror"
#337 "Short Snorter III"

After this raid, it was expected that somehow the Germans intercepted a pre-mission radio broadcast and had their fighters in the air waiting for the American bomber force. For almost a full hour a violent air battle was waged. Fighters of all types hit the formations in wave after wave of head-on attacks, showering the bombers and crews with murderous fire. It was a turkey shoot for the Germans, and virtually every bomber suffered damage. But half of the German factory was leveled in the raid.

There was plenty to report in the post-mission debriefings, but no one felt much like talking about this raid. It was the worst they had seen yet. Six Flying Fortresses and their crews were lost. All of them were from the 401st Squadron in the 91st BG section of the flight. The 306th had it even worse. They lost ten B-17s out of their participating twenty-four on this mission.

The largest formation yet left Bassingbourn for a raid that everyone on base wanted to be a part of—the Focke Wulfe plant, where the famed Fw-190 was produced. The men thought this was a great way of striking back right at the source of exactly what had been causing them so much grief. The thirty-two B-17s of the 91st first experienced enemy fighters early in the mission, out near the Frisian Islands. During the morning briefing the crews were told to expect about 130 fighters on the mission. An estimated 200 showed up for the fight.

One particular Luftwaffe pilot was growing notorious among the bomber crews of the 91st BG. Among the feared "yellow nosed" fighters there was a certain pilot that had drawn the attention of the men on previous missions. He was good, too. Even though no one knew his name, they knew his fighter—the number 14 was painted on his plane, and he was flying this day. Several gunners of the 91st had their eyes on this pilot, as he taunted them just beyond the range of their guns. Flipping and rolling about, trying to show the Americans that he was one tough guy not to be fooled with. Number 14's usual method of attack had him screaming at top speed through the B-17s while he was blazing away on his guns. He usu-

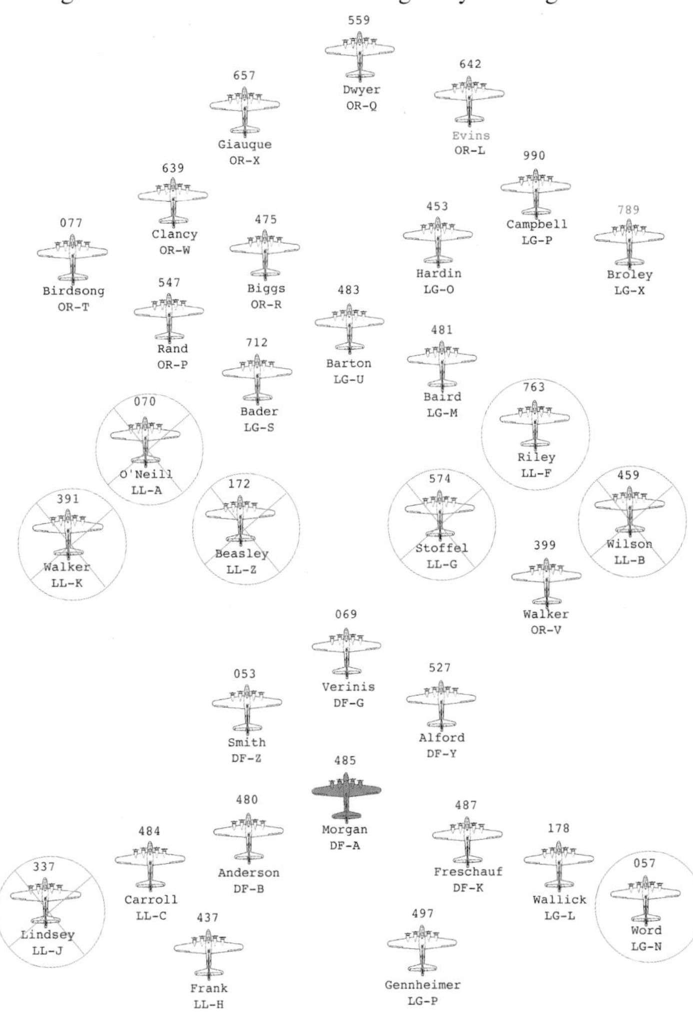

ally did not break off his attack until the Americans could see his face!

The fighter opposition became so terrific that Evans had to fight his fixation and actually tear his hands off the nose guns of the "Memphis Belle" so he could do his job as bombardier. Navigator Leighton, up there with Evans, began spotting for briefed waypoints that would get the "Memphis Belle" and the other bombers to the target. Fighters pressed on their attacks, finding their marks on B-17 after B-17. Crewmen were injured, voices called out "approaching fighters" as bucking fifties roared above the din of the engines. It was a virtual maelstrom of activity from which Evans could not flinch. If the bombs were going to hit the target, he would have to fly "Memphis Belle" through his bombsight on a precise course that would bring the crosshairs of his Norden sight over the factory miles below.

As if the fighters were not bad enough, the flak over Bremen was horrific. The Germans were very sensitive about the aircraft factories, and the best of the German flak gunners were always stationed around high-value targets like this one. The gunners on the ground had the range and the speed of the bombers. Here the formation was rocked with heavy and accurate blasts exploding everywhere. Bob Morgan remembered that he felt like they were flying through steel-filled thunderclouds.

Evans found the target a mere five seconds away from the aiming point and jerked on the bomb release. "Memphis Belle" was instantly taken over again by Bob Morgan in the cockpit and, lightened of its burden, the massive bomber was hauled into a steep climbing turn to throw off the flak gunners. The Bassingbourn Boys placed an accurate and concentrated pattern on the target despite the seamless flak erupting throughout the run. Quinlan in the tail reported that the formation was intact. Evans left the bombsight and returned to the guns protruding through the plexiglass nose of the "Memphis Belle." And guess who showed up—number 14.

Evans poured bullets into the German fighter, and some reports indicate that the plane went down. He was not credited for the kill, though. The German fighters did not follow the bomber stream into the flak this time, as they were waiting to hit the B-17s on the way out. They hit them hard, killing the straggling bombers as they fought their way out to the coast and the safer skies of the North Sea. Quinlan was busy back in the "Memphis Belle" tail trying to get a bead on an attacking fighter, but could not. Like a flagpole sitter in a high wind, he was bounced around his tiny space while Morgan jinked the "Belle" about the sky, trying to throw off the aim of the enemy. Gunners from several bombers merged their bullet streams on this enemy fighter, quickly dispatching it.

The 91st Bomb Group's 401st Squadron was horribly assaulted on the way out of the target area. This raid was going to leave an unbelievable impression on the men of the 91st, when in just a few minutes no less than six B-17s of the 401st Squadron (in the center of the 91st's formation) were shot down. The first was Captain Oscar O'Neill aboard #070 "Invasion 2nd." Three attacking enemy fighters approached his plane out of the 12 o'clock level position. His bomber was already burning from flak hits, and the fighters managed to tear the whole front of his number two engine off. While all ten men on O'Neill's crew survived, they were captured and became prisoners for the duration of the war. No. 070 had no one aboard when she made a decent landing near Oldenburg, Germany. Only moments later, "Hellsapoppin" B-17 #41-24459 was bracketed by heavy and accurate flak bursts that sent torn metal flying through the interior of the B-17. The radio room and the left wing were burning, three feet of the right wingtip was gone, and all the glass in the nose had been blown away. Both pilots were seriously injured and incapacitated. "Hellsapoppin" broke into two pieces just after the last man alive was able to jump and fell to earth roughly ten minutes south of Bremen with five dead and dying men aboard. The other five crewmen were captured. At the same time, B-17 #42-5172 "Thunderbird" was being shot to pieces. Direct hits in both right engines set the wing afire with burning oil. Lt. Beasley had no luck at all with the fire extinguishers, and the moment was growing terminal when the bomb bay and radio room ignited, as well. "Thunderbird" was on her last flight, and her aircrew were leaving as fast as they could. Falling in a tight spin to earth, her two pilots were hopelessley trapped within the burning and dying Fort. The last time her crew saw their plane from their haven in their parachutes, it was nose high in a stall before a pitchover and straight in dive for the ground. The bodies of her pilots Lt. Beasley and Lt. McCain were recovered by the Germans. The rest of the crew were captured and imprisoned.

The fourth Fortress shot down was right on the heels of "Thunderbird." B-17 #42-29574 "Sky Wolf II" was crewed by men of the 92nd Bomb Group. The No. 1 engine was burning on the bomb run only minutes before the attacking German fighters pounced the plane. The nose plexiglass shattered, leaving huge gashes in the bombardier's face. He jumped from the plane hoping to receive medical attention from the Germans, but died in his parachute as he floated down. The tail gunner was either dying or already dead, and a failed electrical system prohibited the ball turret gunner from getting out of his turret. They crashed with "Sky Wolf II" in Ostfriesland, Germany—five men dead, five POWs. Then the Flying Fortress "Rain of Terror" #42-5391 became the fifth victim of the 401st Squadron. Again flak was the culprit, accompanied by horrific sweeping attacks from both Me-109s and Fw-190s. The tail gunner was last seen slumped over in his compartment, and the top turret gunner was badly wounded. As her crew hit the silk, her two pilots managed to stay within the furiously burning bomber and crash-landed the plane on a beach near Norden. Of the ten aboard "Rain of Terror," two were dead and eight became prisoners. Many of those were badly injured.

The final victim of this assaulted squadron was flying in the purple heart corner of the formation. Both of the right engines of "Short Snorter III" (# 42-5337) were hit by flak, and just moments later another burst instantly killed both pilots. The plane made a sickening downward spiral, and some members of her crew were seen to be lying dead inside the plane. The bomber was last seen by the remaining members of her crew, who parachuted. She was headed out to sea where she eventually crashed.

"A hoax from start to finish"

B-17F #41-24485	"Memphis Belle"
Pilot	Robert Morgan
Co-Pilot	C.E. Debaun
Bombardier	Vince Evans
Navigator	J.R. Ehrenburg
TT / Engineer	L. W. Murray
Radio Operator	J. Moore
Waist Gunner	Bill Winchell
Waist Gunner	Scott Miller
Ball Turret	Cecil Scott
Tail Gunner	W. C. Dager

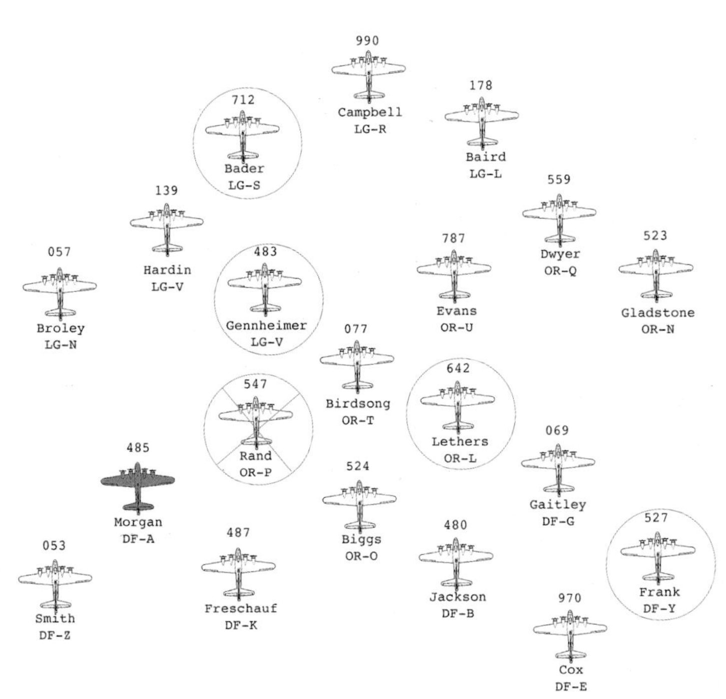

1 May 1943
Target - Submarine Pens at St. Nazaire, France
Alert no.74 - Mission no.35 for the 91st BG(H)

Battle Journal Input
Target: (1) St. Nazaire, France - Tidal Basin Coffer Dam / Locks
Enemy Defenses: Fighters (Single engine) 85 (twin engine) 25
Flak Batteries - 62 heavy guns
Friendly Fighter Support: none
Time Schedule: Sta 0810; taxi 0830; takeoff 0845; leave 0917; zero hour 1030.
Bomb Load: 2 X 2,000 pound High Explosive
Bombing Altitude: 91st BG(H) 25,000 feet
Other Bomb Groups: 306th, 305th, 303rd

A couple of interesting events took place at Bassingbourn during this time. At exactly noon on 21 April 1943, the British officially turned the base over to the Americans in a ceremony that brought top commanders. Bassingbourn would now be referred to as United States Army Air Force Station 121. It was long in coming, and what Col. Wray had been hoping for. He had angered the brass months before when he just moved into Bassingbourn and promised that the American's stay there was temporary. Now it was permanent, and the *Ragged Irregulars* would stay at Bassingbourn for their entire tour of duty. As the American Flag was raised over the base a flight of eighteen B-17s roared over in a review pass for the General's party.

The next day some of the crews started to train for a new employment of the Flying Fortress—zero altitude bombing missions. The pilots experienced a most exhilarating practice mission, where for over three hours the B-17s blazed at 200 mph as low as they could possibly fly, practicing how to fly in a knap of the earth fashion over houses, chimneys, trees, telephone poles, hedges, haystacks, and even flocks of birds. The exhausted pilots were smiling ear to ear when they landed, but they were also wondering if this would be a good way to employ the use of a bomber like this. A B-17 was just not built for this kind of close air support role—it was indeed a bomber built for high altitude work.

This was mission number twenty for the "Memphis Belle" and number twenty-two for five members of the regular "Belle" crew. Morgan, Evans, Winchell, Miller, and Scott were the only members of the regular "Belle" crew that were flying this mission.

As always, the flyers could look out the windows of their airplane when Morgan was starting the engines and see their ground crew, headed by crew chief Joe Giambrone. This was a serious time for the ground crew. Usually they had worked all through the night repairing the B-17 for the next day's mission. Hoping that the cowl flaps, landing gear, engines, instruments, bomb sights, bomb releases, oxygen system, electrical system, hydraulics, intercom, and everything else was in good working order. Serious business, because the lives of ten men depended on Giambrone and his men doing their jobs—and doing it right every time.

Back at Bassingbourn, one of the officers made a comment that the briefings were beginning to feel very different. As they would look around at the pilots and crews, familiar faces were no longer there. The chairs were becoming occupied with newer people as the airmen losses mounted. The veterans of bomber combat flying were becoming fewer and fewer as the war raged on. "The Old Standby" (#41-25178 / LG-L / 322nd Sq) is pictured above at rest in her Bassingbourn hardstand. This bomber was lost to enemy action one year and three days after she was delivered to the Army. She crashed at Flemsburg on 9 October 1943, with one KIA and 9 POWs. -Joe Harlick

Four B-17s of the 91st returned early, and one was shot down in yet another frustrating raid that was flown for no reason. Bombs were not dropped on any German target, but salvoed into the Bay of Biscay because of target cloud coverage.

From the beginning, this mission was going poorly. Two bombers returned to Bassingbourn—more mechanical trouble. They were #712 "Heavyweight Annihilators no.2" and #483 "Spirit of Alcohol." They were followed by #642 "The Vulgar Virgin" and #527 "The Great Speckled Bird."

The weather continued to worsen as the bombers turned to the target. Some bombers in the lead dropped their bombs through an overcast and consequently destroyed some of the civilian area of St. Nazaire. Determined not to repeat this mistake, the bombers that flew alongside "Memphis Belle" that day elected to put their bombs in the ocean.

Frustrations were running high and clouds were growing thicker. The Wing failed to turn to the north towards England at the right time, coupled with an assembly that was quickly becoming scattered like the disastrous error from the mission to Lorient back on 30 December. Again some fliers mistook the Brest Peninsula for Southern England and flew north over occupied France low and scattered. The German pilots had a field day shooting at B-17s.

Lt. Rand and his B-17 "Vertigo" (#547) were lost here in another murderous flak barrage. Five of his crew were killed and five became POWs. The 306th BG lost three or four B-17s, and the 303rd lost another two. Another seven B-17s gone on this day—were we making them as fast as we were losing them? This was the question around all of the bases in the Eighth Air Force.

Lt. Baird in #178 "The Old Stand By" (*see photo above and diagram on preceding page*) broke out of the clouds alone and under attack. Seven Fw-190s followed him down to the waters of the English Channel. One attacking fighter attempting the favored frontal attack pumped some cannon fire into his B-17. As the "Jerry" broke over the top of "Old Stand By" the top turret gunner shot it down into the sea. Baird was running as low and as fast as he could. Enemy fighters attacking from the rear were getting soaked with the spray as the B-17's propeller tips were swirling water and foam into the air from the wavetops! Three crew were injured and another fighter was shot down. Lt. Baird nursed "The Old Stand By" back to England badly damaged. A forced but safe landing was made at Warmwell, and the injured were taken to a nearby hospital.

Baird and his co-pilot were both awarded the Distinguished Flying Cross for their actions, and two members of their crew received Silver Stars. Later that year, on 9 October 1943, "The Old Stand By" was shot down and crashed at Flemsburg, Germany, with nine prisoners and one dead.

"All missions should be like this"

B-17F #41-24527	"The Great Speckled Bird"
Pilot	Robert Morgan
Co-Pilot	V.A Parker
Bombardier	Vince Evans
Navigator	Charles Leighton
TT / Engineer	Harold Loch
Radio Operator	Robert Hanson
Waist Gunner	Bill Winchell
Waist Gunner	Scott Miller
Ball Turret	Cecil Scott
Tail Gunner	John P. Quinlan

4 May 1943
Target - Vehicle Assembly Plants at Antwerp, Belgium
Alert no.76 - Mission no.36 for the 91st BG(H)

Battle Journal Input
Target: (1) Ford Motor & General Motors Plants
Enemy Defenses:
Fighters (Single engine) 175
(twin engine) 90
Flak Batteries - 60 heavy guns
Friendly Fighter Support: 6 sqdns P-47 Thunderbolts/egress
Time Schedule: Sta 1440; taxi 1455; takeoff 1515; zero hour 1800; Estimated time of return; 1934.
Bomb Load: 5 X 1,000 pound High Explosive
Bombing Altitude: 91st BG(H) 24,000 feet
Other Bomb Groups: 305th, 303rd, 306th (diversionary sweep)

Did Not Take Off	Return Early
#787 "Billie K"	#763 "Bomb Boogie"
	#487 "Ritzy Blitz"
	#712 "Heavyweight Annihilators no.2"
	#483 "Spirit of Alcohol"
	#077 "Delta Rebel no.2"
	#511 "Wheel 'N Deal"

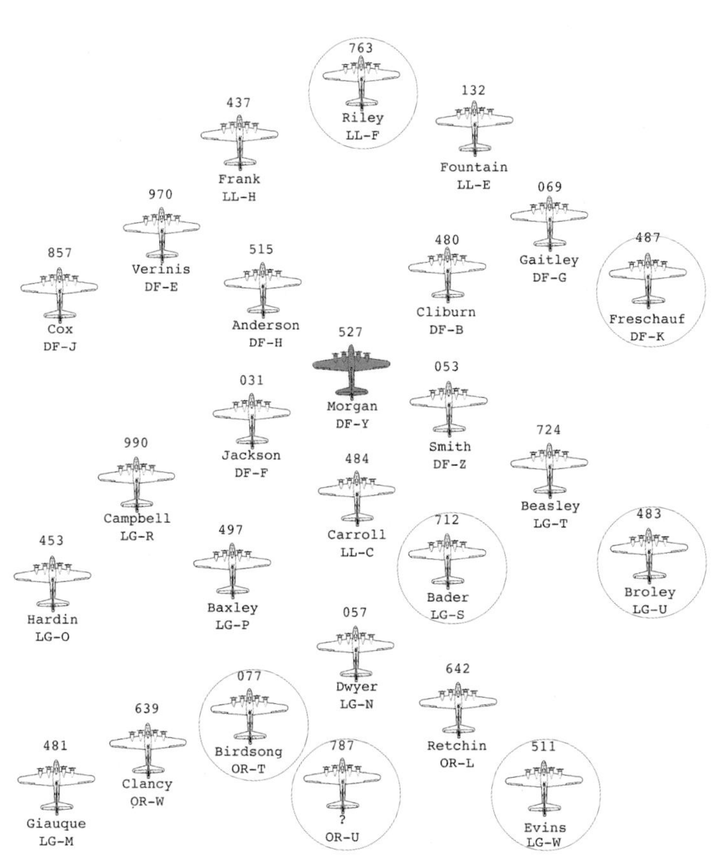

The twenty-third raid for most of the "Memphis Belle" crew. Morgan's men did not fly on their "Belle" this day. Instead "The Great Speckled Bird" was their mount for this mission. This would be the only time that Morgan was assigned to this B-17. Also, he would finish his combat tour on the plane assigned to him before he ever left the U.S. On three earlier missions he had flown a B-17 named "The Jersey Bounce," and on one previous raid it was "The Bad Penny."

The raid was scheduled for a rare afternoon launch, and the bombers were not expected to be back at Bassingbourn until around 7:30 PM. Take off was without incident, with the exception of Lt. Evins in "Billie K." He was unable to make the take off and scrambled to make this raid aboard "Wheel 'N Deal."

As the diagram shows, he was forced to return early because of mechanical problems. Five others, including the lead for the 91st, came home early as well. For once, the weather was beautiful and clear, and the target was bombed with great success. Numerous hits were seen on buildings, and both assembly plants were left in flames.

Three of the P-47 Thunderbolts flying cover for the Fortresses were shot down, but no B-17s were lost. The crews reported excellent performance on the part of the American fighter pilots. There was one report that indicated that the flight crews were not made aware of the planned P-47 escorts. It added that several B-17 gunners mistook them for enemy planes and shot at these U.S. fighters. Evidently the only result from this was fortunately just some really mad P-47 pilots and some angered gunners. For many months, Allied fighter pilots were strictly told to never point their noses toward an Allied bomber. If they were shot down then it was consid-

"Bomb Boogie," lead ship for this raid, with the doors open on a no-flak bomb run. -Joe Harlick

"Memphis Belle" co-pilot holding the only female member of the crew, their mascot "Stuka." Painted on the waist of the "Belle" are two famous names - Wyler & Clothier. -Frank Donofrio

ered to be their own fault. From the front it was very hard to tell if an approaching plane was a Spitfire, Mustang, Thunderbolt, Messerschmitt, or Focke-Wulf!

In early May 1943, the personnel at Bassingbourn were noticing one of their own (Captain Bill "Ace" Clothier) with a motion picture film camera, recording the life and activities around the station—in color! He had flown on at least six missions to film actual combat sequences for a possible film similar to the British "Target for Tonight."

Much activity had been centered around one particular B-17 in the Group, the "Memphis Belle." Clothier was known to be associated with some of Hollywood's best, and even famed Hollywood director William Wyler had been spotted in and around the base.

Even though not many knew for sure what the result of this attention from Hollywood would be, speculation that the focus of the film would be around Bob Morgan's crew and their B-17 becoming among the first Eighth Air Force bombers to make the twenty-five mission mark was brewing.

Wyler himself was rumoured to have flown on at least five combat missions aboard "Memphis Belle." He may have completed as many as sixteen aboard various B-17s. Nonetheless, of those who knew him, no one ever said that he felt like he deserved special treatment because of who he was. A great deal of film was exposed at Bassingbourn and on several B-17s assigned to the 91st by Wyler and his crew. The result was not only a milestone documentary which set the pace for documentary filmaking for the next twenty years, but also some of the most dramatic images ever exposed in aerial combat. Without William Wyler, many historical scenes of American bomber operations in WWII would have been lost to the memories of those who were taking part. "Memphis Belle" for a time carried the names of both Wyler and Clothier painted within the blue field of the right waist insignia (*see* above right).

"B-17 Pilot makes 25!"

B-17F #41-24485	"Memphis Belle"
Pilot	C. L. Anderson
Co-Pilot	D. F. Gladheart
Bombardier	E. M. Bruton
Navigator	W. S. Scovall
TT / Engineer	H. Robbins
Radio Operator	R. E. Current
Waist Gunner	N.W. Kirkpatrick
Waist Gunner	J. W. Carse
Ball Turret	R. W. Cole
Tail Gunner	C. A. Nastal

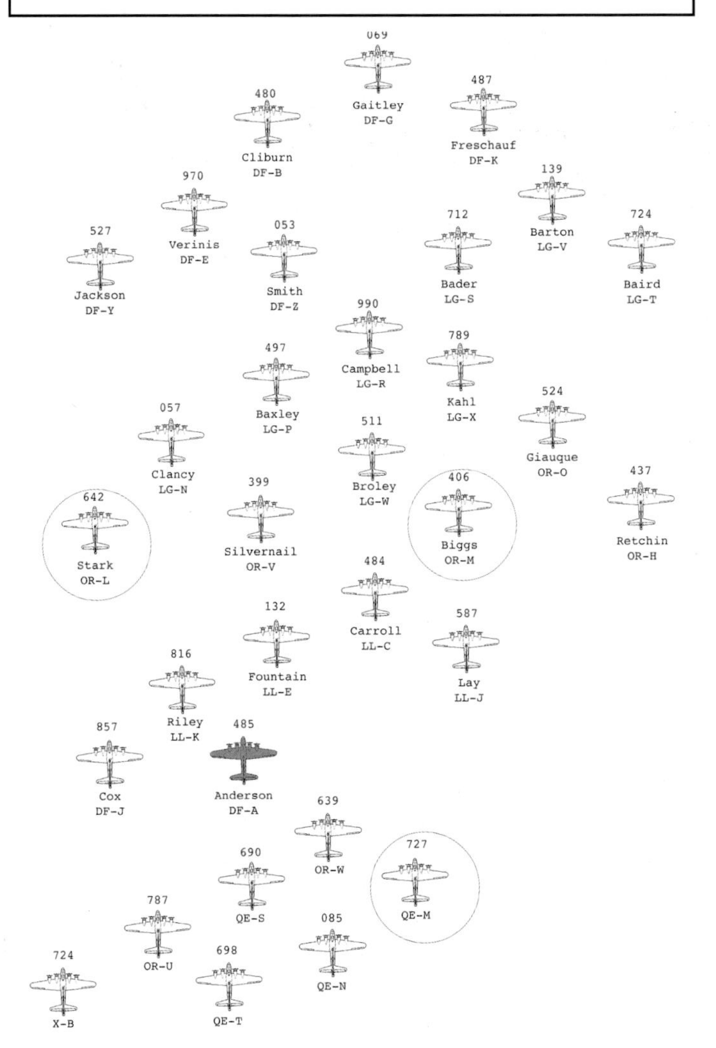

13 May 1943
Target - Air frame works at Meaulte, France
Alert no.78 - Mission no.37 for the 91st BG(H)

Battle Journal Input
Target: (1) Meaulte, France - Aero Frame Works
(2) Abbeville, France - Fighter Aerodrome
Enemy Defenses:
Fighters (Single engine) 210
(twin engine) 65
Flak Batteries - Heavy and Accurate
Friendly Fighter Support: 14 sqdns Spitfires and P-47s
Time Schedule: Sta 1250; taxi 1310; takeoff 1325; zero hour 1600;
Estimated time of return; 1729.
Bomb Load: 10 X 500 pound High Explosive
Bombing Altitude: 91st BG(H) 22,000 feet
Other Bomb Groups: 306th, 305th, 303rd

Morgan and his crew had not yet returned to Bassingbourn from their radio broadcast, and because a raid was scheduled, "Memphis Belle" was assigned to a different crew for this mission. This was the twenty-first mission for the plane, which was rapidly becoming famous and drawing the attention and speculation of all the personel at Bassingbourn. The "Belle" and the other bombers were carrying one of their largest bomb loads yet, with ten 500 pounders each. The math added up to eighty tons of High Explosives that would be dropped on the Nazi war machine from just the thirty-two planes of the 91st—and there were three other Bomb Groups taking part! The Americans would drop well more than 200 tons of bombs on this raid.

The bombs from the 91st planes tracked long and fell just beyond the target, but the 306th in trail hit it dead center. The area was fully involved in flames, and the smoke billowed well above a thousand feet into the sky. The Spitfire escorts did not faze the de-

10 May 1943
Four members of the "Memphis Belle" crew headed for London and something pretty special for 1943—a recorded radio broadcast from England to the United States. This was a big deal for tail gunner Johnny Quinlan, radio operator Bob Hanson, and navigator Chuck Leighton. Of course, no broadcast would be complete without the skipper of the "Memphis Belle," so Capt. Morgan went along, too. As a matter of fact, Leighton and Morgan brought several of their close officer friends along to lend moral support (Must have been a pretty good time!). This broadcast happened regularly once a month, and it is thought that the Eighth Air Force Commanders arranged to include the "Belle" crew because of all of the attention they had been getting in and around the base from the Hollywood film crew, and the fact that they were certainly in line to be among the first to make the golden twenty-five mission mark.

Note: The 94th Bomb Group was brand new, and some of their pilots and crews were getting their first tastes of combat with the experienced crews of the 91st. Five of their bombers and crews took part at the rear of the formation. One returned early due to mechanical problems.

Jim Verinis, "Memphis Belle" co-pilot for the first six missions, was finished. May 13, 1942, was the day he flew his last combat mission. The fighting was over for him. Verinis became the very first airman of the Mighty Eighth to reach his twenty-fifth mission—a mark that was not verified for another fifty-six years. In 1999 The Mighty Eighth Air Force Heritage Museum in Savannah, Georgia, documented that indeed, Jim Verinis was the first to finish throughout the entire Eighth Air Force. The photo above shows some of the "Memphis Belle" crew as they prepared to board the "Belle" and fly back home to the United States. From left to right are: radio operator Robert Hanson; and a pensive co-pilot, Jim Verinis. On the other side of the jeep, pilot Robert Morgan smiles and heaves a B-4 bag up on his shoulder as he heads for the "Belle." Vince Evans, the bombardier, follows Morgan. This was a victorious day for the 91st Bomb Group, as many of the base personnel stood by to cheer the men on. (Mississippi Valley Collection via Dr. Harry Friedman)

termined Luftwaffe fighter pilots. They burned right through the Allied fighter cover, boring in on the bombers that had just caused terrible destruction to their factories below.

"Vulgar Virgin" (B-17 #642) was the first shot down. They were last seen with the number two engine and cockpit enveloped in flames in a gradual spiral descent just before the bomber exploded. Eight airmen were aboard Lt. Starks' bomber were killed, and three more became POWs. One of the survivors was a photographic specialist, 1st Lt. Neill Oakley, who was manning the strike cameras on the flight.

The second B-17 that was lost was assigned to Lt. Homer Biggs (B-17 #406 - an unnamed bomber). Seven of his men were killed, while the other three were taken prisoner and later E & E'd (escaped and evaded).

Another sad note: The composite flight of 94th BG planes hit the German airfield at St. Omer while flying a diversionary attack. Their bombs fell wide and into the nearby town, probably causing considerable collateral damage among the French civilian population.

"Forty Bassingbourn Bombers"

B-17F #41-24485	"Memphis Belle"
Pilot	J. H. Miller
Co-Pilot	V. A. Parker
Bombardier	E. M. Bruton
Navigator	J. R. Ehrenburg
TT / Engineer	W. D. Spofford
Radio Operator	O. L. Stuart
Waist Gunner	Miller
Waist Gunner	R. V. Cupp
Ball Turret	R. H. McDermott
Tail Gunner	N. R. Lane

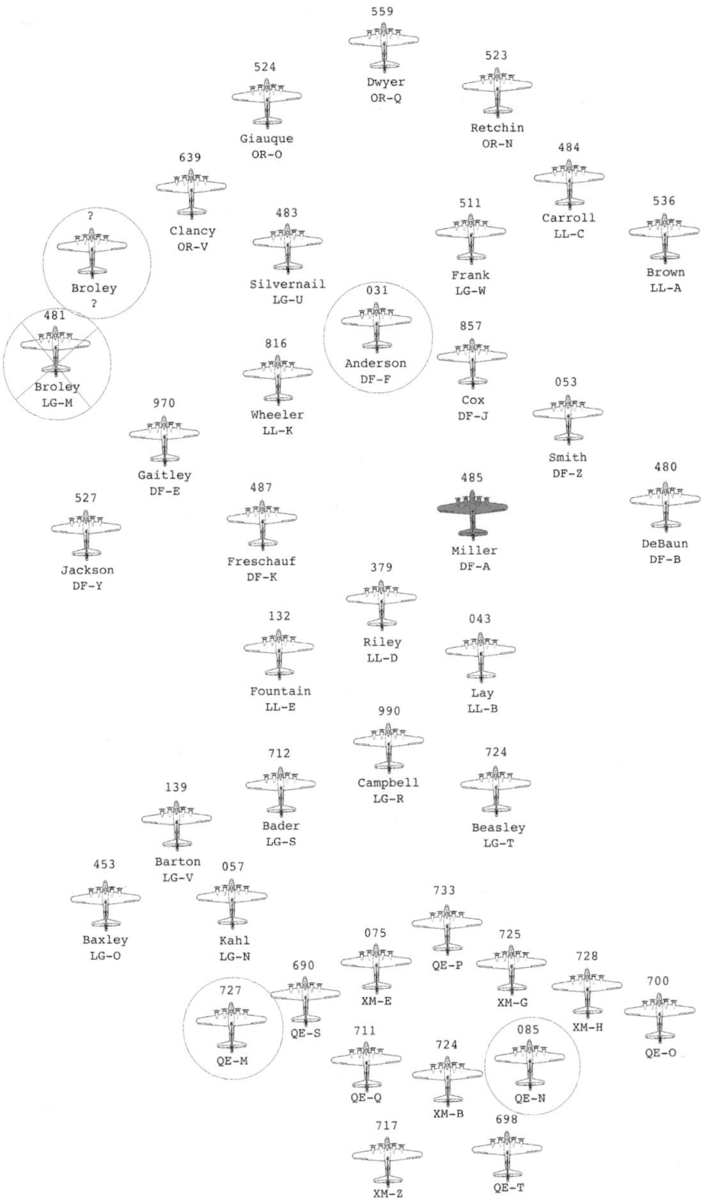

14 May 1943
Target - Shipyards at Kiel, Germany
Alert no.79 - Mission no.38 for the 91st BG(H)

Battle Journal Input
Target: (1) Kiel, Germany - Electric Shops
(2) Courtrai, Belgium
(3) Antwerp, Belgium
Enemy Defenses:
Fighters (Single engine) 100
(twin engine) 235
Flak Batteries - 80-90 in target area
Friendly Fighter Support: for Courtrai and Antwerp only
Time Schedule: Sta 0820; taxi 0840; takeoff 0855; zero hour 1200; Estimated time of return; 1310.
Bomb Load: 5 X 1,000 pound High Explosive
Bombing Altitude: 91st BG(H) 26,000 feet
Other Bomb Groups: 306th, 305th, 303rd

Captain Morgan and the regular "Memphis Belle" crew did not fly this mission, but the "Belle" did. This was the twenty-second raid for "Memphis Belle," and she was assigned to a pilot by the name of Miller, who had been co-pilot on the "Belle" three times before.

This was a tragic day for Lt. William Broley. He was unable to take off in the first bomber assigned to him for this raid. Broley had been getting plenty of experience. On the previous five missions, he had piloted four different B-17s. They were:
#511 "Wheel 'N Deal"
#789 "Golden Bear"
#057 "Blonde Bomber"
#483 "Spirit of Alcohol"

The well-liked and competent crew threw their gear into #481 "Hell's Angels," a well known and battle proven ship. They were last seen struggling to make it across the English Channel. They crashed into the sea only twenty miles short of the British coast. Air-Sea Rescue only found some wreckage floating on the surface. All aboard died.

For the second day in a row Bassingbourn came alive with a very large bomber force preparing to take off. Just before 9 AM some one hundred sixty B-17 engines were creating a thunderous roar that could be heard for miles in all directions as the bomb laden airplanes struggled into the sky. Forty B-17s of the 91st BG and the new 94th BG were scheduled, and only one was unable to take off.

Three returned early, but the rest completed the raid with spectacular results. Post mission strike photos from the 322 squadron (at the rear of the Wing) indicated that the target was already devastated by terrific bombing from the lead element of the Combat Wing, who had appeared over Kiel just minutes before them. Even the few bombs that fell astray and into the nearby city created a great deal of damage. This did not seem to bother too many, as they felt that this was in repayment for a recent German raid on the city of London. The Germans had been bombing civilian targets, while the Allied forces had been concentrating on only military targets. The crews had great concern when bombs went astray over French targets, because friendly civilians might have been casualties. When this occurred over the German homeland it did not hurt the men as badly.

Because the Germans intentionally hit non-military targets in London many innocent people were hurt and killed by these night Luftwaffe raids, and much of London was burned out.

A total of eleven bombers out of the one hundred fifty taking part were lost. The one hundred twenty-five attacking enemy planes seemed to focus on several flights of B-24s also taking part in a different Wing of this mission. The loss of Lt. Broley and his men was felt throughout the ranks of the 91st. They would be missed.

Lt. William Broley's crew before they were shot down and killed on 14 May 1943 aboard B-17F #41-24481 "Hell's Angels." The men had been with the 91st BG (H) since their training and had plenty of experience. One event near the beginning of their combat tour saw them skidding down a runway in a B-17 after they had forgotten to check the tail wheel before landing. It was up, and the entire tail gunner's compartment was broken from the bomber during their landing. The gunner had moved from the compartment before landing, or he would have rolled down the runway inside his tumbling compartment! These men were approaching the end of their scheduled combat tour when they were killed. -Col. Bert W. Humphries

A photograph of a wall in the 324th Squadron Operation's Room in hangar number two at Bassingbourn. Make note of the names that were painted on this wall and compare them to the bomber formations here in this chapter. The "Memphis Belle" artwork here is among the largest, and by coincidence all of the B-17s flown by Morgan and his crew ended up on this wall. -Frank Donofrio

One of the main reasons Col. Bert Humphries was so valuable to the research of this book was the availability of many of the 91st BG formation records.

Written directly from the boards the very morning of each mission, these have been a unique way to almost feel as if we were there with the pilots more than fifty years ago.

As Squadron Operations Officer for the 322 sqdn., Humphries was responsible for documenting the actions of the men of 322 squadron, among many other things.

Humphries was responsible for training "green" crews when they arrived. He helped to develop the thirty pound flak vests. He also worked on new instrument approach techniques to get the bombers back to base during bad weather and at night.

During the morning "briefings" some of the flight crew would hand write the formation stack on note paper to take with them on the raid. It would be referenced to give constant reminders of who should be where throughout the raid. It would also help the flight crews to make adjustments when the inevitable occurred and bombers were shot down or experienced mechanical problems and turned back for home, requiring the formation to make changes in the stack.

Everything from take off, target, and time of return is indicated on the notes. Compare them to the diagram on the next page.

STA 0710 TAXI 0730 TO 0745
Leave 0816
Zero 1100 @Target
MAY 15, 1943
ETR 1331
Land
Bomb load - 10 x 500#
91ST alt - 26,000 Ft

RE: 0950

(101) WILHELMSHAVEN, — Submarines under construction
(104) EMDEN, GERMANY — Trans-shipment pt. of Narvik ore
Enemy A/C 45 SE 100 TE (100 miles)
FLAK - 18 HEAVY GUNS.

MISSION #39 BRIEFING 80

As the crews reached their twenty-fifth mission a buzz job was permitted when they returned to base. This act would generally attract the attention of most of the people around. The jubilant crews almost always put on a spectacular show, mowing the lawn with the props!
There was one instance where a returning B-17 pulled off this stunt with several passes on the field. The bomber had received quite a beating on the raid from which it was returning. This was no matter to the crew, because they were finished and were going home!
After the pilot beat up the field he landed, and as the bomber slowed on the runway, the lift beneath the wings dropped, and so did the wingtips.
The bomber came to a stop with both wings dragging the ground. An enemy shell nearly severed two main wing spars, and the only thing keeping them straight in flight was the lift beneath them! It was evident that the shock of landing finished off the spars, allowing the wings to crazily droop toward the ground. (Joe Harlick)

"Just what in the hell did you bomb?"

B-17F #41-24485	"Memphis Belle"
Pilot	Robert K. Morgan
Co-Pilot	J. H. Miller
Bombardier	Vince Evans
Navigator	Charles Leighton
TT / Engineer	Harold Loch
Radio Operator	Bob Hanson
Waist Gunner	Bill Winchell
Waist Gunner	Scott Miller
Ball Turret	Cecil Scott
Tail Gunner	John Quinlan

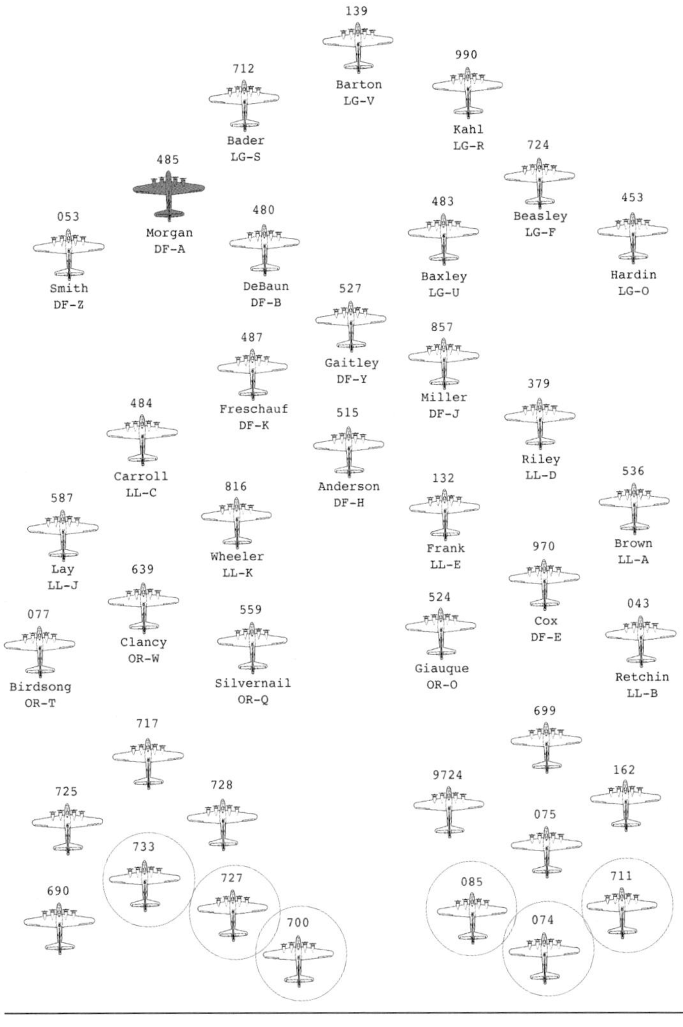

139 Barton LG-V
712 Bader LG-S
990 Kahl LG-R
724 Beasley LG-F
485 Morgan DF-A
053 Smith DF-Z
480 DeBaun DF-B
483 Baxley LG-U
453 Hardin LG-O
527 Gaitley DF-Y
857 Miller DF-J
487 Freschauf DF-K
484 Carroll LL-C
515 Anderson DF-H
379 Riley LL-D
587 Lay LL-J
816 Wheeler LL-K
132 Frank LL-E
536 Brown LL-A
639 Clancy OR-W
970 Cox DF-E
077 Birdsong OR-T
559 Silvernail OR-Q
524 Giauque OR-O
043 Retchin LL-B
717
699
725
728
9724
162
075
733
727
690
085
711
700
074

15 May 1943
Target - Navy Yards at Helgoland, Germany
Alert no.80 - Mission no.39 for the 91st BG(H)

Battle Journal Input

Target: (1) Wilhelmshaven, Germany - eleven enemy submarines under construction
(2) Emden, Germany - transportation and ore shipping
Enemy Defenses:
Fighters (Single engine) 45
(twin engine) 100
Flak Batteries - 18 heavy guns in target area
Friendly Fighter Support: unknown
Time Schedule: Sta 0710; taxi 0730; takeoff 0745; zero hour 1100;
Estimated time of return; 1331.
Bomb Load: 10 X 500 pound High Explosive
Bombing Altitude: 91st BG(H) 26,000 feet
Other Bomb Groups: 306th, 305th, 303rd, 94th composite

Remarks

Briefing at 0430. Conditions—overcast but clear by takeoff. 91st on schedule / 94th BG takeoff was forty-five minutes late. 94th hit the briefed secondary at Emden with very good results and less than half of their strike force. The Primary at Wilhelmshaven and vicinity was obscured and the 91st turned back from the target. On the return to England, Col. Reid (91st lead ship #139 "Chief Sly II") elected to hit a target of opportunity—the heavily fortified German Naval installation at Helgoland. The likely cause of selecting this particular target was that the 91st had flown over the two islands on the way to the primary. Instead of dropping bombs into the water, the 91st would simply hit the islands as they flew back to England. (Sounds easy - it wasn't)

The post mission debriefings brought the attention of many of the Commanders of the 91st BG. Col. Wray was seen walking swiftly into the debrief room with a deeply concerned look on his face. No one believed the crews when they advised Command that the unbriefed islands had been hit.

Strike photos were sent for processing with "priority" stamped all over them. Into the night, genuine concern and aggravation settled over the personnel. While they waited for the strike photos, they went over the mission again and again.

"We selected a target of opportunity."

"We flew over Helgoland on the way to Wilhelmshaven, saw that our target was covered there, and decided to hit the German Navy on the way out."

Mission number twenty-three for the "Memphis Belle," number twenty-four for a majority of her regular crew, and twenty-five for tail gunner Johnny Quinlan and navigator Charles Leighton. For the third day in a row the 91st was launching a tremendous raid, with thirty-nine B-17s taking off to hit the enemy. These were the largest raids yet mounted, and Bassingbourn was taking on the very active appearance of the base that Col. Wray had envisioned when he first set foot there. Again the new 94th BG(H) was taking part in the raid led by the experienced crews of the 91st. But this time six of their element of fourteen bombers returned early for a variety of reasons.

Ninety miles per hour, flaps at thirty, three and green. This baker one-seven is coming over the fence to land at Bassingbourn after a mission. This bomber appears to be in pretty good shape with all four fans turning. Note the woman watching with her small baby in the pram with her. This was a welcome sight to the locals around the bases of the Eighth Air Force. Many British civilians made a habit of counting the bombers as they left, hoping the same number could be tallied in the afternoon when the crews returned. (Frank Donofrio)

B-17 #519 "The Spirit of Billy Mitchell" LL-A of the 401st Bomb Squadron. This bomber was accepted into the inventory on 12 September 1943. German fighters saw to it that this bomber failed to return on April 19, 1944. The bomber crashed near Kassell, Germany, killing four of her crew. The remaining six members became POWs. (Frank Donofrio)

Bottom: Two pictures capture B-17 #23057 "Blonde Bomber." From the 322nd Squadron, this B-17 was first named the "Picadilly Commando." On January 11, 1944, the bomber failed to return from a mission when it crashed near Oschersleben, Germany, with two KIA and eight POWs. "Blonde Bomber" flew some raids with "Memphis Belle" before the "Belle" was sent home. She is shown here well above the clouds and above the early morning sun. The shadow is thrown to the top of the Flying Fortress. (Frank Donofrio/Joe Harlick)

"Our bombs were set to walk across the target area at one hundred foot intervals—more than half of the formation's bombs plastered the target."

"There were almost no wasted bombs."

Frustration was getting high, and Intelligence was unsure of what to report to Headquarters personnel, who were very skeptical about the mission results. One intelligence officer presented a picture of the base from a magazine and positive identification was made. It was Helgoland after all.

This was not an easy target to hit. Enemy fighter opposition was tremendous, and six U.S. bombers were lost. Through an unbriefed bomb run, no recognition maps, and swarms of enemy fighters, the 91st effectively hit the German Navy, causing considerable damage to the installation.

All of this must have added to the tension already on base. You do not want to mess up with the top brass all watching you. Three Generals, nine full bird colonels from other bases, and more than three hundred local guests were at Bassingbourn for a planned 1st birthday party for the Group, with a full buffet dinner in the senior officer's mess, as well as dancing and music.

It must have been the drink that added to the next bit of confusion here. The official 91st Bomb Group historian wrote that the "Memphis Belle" became the very first complete crew to finish their twenty-fifth mission intact. Not only was the "Memphis Belle" not finished, but neither was most of her original crew, and they were not intact! Just like all other bomber crews, Morgan's men had seen the typical personnel changes that were common to all bomber crews. It is probable that there were no "intact" crews at this time in the War.

Right: Posing in full flight gear are just a few of the "Bassingbourn Boys." This picture ended up on the cover of a national magazine. All of the men have their "Mae West" life preservers and parachute harnesses. "Where are my sunglasses?" (Frank Donofrio)

"Memphis Belle Finishes!"

B-17F #41-24485	"Memphis Belle"
Pilot	Robert K. Morgan
Co-Pilot	Hayley Aycock
Bombardier	Vince Evans
Navigator	Raymond Kurtz
TT / Engineer	Harold Loch
Radio Operator	Bob Hanson
Waist Gunner	Bill Winchell
Waist Gunner	Scott Miller
Ball Turret	Cecil Scott
Tail Gunner	R. W. Cole

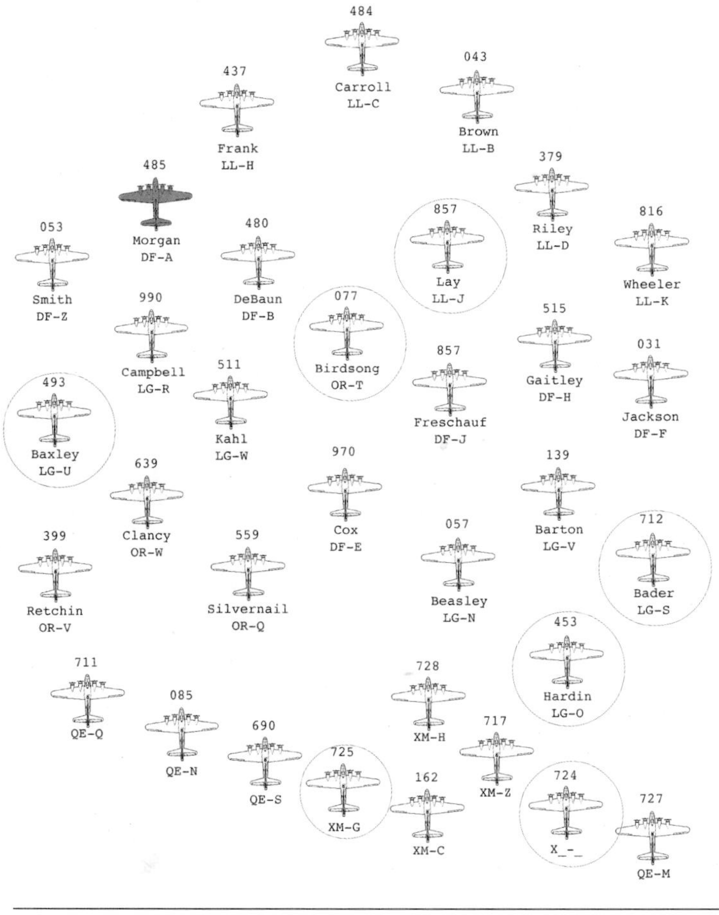

484 Carroll LL-C
437 Frank LL-H
043 Brown LL-B
485 Morgan DF-A
379 Riley LL-D
053 Smith DF-Z
480 DeBaun DF-B
857 Lay LL-J
816 Wheeler LL-K
990 Campbell LG-R
077 Birdsong OR-T
515 Gaitley DF-H
857 Freschauf DF-J
031 Jackson DF-F
493 Baxley LG-U
511 Kahl LG-W
639 Clancy OR-W
970 Cox DF-E
139 Barton LG-V
399 Retchin OR-V
559 Silvernail OR-Q
057 Beasley LG-N
712 Bader LG-S
453 Hardin LG-O
711 QE-Q
085 QE-N
690 QE-S
728 XM-H
717 XM-Z
725 XM-G
162 XM-C
724 X_-_
727 QE-M

17 May 1943
Target - Sub Pens at Lorient, France
Alert no.82 - Mission no.40 for the 91st BG(H)

Battle Journal Input
Target: (1) Lorient, France - Submarine Pens
(2) Lorient, France - Power Station
(3) Brest, France - Dry Docks
Enemy Defenses:
Fighters (Single engine) 85
(twin engine) 25
Flak Batteries - more than 100 throughout
Friendly Fighter Support: Four sqdns Spitfires (Square tips)
Time Schedule: Sta 0835; taxi 0855; takeoff 0910; zero hour 1200; Estimated time of return;1405.
Bomb Load: 5 X 1,000 pound High Explosive
Bombing Altitude: 91st BG(H) 22,000 feet
Other Bomb Groups: (composite) 94th & 92nd attacking targets at Brest

The 91st again rallied their bombers and crews to hit their old familiar target at Lorient. They were going to bounce their 1,000 pound bombs off of the eleven-foot thick reinforced concrete roofs of the sub pens. With the tonnage already dropped on the facilities there should have been substantial damage, but recon photos indicated just superficial damage.

Gripes were heard around Bassingbourn from those wishing that Command would just order the British Halifaxes and Lancasters down there to finish the job with their 22,000 pound delayed fuze "blockbuster" bombs.

Nonetheless, the weather was good for the raid and resistance by both fighters and flak was comparatively light. The bombing results should have been ideal. However, the explosion patterns that were reviewed in the post-mission strike photos showed that the bombs of the 91st fell wide of the target.

The ever-present Luftwaffe did show up, numbering about forty fighters. In what had become almost typical for the crews an air

The final combat mission for the "Memphis Belle" crew. Capt. Robert Morgan and his men were through! Although they were not the first crew to achieve this feat, they were officially recognized as "the first combat crew to complete their required twenty-five combat missions and return to the United States." It is well known that the 303rd Bomb Group at Molesworth, England, produced the first crew to fly twenty-five raids. The bomber they flew was named "Hell's Angels." Available records indicate that the "Memphis Belle" is likely the third bomber of the Eighth Air Force to make the twenty-five missions. The War Department, wanting to boost sagging homefront morale and sell War bonds, was given a directive to find a complete crew that had finished their tour of duty, then send them home to fly around the U.S. showing the public the successes of the USAAF campaign in Europe. The crew of the "Memphis Belle" had already been the subject of William Wyler's Hollywood cameras, and the romantic story surrounding the pilot and his nineteen year old girlfriend from Memphis, Tennessee, was just too much for the brass to pass up. The "Memphis Belle" and her crew would go home to show the United States that we were going to win this War.

This photo was actually a staged shot coordinated by film director William Wyler. The jubilance among the men, however, was not staged. (Frank Donofrio)

"My dearest Darling,
I am writing to you before I taxi the Belle out to takeoff position. You see, this is number 25. The whole tour will be finished. A big load off my shoulders, and yours, too. There isn't a lot to say but, Margaret, you were riding beside me at all times and, darling, I'll finish this later when I have buzzed the hell out of the field. That is what I have longed to do, my sweetheart. I have both your ribbons on this morning and, darling, may God be with us both. Well sweetheart, this is it and so take this kiss and hug and keep me flying, darling angel. Adore you. Bob"

battle ensued, and when one of these fighter pilots thought to attack the "Memphis Belle" the alert left waist gunner Bill Winchell responded by pouring a big burst of well-placed fifty caliber lead into the plane. The Nazi plane exploded and fell to earth, and the "Memphis Belle" would later have the eighth and final swastika painted on her nose.

Morgan pointed the nose of the "Belle" toward England. The final moments of the flight were jubilant. The crew had done it, and now they could go home.

As the formation approached the base, some of the thoughts were on those that would not be finishing with them. On the hell they had flown through twenty-five times. Of what would be next for them. They just kept saying over and over to themselves that it was over and they could not believe it.

Army '485 entered *"a"* landing pattern at Bassingbourn, but Morgan was not lined up on the runway; he was headed for the control tower! The throttles were wide open, pouring fuel into the four hungry engines. Her huge props bit into the air, pulling the thirty-ton B-17 even faster, and "Memphis Belle" sped only a few feet from the grass. Hundreds were standing in awe as Capt. Morgan was putting on his best buzz job yet. Just as he approached the tower, the nose of the "Belle" was jerked into a climbing turn. The tower, the men in it, and the rooftop antennas fell below the windscreen in front of Morgan, and the happiest bomber crew in England came in for a landing.

They taxied up hanging out of the radio hatch and the waist windows, where Winchell was patting his fifty cal. and making a spinning motion with his hand showing everyone that he had smacked another German from the sky.

"Memphis Belle" rolled to a stop, the engines revved, then with her mags flipped to "off" her props slowed into a visible arc as they stopped. Radio Operator Bob Hanson jumped out, fell to his knees, and kissed the ground. It seemed as if a hundred hands grabbed Morgan as they lifted him on their shoulders so he could kiss the leggy pin-up girl that had seen some of the worst the Germans could throw at them. With a little pat on her backside Morgan slipped down to cheer with his warriors.

Some controversy exists over whether or not this was the day that the "Memphis Belle" B-17 flew its final combat mission. It most certainly is the day that Bob Morgan and most of his men flew their last raid. The "Belle" at this time flew its twenty-fourth mission—this is confirmed. It is also where, through time and faded memories, the exact particulars and sporadic records fail to reveal what actually took place.

When the War Department issued the orders to find a complete bomber crew that had finished their required twenty-five missions, "Memphis Belle" was among those in the running. The "Belle" and her crew were considered to be top priority because they would be an invaluable asset to the American public striving to operate the industries that were feeding the War.

Knowing the high value of a twenty-sixth mission to build sagging homefront morale and boost public support of the efforts of the USAAF, it is highly likely that this was also the last day that the "Memphis Belle" flew a combat mission. Some believed that a call came from General Ira Eaker himself to Col. Wray instructing him to ground the "Memphis Belle" from any further combat. Losing the aircraft that was by now so widely promoted and having the crew fly their tour on a different B-17 would not do at all.

If the "Memphis Belle" crew was going home, then it would be aboard the very bomber they flew in combat. The mission of 17 May 1943 is the most probable final combat sortie for this aircraft, even though it was scheduled to fly two days later with another crew for what was to be its absolute final raid.

In 1999 Robert Morgan vividly remembered that when the 91st BG was hitting their target at Kiel, he and some of his crew were airborne in the "Memphis Belle" over England shooting extra footage for the famed William Wyler documentary. Wyler was reportedly aboard with them.

Morgan remembers that "Memphis Belle" was indeed scheduled to fly the 19 May mission, and available records do indeed indicate a B-17 on the formation board in the briefing room with #485 as the tail number—Anderson as pilot—with the squadron code DF-A. That sure is the "Belle," but Morgan believed that what likely happened in the time between briefing and takeoff was a change to a spare B-17 for Anderson's crew had been made, and the records were not altered to document the different bomber now assigned to Anderson.

Understanding the extreme importance of the mission that Command was drumming up for Morgan, his men, and the "Belle," it is highly unlikely that they would have allowed this B-17 to be at such great risk on what was a very tough target to hit. Again, however, formation over target lists indicate Army '485 was over Kiel with the rest of the 91st BG that day. It is therefore assumed that the Belle did indeed fly combat on 19 May 1943. Today it seems rather inconsequential, as her current mission of rememberance is flaw-

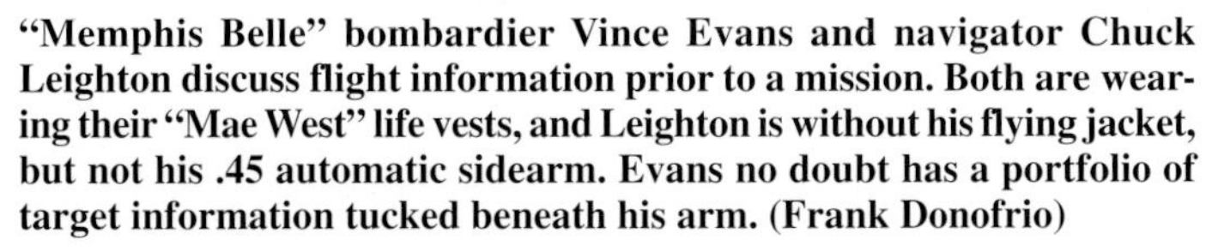

"Memphis Belle" bombardier Vince Evans and navigator Chuck Leighton discuss flight information prior to a mission. Both are wearing their "Mae West" life vests, and Leighton is without his flying jacket, but not his .45 automatic sidearm. Evans no doubt has a portfolio of target information tucked beneath his arm. (Frank Donofrio)

Upper Right: "Memphis Belle" radio operator Bob Hanson in a thankful pose as he kisses the British soil after his twenty-fifth and final bombing raid. (Frank Donofrio)

Middle Right: "Memphis Belle" ball turret gunner Cecil Scott. No enemy fighter was ever shot down from this turret on the "Belle," but there was plenty of shooting from there! (Frank Donofrio)

lessly carried out with all the dignity and honor demanded for the men and women of WWII. Especially since the Belle is the single remaining B-17 of its type in the world that flew this particular tour of duty, and therefore a true treasure in the fraternity of American Aviation artifacts.

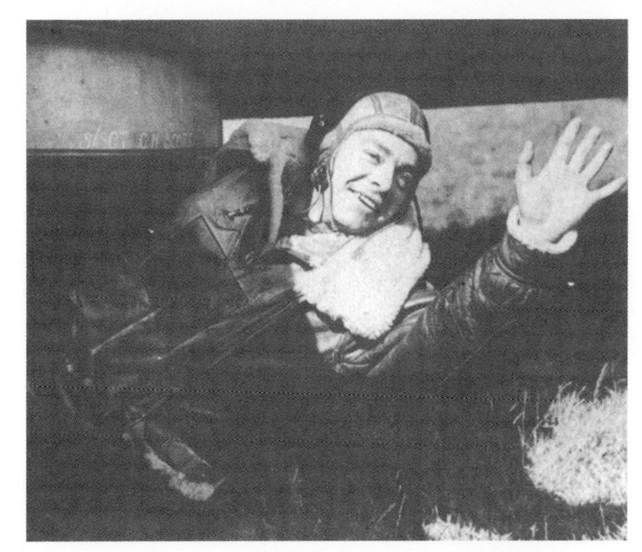

"Memphis Belle" tail gunner Johnny Quinlan pointing out his two victories from the guns that he named "Pete and Repeat." One of his nicknames from the crew was "our lucky horseshoe." John holds up a testimony to this name. (Frank Donofrio)

"Did the Belle fly or not?"

B-17F #41-24485	"Memphis Belle"
Pilot	C. L. Anderson
Co-Pilot	D. F. Gladheart
Bombardier	Eugene Adkins
Navigator	W. S. Scovall
TT / Engineer	A. B. Cornwall
Radio Operator	R. E. Current
Waist Gunner	J. E. Carse
Waist Gunner	N.W. Kirkpartrick
Ball Turret	R. W. Cole
Tail Gunner	E. S. Miller

This mission is included in this book because of the fact that many believe the "Memphis Belle" was participating. As previously stated, there is good reason to believe that, although the bomber was scheduled to fly this raid, a late switch to a replacement B-17 for Anderson and his crew was ordered. (Forty-eight crews were scheduled to fly this raid from Bassingbourn: twenty-six B-17s from the new 94th, along with twenty-two from the 91st BG veterans.)

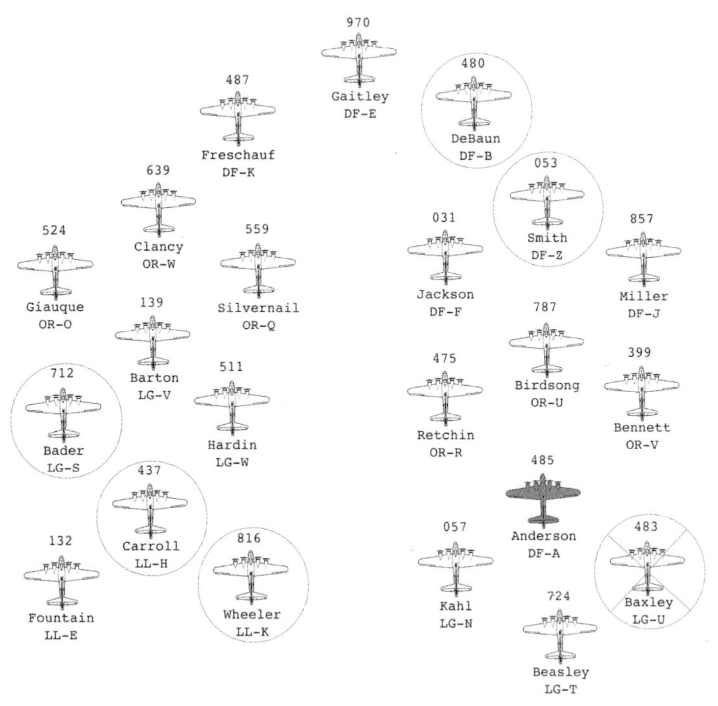

19 May 1943
Target - Sub pens at Kiel, Germany
Alert no.82 - Mission no.41 for the 91st BG(H)

Battle Journal Input

Target: (1) Kiel, Germany - Turbine engine sheds and building slips for submarines.

Enemy Defenses:

Fighters (Single engine) 100
(twin engine) 235
Flak Batteries - Heavy; Smokescreen

Friendly Fighter Support: P-47 Thunderbolts (379th FG) at Amsterdam

Time Schedule: Sta 0925; taxi 0945; takeoff 1000; zero hour 1026; Estimated time of return;1620.

Bomb Load: 10 X 500 pound High Explosive

Bombing Altitude: 91st BG(H) 24,000 feet

Other Bomb Groups: (Diversionary raid) 4th Bomb Wing

The most unusual event on this mission came from a Group which was bombing above and in front of the 91st that day. Their bombers were loaded with incendiary cluster bombs which, upon release, broke apart with some detonations just as they fell from the planes.

This caused a great deal of evasive flying amongst the 91st's bombers, who were desperately trying to avoid hitting these "friendly" bombs as they flew through them. Some 91st B-17s reported damage from the mishap.

One of the longest and determined enemy attacks yet seen was being fought in an air battle that raged for well more than a full hour. The Luftwaffe again attempted bombing the Americans from above their formation, however, no losses were reported from these attempts.

Before the 91st was able to drop their bombs, the melee created by the Group in front of the 91st prohibited the Bombardiers from drawing an accurate aim on their target. The results of this raid were considered unsatisfactory.

Note that the lead ship for this raid was #970 "Connecticut Yankee," the B-17 so christened by Jim Verinis, "Memphis Belle" co-pilot.

Also, two regular members of the "Memphis Belle" crew, Gene Adkins and E. S. Miller, were on this raid. This was Adkins' definite twenty-fifth mission. "Memphis Belle" pilot Bob Morgan was trying to have Miller, one of his regular waist gunners, assigned to fly as many raids as he could so that he could go home with the "Belle" if and when the inevitable War bond tour became official. Neither of these men would be able to come home with the Belle.

When examining the formation diagram (*left*) it is difficult to miss the position of Anderson's crew in the "Memphis Belle," and that they were terribly close to B-17 #483 "Spirit of Alcohol," which was lost.

Only fate really kept the "Memphis Belle" (or Anderson's B-17) from being the ship that met this fate. The disappointing loss of

Lt. Baxley and his crew impacted the men of the 91st. Not only was he well on his way to completing his mission requirements, but his co-pilot was flying his first combat mission.

By June 7th, the 91st had word that the Red Cross had received a postcard stating that Baxley and his men took to their parachutes and were taken prisoner by the Germans.

Whether or not the "Memphis Belle" flew this raid to Kiel that day or was in the English sky shooting extra footage with William Wyler seems insignificant when compared to the fact that the "Memphis Belle" was finished with combat, was now very famous and would be returning to the United States with honors, but more than anything, she would be leaving many other B-17s and their crews in England to face the continued horrors of combat flying in hostile skies.

The feeling around the base, however, was jubilant. The personnel needed a boost to their morale, and "Memphis Belle" and her men gave the boys something to look forward to. In the near seven months that "Memphis Belle" had spent flying combat the 91st BG lost thirty B-17s. The average was more than one bomber for every mission the "Belle" flew—every time a mission was launched! "Memphis Belle" also witnessed the final moments of even more Flying Fortresses and Liberators that flew in accompanying Groups.

Hundreds of American men had been killed or wounded. Some of the men worked hard for promotions, only to find that as they received them, they often were required to stay for further combat assignments. It seemed that a promotion was no longer a desired career objective for a bomber pilot; instead, he just wanted to complete his tour, head back to the States, and get the hell out of this business!

These men had seen too much in too little time. As the "Memphis Belle" gained in noteriety, she would have to carry with her all of those sister B-17s that were lost.

The hellish life of the 91st BG would not end yet. A total of one hundred ninety-seven B-17s were shot down from this group alone—more than any other of the forty-one various Bomb Groups in the Mighty Eighth Air Force.

The chart on the next page shows the B-17s of the 91st BG that were lost while they flew in some of the same formations as "Memphis Belle." One hundred ninety-nine of these brave men were killed before "Memphis Belle" came home.

From the personal notes of Bert W. Humphries
322 Squadron Operations Officer - 91st Bombardment Group
19 May 1943

I was awakened at 3:00 AM, and I sleepily pedaled my bicycle through the pitch black night to arouse the six combat crews that the 322nd would be contributing to forty-eight crews that will be scheduled to fly today!

At the 4:30 AM briefing we found out that the 91st Bomb Group's target was to be Kiel, Germany, and the turbine engine sheds housed near the submarine pens—to be attacked with 10 X 500 pound High Explosive bombs from 24,000 feet.

Simultaneously, the 94th Bomb Group would be attacking Naval installations in the city of Flensburg, Germany, with clusters of incendiary bombs, which was done effectively and left the target area in flames.

However, photographs of the 91st Group's bomb pattern were somewhat amiss due to a very effective smoke screen thrown around the target by its defenders.

Total B-17 losses were six, which included Lt. Edwin T. Baxley and crew in ship No. 483 "Spirit of Alcohol." (Baxley was a classmate of mine and one of the eighteen co-pilots who volunteered for combat with the 91st Bomb Group; there are seven of us left now!)

According to information from various sources, Baxley's ship got hit as soon as it crossed the coast but continued on to bomb the target; and when last seen it was descending, under control, with no engine fires.

I was further depressed when I learned that I had sent the newly arrived first pilot Lt. Breeden along as Baxley's co-pilot (to gain some combat experience before flying with his own crew), and this was his first mission! Needless to say my morale is pretty low this evening.

So as soon as the dining room opened this evening I went in, sat alone, forced myself to eat a few bites, and then slipped out and made straight away to my room, and thence to bed. But I found no solution there, because sleep would not come to me as I kept thinking about everything that had happened during the day.

For example, I kept pondering why our Squadron Commander, Major Fishburne, is suddenly transferred to the 326th Bomb Squadron at Bovingdon, whose mission is Combat Crew Replacement Center (C.C.R.C.). Being unable to sleep, I finally got out of bed and read the mail I received during the day.

The British Broadcasting Corporation

May 19,1943

The very day that the Belle was scheduled to fly her last combat mission, Margaret received a letter from the BBC....

Dear Miss Polk,

We have been notified by our London office that Captain Robert Morgan will participate in the BBC program "Stars and Stripes in Britain," scheduled to be re-broadcast over the mutual system, Sunday May 23, at 7:30 PM Eastern War Time.

The program is broadcast by short-wave from Britain by the British Broadcasting Corporation. We suggest that you check with your local Mutual Station to determine whether or not they will carry the broadcast.

The "Stars and Stripes in Britain" with Ben Lyons and Bebe Daniels as master and mistress of ceremonies, is a BBC service broadcast whose participants are American soldiers, sailors, Marines and Air Force personnel in the British Isles. The difficulties of wartime broadcasting occasionally make it necessary to cancel the appearance of certain men. The BBC will make every effort to bring these men to the microphone again whenever possible.

Sincerely,
Publicity Department
British Broadcasting Corp.

A newspaper photo of Margaret Polk during the summer of 1943.

91st Bomb Group losses suffered Oct.'42-May '43
(during the time frame that the "Memphis Belle" flew combat)

#41-24451 (unnamed)	Oct 3, 1942 crashed Ireland	Salvaged	8 killed / 2 injured
#41-24479 "The Sad Sack"	Nov. 23, 1942 St. Nazaire, France	MIA	10 killed
#41-24503 "Pandora's Box"	Nov. 23, 1942 St. Nazaire, France	MIA	11 killed
#41-24506 "The Shiftless Skunk"	Nov. 23, 1942 St. Nazaire, France	Crashed	5 killed / 5 returned
#41-24432 "Danellen"	Dec. 20, 1942 Romilly, France	MIA	9 killed / 1 POW
#41-24452 (unnamed)	Dec. 20, 1942 Romilly, France	MIA	3 killed / 7 POWs
#41-24449 "Short Snorter"	Dec. 30, 1942 Lorient, France	MIA	10 killed
#42-5084 "Panhandle Dogey"	Jan. 3, 1943 St. Nazaire, France	MIA	9 killed / 1 POW
#41-24482 "Heavyweight Annihilators"	Jan. 3, 1943 Returned England	Salvaged	10 returned
#41-24544 "Pennsylvania Polka"	Feb. 4, 1943 Emden, Germany	MIA	8 killed / 2 POWs
#41-24589 "Texas Bronco"	Feb. 4, 1943 Emden, Germany	MIA	2 killed / 8 POWs
#41-24447 "Kickapoo"	Feb. 26, 1943 Wilhelmshaven, Germany	MIA	10 killed
#42-5362 "Short Snorter II"	Feb. 26, 1943 Bremen, Germany	MIA	10 killed
#41-24464 "Excalibur"	Mar. 4, 1943 Hamm, Germany	Ditched	3 killed / 7 rescued
#41-24512 "Rose O'Day"	Mar. 4, 1943 Hamm, Germany	MIA	7 killed / 3 POWs
#41-24549 "Stupintakit"	Mar. 4, 1943 Hamm, Germany	MIA	8 killed / 2 POWs
#42-5370 (unnamed)	Mar. 4, 1943 Hamm, Germany	MIA	9 killed / 1 POW
#42-29659 "Liberty Bell"	Mar. 22, 1943 Wilhelmshaven, Germany	MIA	11 killed
#42-29537 (unnamed)	Mar. 28, 1943 Rouen, France	MIA	10 killed
#41-24459 "Hellsapoppin"	Apr. 17, 1943 Bremen, Germany	MIA	5 killed / 5 POWs
#42-5070 "Invasion II"	Apr. 17, 1943 Bremen, Germany	MIA	10 POWs
#42-5172 "Thunderbird"	Apr. 17, 1943 Bremen, Germany	MIA	2 killed / 8 POWs
#42-5337 "Short Snorter III"	Apr. 17, 1943 Bremen, Germany	MIA	8 killed / 2 POWs
#42-5391 "Rain of Terror"	Apr. 17, 1943 Bremen, Germany	MIA	2 killed / 8 POWs
#42-29574 "Sky Wolf II"	Apr. 17, 1943 Bremen, Germany	MIA	5 killed / 5 POWs
#41-24547 "Vertigo"	May 1, 1943 St. Nazaire, France	MIA	5 killed / 5 POWs
#42-5406 (unnamed)	May 13, 1943 Meaulte, France	MIA	7 killed / 3 POWs
#42-29642 "Vulgar Virgin"	May 13, 1943 Meaulte, France	MIA	8 killed / 3 POWs
#41-24481 "Hell's Angels"	May 14, 1943 Kiel, Germany	MIA	10 killed
#41-24483 "Spirit of Alcohol"	May 19, 1943 Kiel, Germany	MIA	4 killed / 6 POWs

British Royalty Meets the "Memphis Belle"

26 May 1943

In what was typical fashion for British royalty, King George VI and Queen Elizabeth arrived at the front gate of Bassingbourn at 11:15 AM. After a welcoming by Col. Wray (91st C.O. and other top brass—three Generals and other Commanders), six large limousines carried the entourage around the base through hangars and living areas, then down the perimeter track and taxi strip.

Many of the personnel at the base were suffering the effects of the Group's first birthday held the night before and had to be roused out of bed. The men had flown missions for three straight days, and the party became one that would be talked about for years.

The base was looking its best as the Royals made their way about with the entire Command Corps in trail, pointing and explaining. Several B-17s that had questionable nose art were hidden from view. One of them was "V-Packette" (a phrase that was used to describe the item that was intended for use during the more romantic moments between soldiers and local girls they met.). It simply was not proper for the Royals to see such things.

An experimentally modified B-17 (given the designation YB-40) landed at Bassingbourn that morning. Bert Humphries, Squadron Operations Officer for the 322nd, had rushed out to where it was sitting with his small camera, hoping to sneak some pictures of the "super-armed" Flying Fortress with its extra turrets and twin fifty waist guns. Not many people were briefed on why it was there and did not find out until the King and Queen, along with their entourage, drove up for an inspection tour of the outstanding ships of the 91st.

"Mizpah"s Capt. Hardin and his men ready to meet the Queen.

RAF Officers, along with Col. Wray and General Eaker, escorted the Royals to the crews, who were all assembled in front of their bombers. The 91st presented three distinguished B-17s and their crews for the Royal couple that day:

- From the 322nd Squadron, Capt. John Hardin introduced his men and their bomber "Mizpah."
- The 323rd Squadron presented Capt. George Birdsong, with his crew and the "Delta Rebel no.2."
- And the 324th squadron presented Capt. Morgan, with his crew and the "Memphis Belle."

A dim note: the 401st had no veteran crews left to present to the King and Queen—many had been lost in combat.

A veteran B-24D Liberator was flown to Bassingbourn from the 44th Bomb Group at Shipdam. They too visited with the King and Queen.

As quickly as it started the greeting and review were finished, and the King and Queen went to the main administration building, where they said their official farewells, were given an Honor Guard salute, and then drove away.

Incidently, more than fifty-four years later, in 1997, Morgan, now a retired full Colonel, was invited to tea with the Queen Mother at her private residence at Sandringham. He was in England with his wife Linda for the opening and dedication of the American Air Museum at Duxford. The Queen Mother remembered him from their meeting so many years before when she spoke of her thanks for the many sacrifices of the Americans.

The "Memphis Belle" ground crew answers questions from the Queen. Her Majesty has finished talking with the flight crew lined up on the left wing of the "Memphis Belle."

The B-24D "Liberty Lass" from the 44th BG(H) at Shipdam is presented to the Royals.- Images this page: Bert W. Humphries

"Delta Rebel no. 2" is ready for Royal inspection. The bomber in the distance is the "Memphis Belle."

11

Headed Home

"The Theatre Commander has approved my recommendation that one especially selected crew which has completed its operational tour, be allowed to return to the United States, flying its airplane which also has completed 25 operational missions. You are charged with the responsibility of selecting this plane and crew. You must make it clear to your operational personnel that this is, in no sense, a precedent. It is done because of the beneficial effect it is believed it will have on the Operational Training Units system at home, particularly in the Second Air Force.

It may be that as a result of our continued operations in the coming summer we may allow one especially selected crew from each wing to go back with their aircraft. That has not been definitely determined and will not be until the effect of sending this first crew back has been analyzed.

As soon as you have selected the crew and plane in question, advise me showing particular reasons for its selection with good historical and biographical sketches, including the performance of the crew in combat, so this can be sent back to the United States. In this way we can make sure that the crew will be properly used for morale and training purposes on arrival in the United States."

- Ira C. Eaker, Major General
Eighth Bomber Command USAAF
1 June 1943

(Frank Donofrio)

Captain Robert Morgan renders his farewell salute to the commanders at Bassingbourn just before he took his men back to the United States. This was a landmark day in the story of the "Memphis Belle," because this plane and crew would be going back to represent the many airmen of the Mighty Eighth to the American public. Their job now would be to instruct new crews that were heading into combat, as well as educate America about how the new concept of high-altitude bombing was working. While the crew of the "Memphis Belle" were absolutely happy to be headed home, they all left England knowing the seriousness of what they had spent the past months being involved in. They knew all too well that there was much skill to be thankful for because they were still alive, but they also knew that they were very, very lucky. Forty-nine different men crewed the "Memphis Belle" during her combat days. Because of crew rotation, the "Belle" had seen quite a few different men manning her stations: four different pilots; eleven different co-pilots; five different bombardiers (TT/ENG was a togglier on the 19 May mission); four different navigators; six men flew the top turret; four crewed the radio room; eight different waist gunners; five different men in the ball turret; and four different men flew the tail.

Of these forty-nine combat crewmen that flew the "Memphis Belle" in the war, only three suffered various injuries, and only Johnny Quinlan received the Purple Heart.

All of the bombers flying at this time saw crews rotate from plane to plane, mainly because deaths and injuries broke up the crews, but also because their regularly assigned planes saw so much repair work. Many of these "Memphis Belle" men were injured aboard different B-17s, and some of them were killed.

Miss. #	Date	Target Location	Plane	Pilot	Co-pilot	Bombardier	Navigator	Top Turret	Radio Operator	Waist Gunner	Waist Gunner	Ball Turret	Tail Gunner	Target
1	7 Nov 1942	Brest, France	Memphis Belle	Morgan	Verinis	Evans	Leighton	Dillon	Hanson	Winchell	Loch	Scott	Quinlan	Submarine Pens
2	9 Nov 1942	St. Nazaire, France	Memphis Belle	Morgan	Verinis	Evans	Leighton	Dillon	Hanson	Winchell	Loch	McNally	Quinlan	Submarine Pens
3	17 Nov 1942	St. Nazaire, France	Memphis Belle	Morgan	Verinis	Evans	Leighton	Dillon	Hanson	Winchell	Loch	Scott	Quinlan	Submarine Pens
4	6 Dec 1942	Lille, France	Memphis Belle	Morgan	Verinis	Evans	Leighton	Adkins	Hanson	Winchell	Loch	Scott	Quinlan	Locomotive Works
5	20 Dec 1942	Romilly Sur Seine,France	Memphis Belle	Morgan	Freschauf	Evans	Leighton	Dillon	Hanson	Winchell	Loch	Scott	Quinlan	Luftwaffe Airfield
6	30 Dec 1942	Lorient, France	Memphis Belle	Verinis	J.S Jackson	Evans	Leighton	Adkins	Hanson	Winchell	Loch	Scott	Quinlan	Submarine Pens
7	3 Jan 1943	St. Nazaire, France	Memphis Belle	Morgan	C.E Putnam	Evans	Leighton	Adkins	Hanson	Winchell	Loch	Scott	Quinlan	Submarine Pens
8	13 Jan 1943	Lille, France	Memphis Belle	Morgan	S.J Wray	Evans	Leighton	Adkins	Hanson	Winchell	Loch	Scott	Quinlan	Railroad Yards
9	23 Jan 1943	Lorient, France	Memphis Belle	Morgan	Lawrence	Evans	Leighton	Adkins	Hanson	Winchell	Loch	Scott	Quinlan	Submarine Pens
10	4 Feb 1943	Emden, Germany	Jersey Bounce	Morgan	S.J Wray	Evans	Leighton	Adkins	Hanson	Winchell	Loch	Scott	Quinlan	Submarine Pens
11	14 Feb 1943	Hamm, Germany	Memphis Belle	Morgan	H.W Aycock	Evans	Leighton	Loch	Hanson	Winchell	E.S. Miller	Scott	Quinlan	Railroad Yards
12	16 Feb 1943	St. Nazaire, France	Memphis Belle	Morgan	H.W Aycock	Evans	Leighton	Loch	Hanson	Winchell	Miller	Scott	Quinlan	Docks & Navy Installations
13	26 Feb 1943	Wilhelmshaven, Germany	Jersey Bounce	Morgan	H.W Aycock	Evans	Leighton	Loch	Hanson	Winchell	Miller	Scott	Quinlan	Shipping in Harbor
14	27 Feb 1943	Brest, France	Jersey Bounce	Morgan	J.S Jackson	J.R Ehrenberg	A.B Cornwall	H.Robbins	R. Current	Kirkpatrick	C.B Pope	R.W Cole	C.A Nastal	River Harbor
15	6 Mar 1943	Lorient, France	Memphis Belle	Morgan	H.W Aycock	Evans	Leighton	Loch	Hanson	Winchell	Miller	Scott	Quinlan	Submarine Pens
16	12 Mar 1943	Rouen, France	Memphis Belle	Morgan	C.W Freschauf	Evans	Leighton	Loch	Hanson	Winchell	Miller	Scott	Quinlan	Railroad Yards
17	13 Mar 1943	Abbeville, France	Memphis Belle	Morgan	C.W Freschauf	Evans	Leighton	Loch	Hanson	Winchell	Miller	Scott	Quinlan	Luftwaffe Airfield
18	22 Mar 1943	Wilhelmshaven, Germany	Memphis Belle	Morgan	C.W Freschauf	Evans	Leighton	Loch	Hanson	Winchell	Miller	Scott	Quinlan	Naval Base and Battleship
19	28 Mar 1943	Rouen, France	Memphis Belle	Morgan	J.H Smith	Evans	Leighton	Loch	Hanson	Winchell	Spagnolo	Scott	Quinlan	Railroad Center
20	31 Mar 1943	Rotterdam, Holland	Memphis Belle	J.H Smith	H.W Aycock	Evans	Leighton	Loch	Hanson	Winchell	Miller	Scott	Quinlan	Engine Sheds/Assembly Bldgs
21	5 Apr 1943	Antwerp Belgium	Bad Penny	Morgan	J.H Miller	Evans	Leighton	Loch	Hanson	Winchell	Miller	Scott	Quinlan	Aero Engine repair works
22	16 Apr 1943	Lorient, France	Memphis Belle	Morgan	J.H Miller	Evans	Leighton	Loch	Hanson	Winchell	Miller	Scott	Quinlan	Submarine Pens
23	17 Apr 1943	Bremen, Germany	Memphis Belle	Morgan	J.H Miller	P.T Palmer	Leighton	Loch	Hanson	P.J Shook	Miller	Quinlan	H.W Warner	Aircraft Factory
24	1 May 1943	St. Nazaire, France	Memphis Belle	Morgan	C.E DeBaun	J.R Ehrenberg	Evans	L.W Murray	J Moore	Winchell	Miller	Scott	W.C Dager	Submarine Pens
25	4 May 1943	Antwerp Belgium	Gr.Speckled Bird	Morgan	V.A Parker	Evans	Leighton	Loch	Hanson	Winchell	Miller	Scott	Quinlan	Vehicle Assembly Plants
26	13 May 1943	Meaulte, France	Memphis Belle	C.L Anderson	D.F Gladhart	E.M Bruton	W.S Scovall	H Robbins	R. Current	Kirkpatrick	J.W Carse	R.W Cole	C.A Nastal	Airframe Works
27	14 May 1943	Kiel, Germany	Memphis Belle	J.H Miller	V.A Parker	E.M Bruton	J.R Ehrenberg	Spofford	O.L Stuart	Miller	R.V Cupp	McDermott	N.R Lane	Shipyards
28	15 May 1943	Heligoland, Germany	Memphis Belle	Morgan	J.H Miller	Evans	Leighton	Loch	Hanson	Winchell	Miller	Scott	Quinlan	Ship and Naval yards
29	17 May 1943	Lorient, France	Memphis Belle	Morgan	H.W Aycock	Evans	Ray	Kurtz	Loch	Hanson	Winchell	Miller	Scott	Submarine Pens
30	19 May 1943	Kiel, Germany	Memphis Belle	Anderson	D.F Gladhart	W.S Scovall	Adkins	A.B Cornwall	R.E Current	J.E Carse	Kirkpatrick	R.W Cole	Miller	Shipyards and Submarine Pens

Although the "Memphis Belle" was finished with the War, there was still quite a bit of activity around Station 121. Jim Verinis had been re-assigned to Eighth Air Force Headquarters in London and was there waiting for his new duties. Air crews were flying training and combat missions. Some of the men began to grumble about having to fly the tedious training runs and were actually preferring the combat!

The visit from the King and Queen had come and gone. William Wyler was wrapping up his shooting with the "Memphis Belle" crew, and daily rigors seemed to drone on. Visits to London broke up the monotony but were short-lived. On the first of June Bert Humphries made his way to London and ran into Captain Morgan. The two had some drinks with other 91st Bomb Group Officers that had met up with them. Morgan offered to share his room with Humphries, who considered this to be quite a stroke of luck (rooms were very hard to get) and took Morgan up on his offer. The two split up and Humphries did some sightseeing, visiting all of the big London historical spots. It was very late that night when he finally crawled into bed.

Thursday, 3 June 1943

Humphries did not awaken until after 10:30 AM. Morgan was not in the room, and Major Aycock was somehow sleeping in Morgan's bed. When Morgan came whisking in Humphries was shaving and the phone rang. It was Col. Wray informing him that he was to return to base immediately; he was to start packing his belongings and preflighting the "Memphis Belle," preperatory to returning to the States within forty-eight hours!

The man who had nurtured the 91st Bomb Group from its birth, Col. Stanley was promoted to Commanding Officer of the 103rd Combat Wing and was replaced by Captain Campbell. On 4 June a large party was held for the boys with the new C.O. in attendance.

The next day, presumably due to hangovers, the morning Squadron meeting was canceled, but not before photographers, radio technicians, and correspondents were beginning to set up their things around the airfield. The Bassingbourn Boys were about to receive Captain Clark Gable! He was visiting the experienced crews of the 91st to pick up some pointers so he could incorporate them into his aerial gunnery instruction back in the States.

Although Gable was actually assigned to the 351st Bomb Squadron of the 303rd Bomb Group at Molesworth, he came to the 91st to include Sgt. T.J. Hansbury in a radio broadcast back to America. Hansbury had distinguished himself as one of the most decorated gunners in the Group and had finished his twenty-five missions. He was also the first gunner in the 91st to shoot down an enemy fighter aboard B-17 #41-24482 "Heavyweight Annihilators."

Hansbury and Gable apparently hit it off, and after the broadcast the sergeant brought Capt. Gable around to meet some of the men. Gable and Bert Humphries (who contributed so much to this book) became drinking friends over some beer at the Officer's Club. Gable listened carefully to the combat stories from the crews and seemed particularly interested in the German attack tactics being used then by their fighter pilots against the bombers. The discussions went through the evening, and before retiring the world-famous actor from Hollywood enjoyed an unauthorized late night snack in Lt. Don Bader's quarters—a fried egg sandwich!

Gable was also seen with the "Memphis Belle" and crew. He went to London with Bob Morgan and some others. Gable posed for several pictures in front of "Delta Rebel no.2." The actor was making the most of his stay at the 91st to shoot scenes for the film "Combat America." He reportedly flew several combat missions on B-17s. There were rumors that the actor actually flew some missions on "Delta Rebel no.2," but this is not so. Gable was photographed with the plane because of his participation and involvement with the South in his film "Gone with the Wind." It was just too much for the War Department Public Relations machine to pass up.

He also took part in at least some instruction, and during one of the first attempts to show the "Delta Rebel" crew a thing or two he was interrupted by Captain Birdsong, who politely took Gable around to the nose of the B-17 to show him the row of painted enemy swastikas for each Nazi fighter they had already shot down. The actor looked at the record, then looked at the men and said, "I guess I don't need to tell you guys anything more," then walked off.

Tuesday, 8 June 1943 (from the diary of Bert W. Humphries)

Captain Morgan and his crew took off at noon today in B-17 #485 DF-A "Memphis Belle" for the States—and will be involved in a combination "good will" and "war-bond sales" tour of the United States. Morgan and his crew are certainly deserving of this recognition, for they are outstanding airmen of the first order. After sharing the Mayfair Hotel suite with him on a recent visit to London I feel that I have found a friend worthy of great admiration.

A rare view of the "Memphis Belle" in level flight, wheels up and low to the ground. From this vantage point, no less than four fifty caliber machine guns could be pointed at you. Some Luftwaffe pilots likened an attack run on a B-17 to jumping an angry porcupine. (Frank Donofrio)

Army '485 Ground Crew, led by Joe Giambrone inspects damage to the tail of the "Memphis Belle." Chalk was used to circle damaged areas that would require the attention of the specialists in the sheet metal shop. It is rumored that tail gunner Johnny Quinlan requested that one hole remain to provide him a place to flick his cigar ashes out of the plane while they were on a mission. While he was credited with two official fighters to his guns, he firmly believes that he shot down five enemy fighters during his combat tour. The tail position of the B-17 was critically important and provided a rear-facing set of eyes for the flight crew some sixty feet to the front of him. (Frank Donofrio)

Left: The Crew Chief for B-17 #070 "Invasion II," the ship assigned to Capt. Oscar O'Neill. Above: Col. Gillespie and Capt. Oscar O'Neill of the 401st Bomb Squadron. O'Neill was the lead pilot for the 401st from September 1942 until 17 April 1943, when he and his crew were shot down in their plane "Invasion II" during a raid to Bremen. The entire crew became POWs. Years later, Oscar O'Neil's daughter Jennifer became a very famous Hollywood actress!

M/sgt Joseph M. Giambrone (on the left) makes some small talk with a 91st Bomb Group 1st Sgt.—and probably a little joke. Joe was the original Crew Chief of the "Memphis Belle," as well as "Pistol Packin Mama." In the next photo (right) are the men assigned to Giambrone. These are the men who kept the Belle airworthy. More often than not, they were up all night in the bitter cold, rain, and even snow working on their bomber. It is very likely that these are the men responsible for saving the "Memphis Belle," for if they had not done their jobs so well, then it is likely that Morgan would have had to deal with more than combat damage. When a pilot knew he could count on aircraft systems working properly, then there was less for him to worry about when he faced serious combat damage and a long flight home. When a pilot flipped a switch he did not need the compounded problems of not having something on a damaged airplane work perfectly.

The images on this page are presented courtesy of the 91st BG Memorial Association. Visit them on the web at www.91stbombgroup.com.

B-17 41-24606 "Werewolf" is seen on the active ramp at Bassingbourn. This plane was later purposely landed in a field beside a British hospital by her pilot following a raid. He was trying to get immediate medical assistance to some of the wounded men he had aboard.

A common sight around any Eighth Air Force airfield. The maintenance never stopped.

The waist code DF-G of #069 "Our Gang" is clearly seen as the bomber tracks down the Bassingbourn runway.

Actor Clark Gable is seen donning his leather A-2 flying jacket at the tail of Capt. George Birdsong's "Delta Rebel no.2." Gable was well known around USAAF Station 121.

Just after landing in Memphis for the tour, Bob points out the battle awards painted down the nose of the Belle to Margaret.

In his first letter of 5 October 1942 to Margaret after arriving in England Morgan wrote:

> My dearest Darling, I sit here so many miles from you and write you my first letter from England. It is quite hard to write, darling, for there is so much on my heart, and it is quite hard to put it on paper. You must always have faith that I'll be back one of these days. You and I have a wonderful life ahead of us and I'm sure we will both dream of it much in the near future. Our life was meant to be and it will be, my love. I shall cable you as often as possible to give you assurance of my well-being. I read your old letters over and over. They are such a great help, as is our record, "Just as though you were here." I play it over and over, my darling. The "Memphis Belle" will always stick by us and it will make our future secure.
>
> Your Bob

Margaret and Bob caught in a moment of passion when she surprised him during the Belle's tour stop in Cleveland, Ohio, in 1943. This picture was a favorite of editors across the country, and even ended up in Life magazine!

Bob never lost his enthusiasm for Margaret during his days in combat with the "Memphis Belle." He also never lost his enthusiasm for a good date with a pretty girl. He and his Bombardier Vince Evans often had enough time to meet a girl or two at area pubs when they were in England. But he never failed to notify Margaret every time that he finished a mission that he was safely back on the ground. He always told her that he was flying for her and that she was his inspiration for getting through it alive.

Bob told her again and again how proud he was for flying *her* plane for them. He told her as much as the censors would allow about how the Germans would attack and shoot flak at them, but that the "Memphis Belle" always brought them home. Their love flourished even though they had rarely been together since they met in Washington a few months before. The Belle was making headlines by now in the International News Service, and it seemed that everyone wanted to know all they could about this great bomber. Margaret was being invited to social events, and Memphis newspaper readers devoured every story printed in the Commercial Appeal about her and the B-17 that was named after her. Mission after mission went by, and soon it was all over. Bob would be bringing the Belle home. In one of his last telegrams to her from England Bob wrote:

Safe. Tour of duty completed. Fingers crossed. Adore you. Bob.

Soon he could be in her arms again. He just did not know how soon. Within the next thirty days, the "Memphis Belle" and her men landed in Washington, D.C. Their next stop was Memphis, Tennessee, and Margaret Polk. Her telegram from Bob read:

> Memphis Belle will arrive tomorrow. Don't forget there are nine other men. You are going to have one heck of a busy time. Am so tired I hope there will be some rest. Won't leave until Tuesday morning. Adore you, Bob.

Almost nine months to the day after the "Memphis Belle" landed in Memphis for the first time she was back in her namesake city. It was Saturday, 19 June 1943, and Bob was bringing the Belle in for a show over the airport. Even he did not know what he was going to do. But when the two P-47s escorting the Belle into the city began looping and rolling, Bob felt it was time to outdo them. After all, Margaret was watching on the ground, and Bob could not let her see a couple of fighter jocks beat him out of his glory.

He pointed the nose of the great bomber into a shallow dive and pushed the four throttles all the way to their stops. When the Belle gained sufficient speed, he pulled back on the yoke and at the same time turned the wheel hard over all the way. On the ground, jaws dropped and gasps were heard as "Memphis Belle" began a series of climbs, dives, chandelles, and high speed turns. An Army General gave him permission to fly as low as he wanted wherever he took his plane. Many thought he was a foolish stunt flyer and that he should have been court-martialed for flying that way, but who really knew the B-17 better than him? Morgan was untouchable and he knew it.

Margaret had goose bumps as the Belle taxied up and parked. The four engines whispered into silence, and out jumped Bob, who threw himself into her arms. Cameras flashed. Reporters craned to get a better view and ask their questions, and the War Department learned that they had a media coup on their hands. This was the most promotable thing they could have imagined.

Margaret poses in the middle of her guys and in front of her plane just after the Belle landed at Memphis airport, June 1943.

The Belle is caught in her famous power dive over Memphis airport just before landing. To this day, almost no one believes that Morgan could fly the huge B-17 so bravely in his display to the Memphis audience on hand for their arrival.

Another staged photo of the young couple. Note the photographer in the waist of the Belle. Stories of Margaret and "her" B-17 were devoured by Memphis newspaper readers for almost a year.

Bob and Margaret again pose for a shot in front of her home.

Everyone wanted to hear what Margaret had to say, so here she is being interviewed live on the radio on station WMC and the NBC radio network. Note Bob standing at the far right of the picture.

The two are seen years later during a reunion in the 1970s.

When Bob and Margaret told them that there would be a marriage, the War Department asked them to postpone their wedding, and they did so. After all, they could not see flying a "wife" around the country in what was a new and modern strategic heavy bomber. It was not long before the tide began to turn for them. During a surprise visit to Cleveland during the tour of the U.S., Margaret noticed that Bob was with another girl. Then came the shock that he was with a woman in his hotel in Denver, and the marriage was off. Bob was despondent and did all he could to get her back. She was unmoved, and not long after that Bob was married to another woman.

Margaret was ok with that. She went on to a career as an airline stewardess and even tried marriage herself. It only lasted about five years. She fought a terrible battle with alcoholism that she eventually won, and even volunteered her time with a local recovery program.

Margaret Polk spent the rest of her life alone but happy in her small East Memphis home. She was close to her airplane, and often donated her time to the cause of raising money for its preservation.

Miss Margaret, the real Memphis Belle, died quietly in her home among her beloved squirrel friends and her swimming pool in 1990.

Captain Robert Morgan says goodbye and congragulates the outstanding efforts of the "Memphis Belle" groundcrew while co-pilot Jim Verinis and navigator Chuck Leighton look on. Left to right are: Cpl. Oliver Champion; S/Sgt. Max Armstrong; Sgt. Wayne Lipscomb; Sgt. Leonard Sowers; Sgt, Charles Blauser; M/Sgt. Robert Walters; and Crew Chief M/Sgt. Joe Giambrone. These men often toiled throughout the night, sweeping over every inch of the Belle to get her ready for a raid the next day. Conditions were not the best for these guys. They worked in the open no matter the weather, and Joe Giambrone was even blown off ladders during engine tests from turbulent propwash. (Frank Donofrio)

One of the many publicity photos taken by War Department photographers to promote the War bond and good will tour. The plexiglass nose piece shined to perfection is in contrast with the battle worn skin and paint of the "Memphis Belle." It is easy to see from this perspective how difficult it must have been for a bombardier to concentrate on his job with his head only inches below the fifty caliber machine guns. When a B-17 was assigned to fly in the lead position for the Group or the Combat Wing, an additional gunner was often assigned to operate these foward guns. The Bombardier had to focus his attention on bombing and could not be distracted by having to jump on the guns away from his sight. (Frank Donofrio)

Jim Verinis poses with the only "authorized" female to ever fly aboard the famous bomber. Stuka was a Scottish Terrier who was purchased in a London shop by Verinis while he was there on R&R. While the dog never flew a combat mission, she did fly aboard the "Belle" across the Atlantic with the crew for the War bond and good will tour. She accompanied them to all of their thirty promotional stops across the country, and was a special part of the parades that the men took part in. Some of the crew remembered how Stuka used to run back and forth through the "Memphis Belle" during engine start before settling down to enjoy the ride in the plexiglass nose of the plane. (Frank Donofrio)

The International News Service issued a story written by a reporter named Dixie Tighe. It covered the December 6, 1942, mission flown by the 91st BG. This was a significant raid, as it occurred at the one year anniversary of the bombing of Pearl Harbor and America's entry into WWII. It was the fourth raid for the "Memphis Belle," and their first shot at a target that was not on the German occupied coast. This time they would focus their weapons on targets inland at Lille, France. The Group had suffered terrific losses the last time they flew on 23 November, and they were determined to complete the mission on this important day. The story proudly stated that "every man and every plane returned safely to base" (even though there were three injuries among airmen in the Group). The article focused on the names of the B-17s taking part in the raid, specifically naming "Bad Penny," "Jack the Ripper," "Delta Rebel," and "Memphis Belle." It went on to note that the latter bomber was named in honor of the sweetheart (Margaret Polk) of the pilot, who lived in Memphis.

The story was picked up by the Memphis Commercial Appeal locally, and the copy editor for the afternoon paper, The Press Scimitar, made note of the mention of the sweetheart's name. A young

(Frank Donofrio)

One of the treasures from the scrapbook of Col. Bert W. Humphries (322 Sq. 91st BG). Here Bert is seen standing on the perimeter track at Bassingbourn prior to a launch to hit a target in Emden, Germany. The "Memphis Belle" is staged and is the bomber third from the left. The March English weather is typically overcast and cold, a condition that the crews did not enjoy but became accustomed to. (Col. Bert W. Humphries)

Some "Memphis Belle" snapshots taken during the war. Top left to right: The King and Queen of England arrive at USAAF Station 121 HQ. Then reviewing the B-17 "Delta Rebel no. 2" and her crew led by Capt. George Birdsong. For many years it went forgotten that among the B-17s reviewed by the Royal Family was the B-24D "Liberty Lass" of the 44th Bombardment Group (H) at Shipdam. Here the party passes at the front of that plane on their way to their waiting limousines.

reporter named Menno Duerksen was assigned to find out who she was and write a story about her. Suddenly, there was terrific interest locally in the big bomber that was fighting the Germans. It seemed that sometimes readers in Memphis could not get enough news about "their" plane. One year to the day after the dreadful attack on Pearl Harbor, Memphians knew about a special young girl, her brave pilot, and the special Flying Fortress that would share their world for the rest of their lives.

It is important to note that by sheer fate, it seemed that from her first day the "Memphis Belle" was going to be a very special airplane. From a small mention in a news story on the other side of the world, a review by the King and Queen of England, and a Hollywood movie star, this Flying Fortress out of more than 12,731 built would become endeared to millions of Americans for all time. Again, all of the bombers that flew in these tragic times can today be represented by this single aircraft. Recently, a curator and historian at the Smithsonian Institute told the author that the "Memphis Belle" is considered to be among the top five American aircraft ever built. He mentioned the Wright Brothers Flyer, the Spirit of St. Louis, General Chuck Yeager's Bell X-1, the Enola Gay B-29, and the "Memphis Belle." Certainly this is quite distinguished company, and it remains the only one of these historic aviation artifacts that is not housed by this fine institution.

Vivid in my memory is the day June 6, 1943, almost 44 years ago, when General Devers, Commander of the European theatre, and I bade you farewell at Bovingdon...the first Eighth Air Force bomber to be sent home upon completion of 25 missions...the "Memphis Belle" shall remain a living memorial.

A letter from General Ira Eaker USAAF (ret)
May 1987

A fitting British farewell to the "Memphis Belle" and crew. In a ceremony that occurred at Bovingdon, the veteran warbird and crew are at full attention as Generals Devers and Eaker prepare to send her back to the zone of the interior. The Belle would arrive in Washington, D.C., almost exactly one month after flying her final combat mission. Pilot Bob Morgan did not know of his new mission until some 48 hours before they were winging back across the Atlantic Ocean.

"Memphis Belle" Returns to the United States! 16 June 1943

The trip back to the United States did not produce any drama as the previous eight months had. They made their usual fuel and maintenance stop in Greenland before heading west again for America. The "Memphis Belle" was above a cloud deck when the men were getting close to the American coast, and Morgan asked his Navigator to make some plots to begin to let down through the weather, because he did not want to make an instrument approach. The thick layer of clouds was making them a little nervous, and as the altimeter wound through five hundred feet, Morgan lightened the moment by telling Leighton that he hoped there were no icebergs down there. As they were still descending, Leighton shot back that there wasn't a damned thing any navigator could do about that! Morgan got the Belle down on the water so that they could make their approach under visual flight rules and made the landing at one of the New England military bases to rest before the flight down to Washington, D.C., in their dress uniforms. Among the dignitaries that were scheduled to meet them were Generals Arnold and Spaatz, the Secretary of War, and even President Roosevelt. Roosevelt could not make it, but the welcome home speeches were all given without delay. The date was 16 June 1943.

Nearly one month to the day after flying their 25th mission, the crew of the "Memphis Belle" returned to America with their bomber to begin an unprecedented 26th mission. It was also nearly exactly one year to the day after Margaret Polk went to Washington state with her sister, where fate saw to it that the storybook romance for her and Robert Morgan would begin. The Belle is seen parked at Washington D.C.'s National Airport just after arrival ceremonies.

"Memphis Belle" Returns to America

June 16, 1943 - Washington D.C. National Airport
After three days traversing the North Atlantic Ocean, the B-17 Flying Fortress "Memphis Belle" again rests on American soil. Capt. Robert K. Morgan, a twenty-four year old from Asheville, North Carolina, took this bomber and her crew to England only 264 days ago. Today, they return in an airplane which has been gouged, holed, and burned by Nazi anti-aircraft and enemy fighter shells. With their requirement of twenty-five combat missions behind them, they are now the most well-known bomber crew in the world.

Following an eighteen hour flight with a fuel stop in Greenland, the massive Flying Fortress, which will soon become a William Wyler star of the big silver screen, quietly landed within the safety of the Zone of the Interior last night at Westover Field, Massachusetts. After showering and getting a good night sleep, the pilot made sure his combat-proven crew had coffee and their class "A" uniforms as they flew off for the celebratory return ceremonies in our Nation's Capitol.

A few hours later a four-engined speck appeared over the horizon of Washington, D.C., and the thousands of spectators gathered there to welcome home America's newest heroes—the crew of what is sure to become a nationwide darling—the "Memphis Belle."

From one of the most prominent guests on hand came a direct order to Morgan in the cockpit. General Arnold ordered up a buzz job, and Capt. Morgan did not let him down. The men on the Belle could see the main reviewing stand and headed right for it. The sight and sound of a flying B-17 was witnessed firsthand by those who were looking for cover as the huge plane roared just over their heads.

The "Memphis Belle" then landed and slowly taxied up to halt while the sound of her brakes squealing was overcome by the throngs of those yelling and applauding. Flashbulbs popped and dignitaries strolled up to the bomber for this heroe's first official welcome home.

From here, the famous plane and crew will begin a nationwide publicity trip visiting War production facilities while they embark on their morale-boosting tour of the United States.

The following article was printed in "Air Force," the official service journal of the U.S. Army Air Forces, in September 1943.

In this article, it should be noted that not once is the "Memphis Belle" mentioned. The only indication that the Belle is involved is through the drawings. Also make note that the Belle is releasing bombs, and all twenty-five mission bomb symbols are present on the nose!

Bombs away—and the bombardier keeps his mind on his business despite flak and other distracting elements.

As Allied invasion forces moved in on the European fortress last month, round-the-clock bombing of continental targets by American and British aircraft based on the British Isles continued with increasing tempo.

Major responsibility for daylight bombing missions remained the every-day job of AAF four-engine bombers manned by crews of the Eighth Air Force. As the bombings increased-and more and more industrial centers and shipping points felt the blows, so increased Axis opposition. Fighter planes, the best the Nazis had to offer, and flak, often as thick as a blanket, tested the mettle of our airmen.

The Germans were trying every trick in the book—new tactical maneuvers with their fighters, air-to-air bombing, variations in flak concentrations and patterns.

Our formations and tactics were constantly being changed to meet the enemy's new techniques.

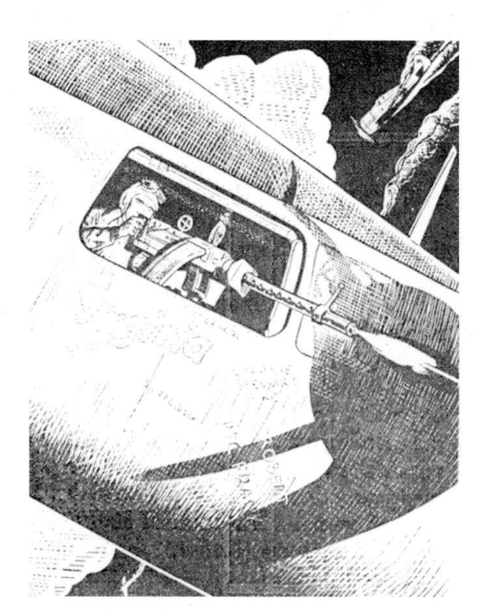

A waist gunner can watch his tracers plow into an enemy fighter. This gunner has already chalked up a swastika.

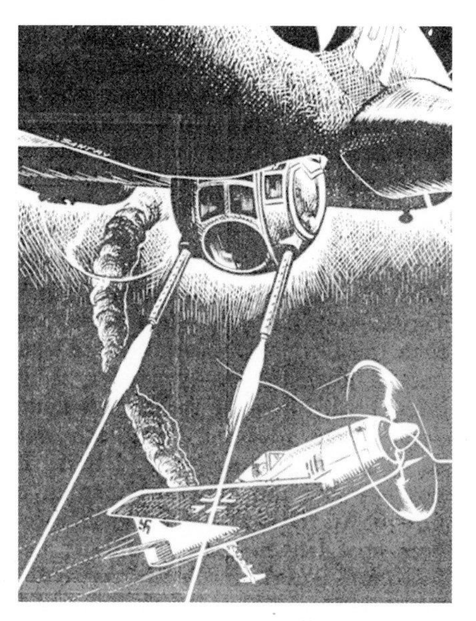

The belly turret gunner thinks he has the best spot on the ship, mainly because he gets a crack at plenty of them.

From the top turret the gunner frequently gets an eyefull, too. He takes on attacking planes from many angles.

OVER EUROPE

Despite sterner opposition our crews and our planes are more than holding up. In a report to Headquarters, Maj. Gen. Ira C. Eaker, the Eighth's commanding general, commented:

"None of the crews has a feeling that they are overmatched. The bomber crews have a complete confidence in their ability to take a heavy toll of German fighters. It is not necessary to drive these men to their tasks, as they are enthusiastic about it.

"We employ all possible deception to avoid fighter concentration and radar detection. This is done in order to prevent interference with the bombing by enemy fighters. However, when a hot air battle results, we do not count the mission lost but consider it a victory when we destroy a large number of enemy aircraft."

As always, the primary job was the destruction of production facilities of the Axis war machine, with shooting down enemy fighters a defensive sidelight of the main task.

Coinciding with the acceleration of Allied air operations over the continent was War Department recognition of the first anniversary of American aerial participation in the Battle of Europe. On July 4, 1942, AAF crews, manning six A-20s, got their first taste of war over the occupied Lowlands. One year later, our airmen celebrated by taking over several hundred heavy bombers to paste targets at Le Mans, Nantes, and La Pallice with 544 tons of bombs. They shot down 46 German planes, scored 36 probables and damaged seven more. Eight bombers were lost.

During that first year, the War Department reports B-17s and B-24s of the Eighth Air Force destroyed or damaged 102 industrial targets, naval bases, and rail centers with 11,423 tons of bombs on 68 daylight precision bombing missions.

They shot down 1,199 enemy planes, probably destroyed 525 more and damaged 501. We lost 276 bombers.

Eighth Air Force 17s and 24s flew 7,067 sorties against Axis targets during the year and averaged only 3.91 percent losses. Enemy planes destroyed by these lost American aircraft in fighting before they were shot down are not included in the tabulation of enemy planes shot down, probably destroyed and damaged.

The accompanying drawings represent an artist's conception of a typical bombing raid over Europe. Key positions in the crew of a B-17 are played up in the in-dividual sketches, which were done by Phil Santry of the AAF Training Aids Division on the basis of reports obtained in Washington.

Action as portrayed in these drawings is the type of action that has become al-most routine for our airmen engaged in daylight runs over enemy target areas on the European continent.

Navigators have to train themselves to keep at their navigation; a tough assignment when the fighting is the heaviest.

In some tight spots, it takes the combined strength of pilot and co-pilot to kick the B-17 around in a vitally necessary manner.

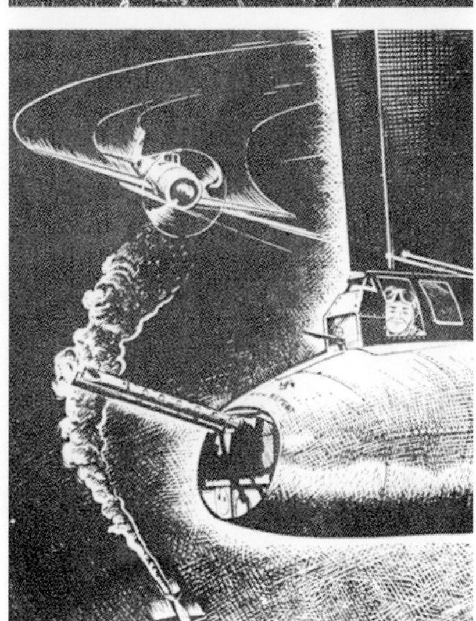

If a tailgunner gloats over one he shot down, another tough enemy might be looking him in the face before he knows it.

With their B-17 parked in the grass, the crew poses for a picture during their visit to Bridgeport, Connecticut, and the workers of Hamilton Standard. This was an important place to visit, because the company manufactured the propellers that pulled the thirty ton bomber over enemy territory. The company produced literally hundreds of thousands of these units. The props themselves had an arc of thirteen feet, and each weighed just over one thousand pounds. "Memphis Belle" maintenance records indicate that during her military life, the Belle had more than forty propellers that needed to be replaced.

While the Nation's capitol was the location for the welcoming ceremonies for Morgan, his crew, and their plane, Memphis, TN, was the first official stop for the tour of America. After spending two days in the city Belle left for Nashville. Margaret was there to see her man off, and he gallantly climbed into the plane to start the engines. One by one the mighty engines caught up the song of battle until all four were harmonizing in the full-throated hymn which had struck terror so often in the hearts of the herrenvolk of Europe. Her nose, artistically decorated with 25 miniature bombs and 8 trophy swastikas, met the wind, and the great dun-colored creature lumbered down the field.

With increasing speed, she seemed to shed her awkwardness; and when her pinions caught the air she became a thing of grace. Set free from the earth, and in her element once more she lived up to her name, Memphis Belle.

Reduced to a silhouette, the Fortress circled the field. Then came the dreadful roar again, and she swooped low over the wildly waving knot of well wishers in tribute. She was off again, this time to make a wider circle, a circle of the entire country, then Memphians hoped to return home to Memphis, TN.

Before the War bond and good will tour could begin, the crew of the rapidly famous "Memphis Belle" were guests at a special luncheon hosted by Congressman and Mrs. Snyder in the House of Representatives' dining room. They were joined by members of Snyder's Army Appropriations Committee, as well as General Arnold and a whole string of top brass. Shown here around a rather crowded table are the crew and dignitaries just before coffee is served. This was a photo opportunity that the media could not pass up, and it was also a hint of the attention that was to come. Over the next three months the men were wined and dined in every city they visited. Some of the crew would later comment that in many ways, the tour of the United States was more difficult than combat. Even though the War Department asked them to extend the tour for another three months, they all quickly agreed that it was time to go on to future assignments.

"Memphis Belle" is seen here in Detroit, Michigan, during the nationwide tour on 3 July 1943. Like all other cities they visited, the crew would focus their attention on War production plants and workers.

Good Will Tour Dates

Washington DC	June 17, '43
Memphis, TN	June 19, '43
Nashville, TN	June 22, '43
Bridgeport, CT	June 25, '43
Hartford, CT	June 26, '43
Boston, MA	June 28, '43
Mifflin, PA (Pittsburgh?)	
Detroit, MI	July 3, '43
Akron, OH	
Cleveland, OH	July 8, '43
Dayton, OH	July 9, '43
Las Vegas, NV	
San Antonio, TX	July 17-20 '43
Harlingen, TX	July 21s '43
Laredo, TX	July 23 – 25 '43
Oklahoma City, OK	
Wichita, KS	late July / early August
Cheyenne, WY	July 27-29 '43
Denver, CO	Aug. 1, '43
Mobile, AL	Aug. 3 '43
Tyndall, FL	Aug. 5, '43
Ft. Myers, FL	Aug. 6, '43
Orlando, FL	Aug. 10 '43
Asheville, NC	Aug. 11 '43
Columbus, OH	Aug. 14 '43
Ogden, UT	(blown engine)
Hollywood, CA	
Chicago, IL	
Camden, SC	
New York, NY	
Spokane, WA (refurbish)	Sep. 29, '43

Some records indicate stops in Yuma, AZ, and Roswell, NM.

Note: A study of the maintenance forms 1-A for B-17 #41-24485 ("Memphis Belle") show an engine change on 17 August 1943—the same day that the Mighty Eighth suffered huge losses on the other side of the world during the raid to Schweinfurt and the ball bearing plants there. The forms state that by this date, the Belle had accrued 498:45 hours of time in the air. Had the Belle not been sent home, it is quite likely that she would have participated in the raid to Schweinfurt, and possibly be among the 60 bombers shot down that fateful "Black Thursday."

Top: Jim Verinis is seen here holding the "Memphis Belle" mascot. The Scottish Terrier is named "Stuka," and she was a favorite to crowds all along the tour of America.

Middle: The crew is pictured at yet another tour stop among throngs of excited visitors.

During the Tour, the crew was expected to make patriotic speeches to boost the efforts of the workers in the huge grinding war machine. Here (right) the "Memphis Belle" ball turret gunner Cecil Scott looks over the shoulder of a plant worker installing control cables in a P-47 Thunderbolt. Leaders felt that if the workers saw the actual men that flew their handiwork in combat they would do their jobs that much better. Almost everyone knew that alongside the bravery of combat flying and fighting, the real weapon was the War production factories. The aircraft industries alone turned out more than 296,600 airplanes of all types. Of these more than 52,000 were classified as bombers. Literally billions of bullets were made, bandages by the millions, and tens of thousands of aircraft parts and supplies. A Jeep rolled off assembly lines every six minutes and a rifle every minute. The Boeing, Vega, Douglas co-op, at the height of production, delivered more than eighteen flight-ready B-17s every day!

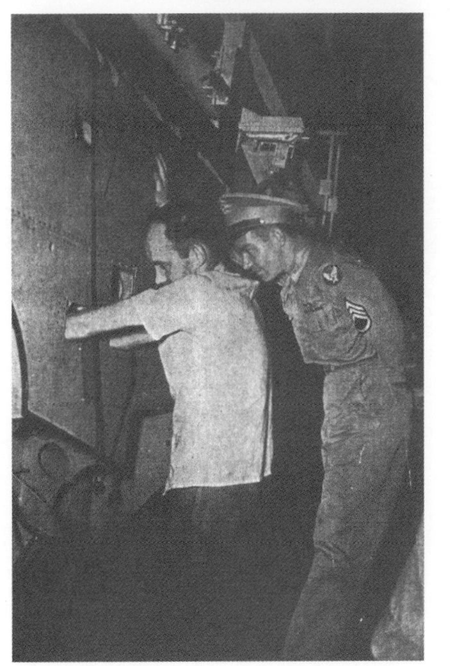

The "Memphis Belle" at Wright-Patterson Air Force Base, Ohio, during the big War bond and morale tour 9 July 1943. Being

The "Memphis Belle" at Wright-Patterson AFB, Ohio, during the big war bond and morale tour, 9 July 1943. Being one of the biggest military bases in the country, her reception here was attended by thousands. The rigorous schedule for the plane and crew only lasted three months, and was blamed by Bob Morgan for destroying the relationship between he and Margaret. The Army wanted to extend the tour for another three months, and even though the crew was getting the very best treatment of their lives, they decided that they had had all they could take of the wine, women, and song. After leaving their famous bomber in New York City, the crew split up and asked for reassignments. Many of the men actually went back into combat. Evident here is the mottled camoflauge that was popular among the very first B-17s to fight from England. A tow tractor is being used to position the "Memphis Belle" for the ceremony using the tail wheel towing point. At the front of the plane the VIP platform can be seen, as well as the loudspeakers to the left and right. One man near the right horizontal stabilizer seems to be directing the tug driver in the delicate hookup of the tow bar.

"MEMPHIS BELLE"—Buxom heroine of 25 missions over Nazi Europe is undergoing a thorough overhaul at SPAD for return to enemy skies and new adventures. "Keep 'Em Flying" repair crews will put the battle-scarred Amazon into shape for her return to battle.

When this photo and caption appeared in a Spokane, Washington, newspaper, it was clear that the Army believed that "Memphis Belle" would return to combat. The bomber was ferried to Spokane for a complete overhaul on 8 October 1943. She was ready again for fighting by 7 December 1943 after most of her battle scars were professionally repaired. Since by this time in the War the much improved B-17G was arriving on the scene, it was determined that this famous plane would be re-designated as a TB-17F and assigned to the 326th Bomb Group, and then the 815th Bomb Group down at McDill near Tampa, Florida, in a training role.

It was here that her flight logs swelled with the hundreds of names of the men who plied through the Florida skies learning how to operate the complicated systems of the B-17. Many reported years later that "Memphis Belle" had grown into a terrible airplane to fly, with misaligned throttles and supercharger levers, and had difficulty in holding formation. One evening during a night training mission the electrical system supplying power to her radios and exterior lights failed. Her pilot, Andrew Kelleher, said that he saw another B-17's lights and took up a position in trail so he could follow that plane back to MacDill. What he did not know was that yet a third B-17 was now behind them. When Kelleher landed the Belle, he was forced to turn off the runway and missed the taxiway. The main wheels dug into the soft turf, leaving the entire tail of the "Memphis Belle" on the runway. The men were quite surprised when the landing lights of the plane trailing them lit them up like the sun, and they jumped from the Belle and ran like hell. The pilot landing his bomber cobbed his throttles, made a low evasive turn, and cleared the vertical stabilizer of the stuck Belle by only inches! "Memphis Belle" was still a very, very lucky ship.

Left: "Memphis Belle" waist gunner and radio operator Bill Winchell and Bob Hanson pose for a shot in front of a B-17 (not the Belle!) at Yuma, Arizona, during their Officer training days after coming back to the United States.

Captain Robert Morgan talks about War production with workers at the Fischer Body plant in Detroit, Michigan.

The crew poses with one of the many mayors they met during the War bond and good will tour.

NOT A PASS · FOR INDENTIFICATION ONLY

PROPERTY OF U.S. GOVT.

WAR DEPARTMENT
THE ADJUTANT GENERAL'S OFFICE
WASHINGTON, D.C.

IDENTIFICATION CARD

Robert J Hanson
NAME

2nd LT AC
DESIGNATION

SIGNATURE

COUNTERSIGNED W.D., A.G.O. 65

HANSON
ROBERT J
O 584938

DATE ISSUED SEP 1944

WIRE A.G.O. IN CASE OF EMERGENCY

LOSS OF THIS CARD MUST BE REPORTED AT ONCE

Bob Hanson's identification card.

"Memphis Belle" co-pilot Jim Verinis holds the only female ever allowed to fly on the famous bomber—Stuka! Margaret Polk, for whom the plane was named, never got to fly on "her" B-17.

The crew poses for yet another shot during the tour.

RECONSTRUCTION FINANCE CORPORATION
Office of Surplus Property—Aircraft Division—Educational Disposal Section
WASHINGTON 25, D. C.

PURCHASE ORDER

Agreement No. .... Order No. ....

Date ..... March 30, ... 19 46

Ship to: City of Memphis
Full Name of Institution

Courthouse, Memphis, Tennessee.
No. Street City State

....
Signature

Catalog No.	Quantity	Nomenclature	Type	Disposal Cost Unit	Disposal Cost Total	Quantity Shipped	Date Shipped	
RFC No. 84	1	Airplane, Serial No. 41-24485, known as "Memphis Belle"	B-17	$350.00	350.00			

For instruction, education, and memorial purposes.

World War II was finally over. The Axis nations had all capitulated, and America was putting the brakes on every part of the military. The churning war machine had gained historical momentum, and it was going to be very hard to slow it down. If brand new planes intended for combat throughout the world were flown from the factories directly to aviation burial grounds, then it was assured that the government would have no use for a five-year old battle-scarred bomber, no matter how famous. Only two days after the bomb was dropped on Nagasaki, the "Memphis Belle" was declared to be surplused and ordered to be transferred to the custody of the Reconstruction Finance Corporation for disposal.

The R.F.C. had been partially cooperating in a program to see valuable assets from the war transferred to various communities and associations throughout America. It was found that airplanes were popular war memorials, and several noteworthy aircraft were transferred to the care of business operators who wanted them, as well as various city administrations across the country. One businessman wanted to operate a string of gas stations throughout the U.S. and purchased several noteworthy B-17s. Among these was even the famed "Hell's Angels," which was last seen some time during the fifties in Oklahoma being prepped to be placed on stands over his business!

The Mayor of Memphis, TN, wanted the "Memphis Belle," but not for a gas station. He wanted an appropriate war memorial in his city and did everything he could to get the plane. After the Belle was located, it was determined that the plane could either be purchased outright for a mere $13,750.00, or it could be transferred to the city of Memphis for a procurement fee of just $350.00. The city opted for the latter, and the paperwork was started. It is not known why R.F.C. would write the "borrowing" of the plane from the government on a purchase order. This was a cause of great confusion over ownership that would stay with the Belle for the next fifty years.

12

Out of the Smelter

They were on a mission to bring to Memphis the world's most famous B-17. Chosen from literally hundreds of applicants, these men from every branch of the U.S. military assembled for what would be the most important flight of the "Memphis Belle." Whisked from the mouth of the smelter, the famed bomber was crewed for the final time and flown to her home. A Flying Fortress which had a reputation for bringing her crews safely back from the drama and danger of combat, "Memphis Belle" would do it again. This war-weary veteran would command the attention of her crew on this historic mission. Through failed radios, marginal landing gear, and smoke in the cockpit, their story unfolds....

When they left the boneyard at Altus, Oklahoma, they were simply on a test flight. The Aircraft Commander was convinced that they had just made their final takeoff and told his crew that they were headed directly for Memphis, Tennessee. They flew over the Oklahoma airfield once more just to say goodbye and pointed the nose of the great bomber east and into history. Aboard the "Memphis Belle" for her final flight ever were: Robert Little and Hamp Morrison as her pilots; navigating was done by Robert Taylor and James Gowdy; the radio was manned by Stuart Griffin; Crew Chief was Charles Crowe,; and the Flight Engineer was Percy Roberts, Jr. These men had saved the great "Memphis Belle" from oblivion.

Altus, February 9. — The six P-61s came in low. Below them on a field a mile and a half square were sitting ducks, 2,621 of them jammed wing tip to wing tip. It was a fighter pilot's strafing dream.

The squadron leader peeled off and made a run over the target, and the other five followed. They could not miss. This was fun—the best target they'd had in the last five years.

Then up into the clear blue sky they went, over and over dodging, diving, and zooming, but always in close formation to keep the "enemy fighters off." Then, the fun over, they swooped in for perfect landings.

Final Flight Crew 17 July 1946: "There's smoke in the cockpit, the radios are busted, and we aren't sure if the wheels are down and locked!"

The "Memphis Belle" as she appeared when first approached by the men who flew her home to Memphis. Heavy sighs were heard as they realized they would have to pick up the tools and get to work on the plane before any flight would take place.

Stafford Rogers, Reconstruction Finance Corp. Supervisor in charge of the Altus Army Airfield, walked out of the control tower and greeted the six pilots. There was an exchange of papers, some signature formalities, and six more planes were added to the Graveyard of the Eagles.

This aerial graveyard is an impressive sight. The 2,600 plus planes resting there now are worth half a billion dollars of U.S. tax money. Only God knows what the planes they shot out of the European skies cost the Axis.

Rogers drives you down the lanes of planes and points to individuals. "Now there's the Memphis Belle...see those German crosses and the purple hearts? Some gal that Belle.....and there's Pistol Packin' Mama....more than a hundred missions over Europe....and there's Suzie; see those setting suns and those camels? Ten of them. Each camel means a trip over the hump....and then there's old Thunderbird....60 missions over Europe....and the Lonesome Baby....across the Channel 125 times, and every time the Nazi's oil supply shrank. Hear that one coming in? B-24."

And so it was, with a luscious Varga Gal painted on the nose and a long row of fading setting suns behind the pinup.

Rogers drives farther down the line and turns a corner, spouting information and statistics. Since last July the field mechanics

Next to the Belle at Altus is none other than "V Grand," one of the more well known B-24s of the War. She did not escape the yard.

The Belle arrived at Altus on 2 July 1945. On 9 September she was placed on storage status, then released to the Reconstruction Finance Corporation on 18 October 1945. It is believed that the Belle was parked on the section of the airfield pictured in the lower right in this image at the time this photo was taken.

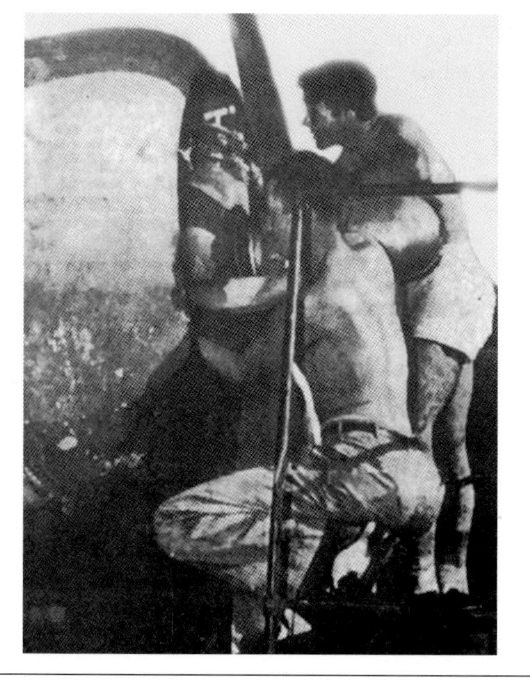

Percy Roberts and Charles Crowe endure the extreme Oklahoma sun as they operate on the #3 engine of the Belle.

This article was published in the "Daily Oklahoman" newspaper in February 1946.

have stripped 3,670 engines from the weary planes. The engines were good, too; they brought the planes in and will still spin the props. Now, they are for sale.

But who wants them? Nobody. The Army has new ones for the planes it needs, and these are combat jobs, not suitable for commercial or civilian flying. Eventually they will go to the scrap pile.

Close by the acres of engines are the props, those thin metallic blades that pulled the eagles through flak-riddled skies in Europe and the Pacific. They have been stripped from B-17s, B-24s, B-25s, P-38s, P-40s, P-47s, and P-51s. There are some from even the Navy's lumbering old workhorses—the PBY.

In another section of the field are 300 brand new combat planes delivered direct from the factory in the first hectic hours after VJ day. They are maintained in fly away condition, and the engines are fired up and run for a few minutes every day just to keep them operational. If, through some unconceivable mischance, the Army needed combat planes in a hurry, all that would be needed would be to put the crews aboard. But nobody here knows how long that status will be maintained.

Here are some planes that have been stripped of engines. "Notice" says Rogers, "how they three-point themselves with the weight of the motors gone. Zombies. I hope their crews never come by here."

You pass a fire truck on patrol. Good care is taken of the graveyard denizens, and meantime, minor routine salvage operations go ahead. This autumn and winter the field sold more than 3,000 gallons of anti-freeze drained from the in-line liquid cooled engines.

Gasoline left in the tanks after they make the last landing is drained off and sold. All usable parts are taken off.

Rogers will be glad to sell you any of the planes, and at bargain prices, too, but he does not expect to make any sales. The famous old "Memphis Belle" is listed at $13,750.00, and a slightly worn B-29 that cost close to a million dollars new is ticketed at $32,500.00. Some fast P-38s with hours against the Japs are on sale for $1,250.00, and some P-61s go for a flat $6,000.00.

At those prices any of those planes is a rank steal. Except that there aren't any Axis planes to fight any more; you can't get a civilian license to fly them; and they are too hot to handle on commercial airports. You could not get them in and off any but an Army field, and Army fields are being folded up rapidly.

So the old heroes of the great raids on Cologne and Berlin and Ploesti; of the battle of Midway and Guadalcanal and Iwo and Tarawa, sit out in the sun and wind and gather rust. One of these days a man with a sledge, torch, and cutting gear will come along, and they will disinegrate into smaller chunks of metal that will go back to the furnaces.

They did their job. Now they are expendable. They have come to the place everything finally does - the *graveyard.*

"We didn't do anything huge, we just went over there and got it!" -Percy Roberts, Jr. - 24 September 1998

They played down their vital roles in saving the aircraft and spoke highly of those that flew with them that day. Bob Little and Percy Roberts, Jr., were visiting the "Memphis Belle" in downtown Memphis almost sixty years after they took part in flying the famous warbird from the reclamation center in Oklahoma to Memphis. "The old gal gave us some trouble with the spark plugs. Crowe and Roberts were working on the engines and making sure they would get us here," commented Little. "The people at Altus asked us if we wanted to take a test flight. I told them no, when we get it up there, we're going all the way."

The "Memphis Belle" arrives in Memphis, Tennessee, at last. Had this flight occurred during a combat mission, it would have been scrubbed. Seen here, she is greeted by hundreds of well-wishers who were on hand when her engines were switched off for the very last time and whispered into permanent silence. (Photo courtesy of the Memphis Commercial Appeal)

And all the way they came. "When we tried to raise the gear, it wouldn't come up. So Percy got in the bomb bay and cranked it up as far as he could," said Little. Roberts added:

"I was crankin' the hell out of the thing and got it about halfway, and it wouldn't go anymore. So we came all the way to Memphis with the wheels part of the way up. On the way, Little asked me if I wanted to fly it for a while. I told him sure, but it wasn't long before we had climbed a few hundred feet. Then he told me 'Percy, look out at that left wing. Now look at the other.' I was in a pretty good turn and didn't know it. Wasn't the prettiest thing, but I flew the 'Memphis Belle' for about fifteen minutes."

Then there was an event that got the attention of all those on board. And while it turned out to be nothing, in a bomber with that many hours on it that had been sitting for so long, smoke in the cockpit will get your attention. "We smelled something burning and couldn't find it. Looked all over until we found that someone had laid an oily rag on one of the inverters. They get pretty hot in flight, and that rag was smoldering and sending some smoke up," Roberts remembered.

They pressed on. Even when over Little Rock, Arkansas, the radio stopped transmitting there was no turning back. Almost there. After all, they were not going to bomb the German sub pens that day.

By the time they arrived over Memphis, thay had a plan to contact the tower by flying by and wagging the wings to signify that there could be no radio transmissions. Percy had already gone back into the bomb bay, and the wheels were cranked by hand back down. Now there was another problem. The down and locked indicator lamps were not lit up. They could not be sure that the landing gear had locked into place.

Roberts even climbed down into the ball turret to see if he could visually confirm that the wheels of the "Memphis Belle" were indeed down *and* locked! "She could have settled onto her belly as we put the tires on the runway" said Roberts. "There were a bunch of fire trucks all over the place. I guess they thought that we were going to have some real trouble when we landed. But they held up!"

"Memphis Belle" rolled to a stop. Her engines whispered into silence as Little and Morrison moved her magnetos to "off" for the last time ever. Morrison had taken off in Oklahoma, and Little made the final landing of the Belle at Memphis airport. Everyone had done their jobs, and now the world-famous B-17 was in her adoptive home.

The Mayor of the city tried to get the wartime pilot Robert Morgan to make this final flight, but business prevented him from coming to Memphis to fly the very plane that had saved him twenty-five times over Europe. Mayor Chandler wrote a letter to Morgan:

This morning at 11:30, the ship took off and came directly to Memphis without stopping, arriving here at 2:45 PM. A large crowd was at the airport to greet her, and there was great rejoicing. She looks every inch the great lady she was when you and your magnificent crew flew her so successfully on those missions, and we hope that it will not be long before you feel the urge and fly to Memphis yourself to give her your own affectionate greetings.

Many qualified men had been attempting to volunteer for the historic final flight of this great plane. The Mayor's office had the task of deciding who would be selected for the crew. Among them was one significant Memphian named Calvin Swaffer. He had flown a B-17 that he named after the musical heritage of his home town, "Memphis Blues." He had phoned and written the Mayor as early as February 1946. He assured the Mayor that during his time with the 303rd BG(H) at Molesworth, England, he had accumulated enough hours in the B-17 to fly the "Memphis Belle" home. He had been through enough emergencies and knew how to handle the huge

"Memphis Belle" is seen here in a temporary living space at the Memphis Airport. Were it not for a caring Air Guard Commander she never would have been moved indoors for her protection. Sadly this could not last, and the Belle was soon moved back out into the elements. The officer inspecting the number four engine is not identified. The photo was taken some time in 1946.

After nearly four years of sitting at the Memphis airport after landing there in 1946, the "Memphis Belle" was taken apart and moved to the corner of Hollywood Avenue and Central Avenue. This was the first time the bomber had been disassembled and relocated for want of a better place. Under the control of the American Legion, the plane was placed atop a huge concrete pedestal, where she would remain for the next twenty-seven years. The idea was to keep the Belle out of reach of prying hands and treasure hunters. It did not work.

plane if something should go wrong. Swaffer was shot down and forced to ditch the "Memphis Blues" in the Channel on his 24th mission. In one of his letters to Mayor Chandler he stated that his "Memphis Blues" had actually flown several combat missions alongside the "Memphis Belle."

So keen was the interest among those who were attempting to be selected to crew the Belle on her final flight that even the Twentieth Century Fox film corporation wrote to show their interest in documenting this last "mission." In their letter, they wrote that they hoped to fly along with the Belle to capture film of the flight for Movietone News! For reasons unkown, they did not make the trip.

The Mayor had been trying to get the Belle moved to Memphis since well before the War came to an end. Initially he was informed that the plane was still needed for military service, and until Chandler saw the 1946 newspaper article from the Daily Oklahoman, he was not even sure that the famous B-17 existed any more. In all there were more than seventy letters written between 1943 and 1946 trying to locate the Belle.

The focus of these letters ranged from asking the War Department where the "Memphis Belle" was through absolutely identifying that it actually was the Belle after the aircraft was found languishing in the boneyard. Then, how much it would cost the city of Memphis to get the bomber. After making the necessary arrangements to legally have the Belle transferred to the city, a check in the amount of $350.00 was written to the Reconstruction Finance Corporation to cover the cost of making the aircraft operational one more time. The "Memphis Belle" had been procured for the price of a second-hand car!

Letters continued. They stressed that maintenance was required on the plane before it could safely fly. They assured that a competent crew had been selected and was ready to go. Some were to Morgan asking over and over if he was sure he could not make the

Here it can be clearly seen that the barbed wire fence did not stop treasure hunters. There is a broken window above the navigator station, and the open oil filler door on the number four engine. The missing crew entry door in front of the right horizontal stabilizer became an open invitation to anyone who managed to climb the fence. Often these were vagrants looking for a place to sleep. The absence of cockpit glass invited all of the harsh Memphis weather, as the flight deck slid further into a nearly impossible state of restoration.

Her markings are of interest here, where the painted-over stars and bars insignia is well noted in the floodlights. On the nose, the "J" is still visible from her days as a training B-17 at McDill AFB, Florida. About the only accurate markings left were the Petty Girl and the mission symbols. With no paint on her props and the light thrown from the ground the Belle gives a somewhat ghostly appearance when compared to her former days of glory. Make note of the memorial wreath at the base of the pedestal.

flight. Others were from Morgan saying that indeed he would command his B-17 to Memphis. He expressed his approval after reviewing the crew selected to man the bomber. Even how the final flight should be publicized was written about. Among one of the more interesting communications was the letter from the Acting Chief of The War Assets Administration to Mayor Chandler dated 18 June 1946:

> On 10 April 1946, this office notified you to accept delivery of B-17F 41-24485 aka "Memphis Belle," but to date we have received no response. This airplane has been prepared for over two months now and it would be greatly appreciated if you could pick it up at your earliest convenience.

Here (below) the Belle is shown in Detroit in 1943. Only a few short years later, she was unflyable and in horrible shape, mounted like a hunting prize atop a high concrete pedestal. The once-famous "Memphis Belle" was now in a very different battle in Memphis, as she suffered far worse here than she ever did in combat. Over time, the Flying Fortress was thoroughly stripped piece by piece by those who desired a part of the plane to display in their living rooms. During the sixties, with the social stance on the negatives of the Vietnam War, the B-17 slid even further downhill, even through the attempts of only a few caring individuals who would wash her down and apply new coats of paint. This action provided a very temporary respite from the harsh elements, and the "Memphis Belle" became an eyesore and an embarrasement to the city of Memphis, Tennessee. Those who realized the historical significance of this plane complained to city leaders whose ears were deaf to their needs and wants. Even letters from her wartime pilot, Robert Morgan, went ignored and unanswered. It seemed that the only ones that really cared about the Belle were the vagrants and birds that were making her their home. By the time the seventies rolled around, empty beer cans occupied her floors where once spent fifty caliber shell casings gathered. The din of battle was long gone, but the voices of those who crewed her in combat were still heard by the few civic-minded people who tried to take care of her as best they could.

More and more paint was applied in failed attempts to make the Belle look good. This only resulted in inaccurate markings and

nose art. Each time this was done more of her battle scars were hidden, and the once proud Petty girl that graced her nose was buried under improperly and poorly applied paint. The plexiglass nose piece was now broken, not by enemy gunfire, but by vandals, and had to be fiberglassed over. Wooden rods now replaced her mighty twin fifties that once valiantly protected her from German fighter attacks. A fence ringed the perimeter around the plane, and even though it was topped with barbed wire people still made their way through it and into the plane, where they took everything that was not bolted or riveted down. Weather, vandals, and even birds were going to be successful where the potent German enemy could not—the systematic destruction of one of the most powerful and historical American airplanes ever built. Even the urging of the United States Air Force could not motivate the city of Memphis to protect the very plane that they lobbied so hard to adopt. To historians, it seemed that the "Memphis Belle" became the bastard stepchild, and by 1977 the Mayor of the city was prompted to take action that proved to be very controversial. In a letter to the United States Air Force Museum Program at Wright-Patterson Air Force Base, Mayor Wyeth Chandler stated the following:

> The City of Memphis wishes to relenquish any claim it has on the B-17 known as the "Memphis Belle." It does this so that the airplane can be placed on permanent loan to the Memphis Belle Memorial Association for display in a suitable museum here in our city.

Indeed, this Flying Fortress had become a white elephant that the city knew it could not care for or protect. The future looked dim, and if not for the efforts of one local business leader, the plight of the Belle would have continued on its downhill slide. Frank Donofrio, who had successfully run five industrial companies in Memphis, would drive by the famous plane every morning on his way to work. A very patriotic American who also lost a brother at Iwo Jima, Frank grew tired of seeing this symbol of American power and might slowly torn apart. He formed the Memphis Belle Committee in 1967, and for a long time he was the only member. Through his determined efforts, he began to draw the attention of city leaders in business and government.

The plane was actually evicted from her home of twenty-seven years in midtown Memphis, and again, for want of a better place, was taken apart, placed on trucks, and carted in disgrace back to Memphis' airport, reassembled, and tied down in the weeds next to a restaurant. Here she spent the next nine years falling apart. By 1986, the Air Force had grown very impatient with the lack of progress in the attempts to enclose the plane in a proper building and wrote Donofrio a letter of ultimatum:

> If the City of Memphis cannot come up with a viable plan to properly house the "Memphis Belle" B-17, then proof would be evident that the City does not have the desire to keep this historic airplane, and steps will be taken to reclaim the bomber for the United States Air Force Museum progam located at Wright-Patterson Air Force Base, Ohio.

That was all it took, and immediately citizens of Memphis rallied to save the "Memphis Belle." In only three months more than $576,000.00 was raised to restore the Belle and construct a proper and fitting building in downtown Memphis, Tennessee.

After her grand homecoming in July 1946 the Belle was quickly forgotten. The main reason for this was because the very patriotic Mayor Chandler was not re-elected, and his replacement was not as willing to complete the promises Chandler made to the Air Force. By November 1946 an article appeared in the local newspaper, The Commercial Appeal, which stated, "The plane which led a blazing trail of invincibility through the flak and fighter filled skies of Nazi occupied Europe has been forgotten but to the weather, vandals, and birds." Throughout the next sixty years, reams of paper would be used to record the complaints of all those who cared. They seemed to say in unison "Put the Belle indoors!"

Over the years the Belle had been relocated three times. Each time the disassembly was managed by Nute Paulk of the Tennessee Air National Guard. The first move in May 1950 was undertaken

By the mid-seventies, a more accurate paint scheme was applied, and the fabric control surfaces had been removed for restoration. Still, it was evident that much work was still needed. Both powered turrets had been removed, as well as the wing root gap covers.

Sporadic attempts at restoration through the '70s were just not enough to ever quite bring the "Memphis Belle" back to her former glory.

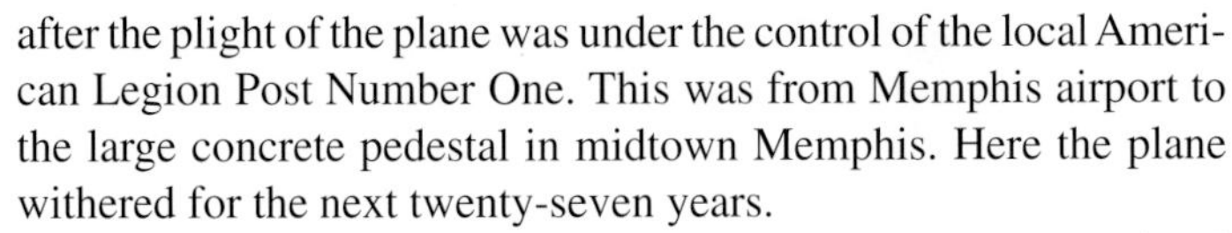

after the plight of the plane was under the control of the local American Legion Post Number One. This was from Memphis airport to the large concrete pedestal in midtown Memphis. Here the plane withered for the next twenty-seven years.

The years were not good to the plane. And since the National Guard was being relocated and the government owned ground upon which she sat, that place would no longer be available for the Belle. The Memphis Belle Memorial Association (who gained control of the bomber in 1977) was served with an eviction notice from the new owners of the property. She was lifted down from her perch, taken apart, and carted in disgrace back to Memphis Airport, only to see her condition slide further downhill.

Over the next nine years, sporadic restoration efforts were undertaken by volunteers from a local aviation school and the Tennessee Air National Guard, and she went from site to site around the airport while group after group tried to restore the great bomber. Even the best attempts soon fell from interest, and it seemed that the Belle would never be restored properly. Until the efforts were noticed by Hugh Downs of ABC News, the work failed to gain any real momentum. In a twenty minute story on the national news program 20/20 the plight of the "Memphis Belle" was well documented, and individuals from all over the world began to send their "Bucks for the Belle."

Yes, its really the "Memphis Belle." By July 1977, the restoration had taken on the aggressive task of entirely stripping the many layers of paint that had been applied over the years. Devoid of her top turret, wing root gap covers, and waist window covers, the Belle, at a glance, almost looks like a new B-17G. It did not require much to see, however, that there was a great deal of work left.

When the time came to remove more than eight layers of paint, the process revealed the original Petty Girl nose art that had been unwittingly protected beneath it.

This worker carefully tends to the famous Petty Girl during the extensive stripping process. Efforts were made to preserve the historic painting, but sadly, it was impossible to do so.

Moving day for the Belle saw her trucked in a convoy that actually rolled right in front of the home of her namesake, Margaret Polk. The procession stopped briefly to honor Miss Polk before moving on toward Downtown Memphis.

Construction of the best home the Belle ever had. After being reassembled on a parking lot, the great B-17 was pushed by hand into the dome's perimeter before the steel ribbing was erected. The plasticized canvas was then pulled tightly over the structure and secured.

New paint just before disassembly commenced.

You just don't see this every day!

Taking the right wing off for transportation.

The left wing is lifted over the Mud Island guard house. It was too big to go through the gate.

Under the careful eye of Nute Paulk of the Tennessee Air National Guard, the airframe had been carefully reassembled before the steel of her new home began going up. The "Memphis Belle" now had a place of honor in the center of a Memphis, Tennessee, riverfront park named Mud Island.

Like putting the biggest puzzle ever together, the right H-Stab is moved into place and then bolted.

It does not take much to lift the horizontal stabilizer.

The original "Memphis Belle" nose artist, Cpl. Tony Starcer, applied the first Petty Girl to the bomber. Arrangements had been made to have him reapply his work some 45 years later on the very same plane. Since he had passed, his nephew Phil Starcer came to paint the new Petty Girl in honor of his uncle. At the bottom of the page, Margaret Polk studies the nose art.

Bill Winchell, Gene Adkins, and Scott Miller catch up during a reunion at their old plane.

Below that, Bob Morgan signals that he is still ready to fly his "Memphis Belle."

Margaret flashes a smile as she waits for her former beau at Memphis Airport.

Together again! Margaret and Bob share a laugh and the well wishes of those who went to the airport to see them.

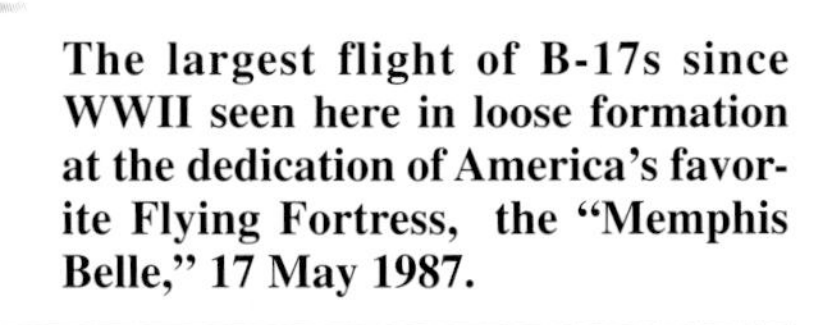

The largest flight of B-17s since WWII seen here in loose formation at the dedication of America's favorite Flying Fortress, the "Memphis Belle," 17 May 1987.

The flight deck of the "Memphis Belle" during restoration.

Even though he never manned a gun when he flew the Belle, Bob could not pass up the chance to study the right waist fifty caliber machine gun. Was he thinking of German fighters, or hunting squirrels back in Biltmore?

Bob and Margaret cut the ribbon at the opening ceremonies of the brand new Memphis Belle Pavilion in downtown Memphis, Tennessee.

Continuing Restoration

After nearly sixty years, the "Memphis Belle" rests at a level of restoration of nearly sixty-five percent or so. The Belle will never fly again as per the arrangement with the United States Air Force Museum program located at Wright-Patterson Air Force Base, Ohio. "Memphis Belle" is the property of the Air Force and is on loan to the Memphis Belle Memorial Assoc., Inc. in Memphis.

Throughout the years sporadic efforts have sometimes resulted in less than desirable restoration efforts to a degree. Many companies and volunteers have crawled for years throughout the entire airframe sanding, painting, and repairing various parts of the Belle. But much of the enthusiasm does not seem to last. Volunteers grow tired and funds run low. But the MBMA never seems to run out of steam.

One of the larger problems is the current environment where the Belle resides. She is beneath a pavilion of sorts that does not provide much more protection from the weather than during the years she spent near midtown Memphis.

Looking back at the various groups which have taken an active role in the restoration of the Belle, we can see the USAF crews from Blytheville AFB, Arkansas, the instructors and students from the Tennessee Technology Center, the intense support from the Tennessee Air National Guard in Memphis, and, of course, Federal Express.

Overall the "Memphis Belle" is in very good static display condition, despite the environmental concerns. Soon this bomber will reside in honor within the Memorial now being planned.

This effort is now being headed by a brand new non-profit group which is known as the Memphis Belle War Memorial Foundation. Various prominent civic-minded Memphis corporate leaders are working hard to fund and build the structure that has been promised for almost sixty years.

With only a few good-weather months available throughout the year, the MBMA does not have a large window of opportunity to carry out large-scale restoration to this plane. The aircraft needs to be in a good condition of display through the summer tourist

Volunteers from the Tennessee Technology Center inspect the Belle during their annual trek to help out. The maintenance director for the Memphis Belle Memorial Association, James Webb, Sr., is seen peering into the number three engine cowling.

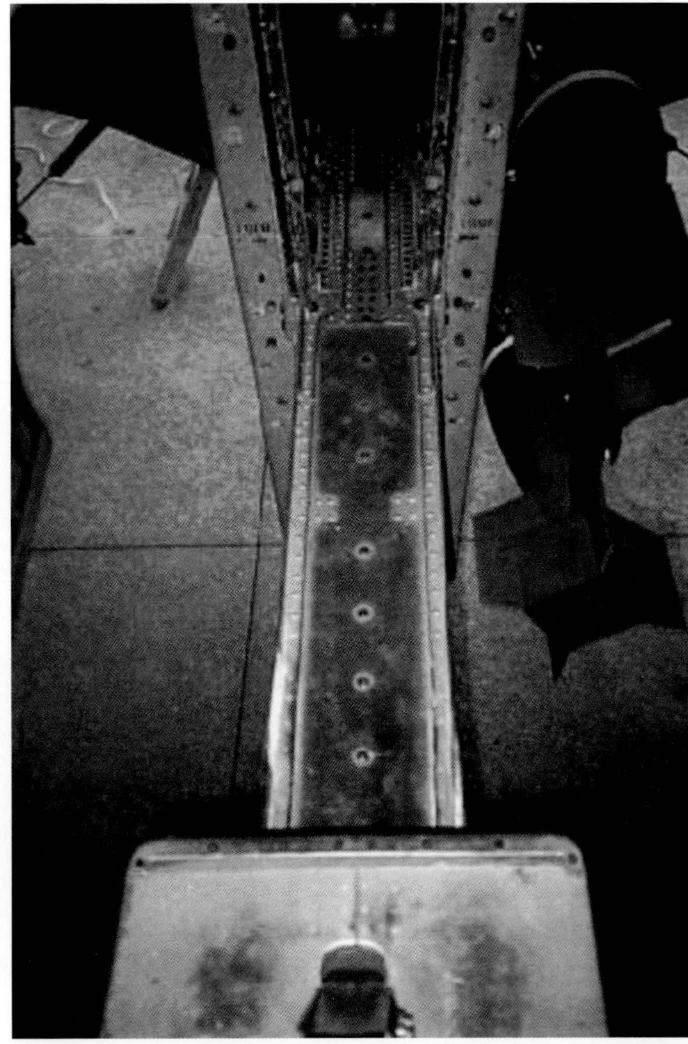

Top Left: With number four engine cowlings removed, the engine is inspected for signs of stress to the metal supports, rust, and even bird nest removal. Top Right: The catwalk through the bomb bay in the Belle is seen. This structure more or less acts as some of the backbone of the B-17 and is critical to the stability of the Fortress. Bottom Left: Just some of the oxygen bottles rest in their harnesses below the floor of the top turret. The redundant system provided the only means of breathable air to the crew at altitude.

season, and this leaves just several weeks between the operating season of Mud Island closing date and the onset of the winter cold to allow for repair work. Then another few weeks in the spring are available to prepare the B-17 for the next year's tourist season.

Each year the interior is thoroughly sprayed with a museum approved metal preservative. The bomber is washed very often, and light restoration is carried out through the summer, sometimes before the eyes of visitors.

The much needed large-scale work cannot be accomplished until a fitting structure is completed to allow various large components to undergo disassembly, cleaning, and repair. The MBMA has been working for years to gain the attention of city fathers and various corporations in Memphis to provide assistance. There has been some success, and at this time plans are afoot to fund construction of the world class and interactive environment for the Belle.

The future will be different for the "Memphis Belle" than the level of attention she has endured over the years. A Memorial will surround the bomber, giving tribute the fallen Veterans of WWII. Measures are being taken to see to it that all branches of the military will be honored beneath the sturdy wings of America's favorite Flying Fortress.

13

All the "Belles"

Above and following: Boeing B-52 "Memphis Belle" III

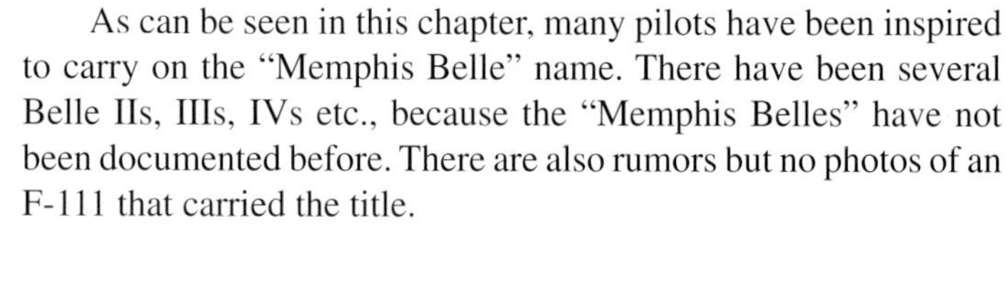

As can be seen in this chapter, many pilots have been inspired to carry on the "Memphis Belle" name. There have been several Belle IIs, IIIs, IVs etc., because the "Memphis Belles" have not been documented before. There are also rumors but no photos of an F-111 that carried the title.

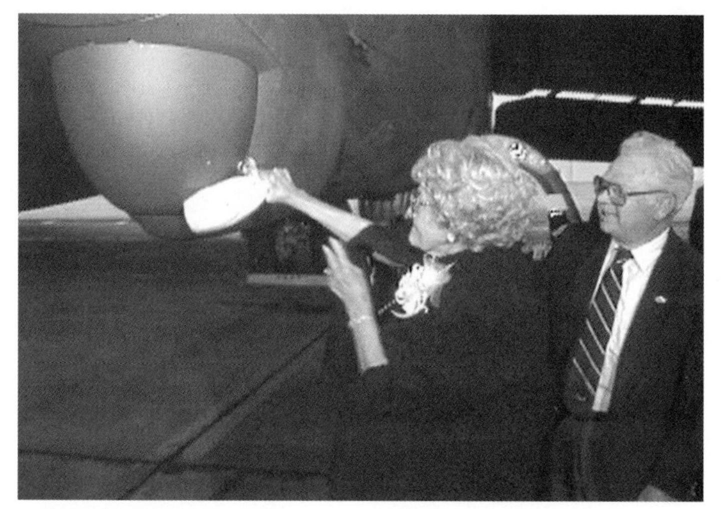

The Memphis Belle F9F Panther jet, which flew missions in the Korean War off the Aircraft Carrier *Oriskany*.

The Memphis Belle C-130 Hercules was attached to the 155th A.S. of the Tennessee Air National Guard in Memphis, TN. She was written off when the unit changed their mission to the huge C-141 Starlifter. She is seen here just after arriving at her final resting place in the Arizona boneyard, where she spent her final days.

The Memphis Belle F-100 Super Sabre Jet completed more than 250 combat missions in the Vietnam War.

The first "Memphis Belle" B-52 Stratofortress was based at Eaker Air Force Base, Arkansas, near Memphis, TN, and was named in honor of Miss Memphis 1967.

Not much is known about the Memphis Belle F-4 Phantom that flew with the U.S. Marines. It is believed that the jet was named by her crew chief, who was from Memphis, TN.

The second Memphis Belle B-52 just barely shows the famous Petty Girl (highlighted) beneath the preservative as she awaits the cutter's torch at Davis-Monthan Air Force Base, Arizona. The nose art survived and was removed from the bomber, then taken to the Air Force Museum in Ohio.

The third B-52 to carry the name Memphis Belle is part of the 20th Bomb Wing of the USAF and is based at Barksdale AFB near Shreveport, Louisiana. She was built in 1960 and still flies more than forty years later. She is almost double the age of many of the crews that fly her!

The Memphis Belle F-105 Thunderchief was flown by Major Buddy Jones of Memphis, Tennessee, and scored two MiG kills in the Vietnam War.

The Memphis Belle B-1 lancer #86-0133 is attached to the 116th Bomb Wing at Warner-Robins Air Force Base near Macon, Georgia, and is the first to make the famous Petty Girl supersonic! The wartime pilot of her namesake WWII B-17 poses for a picture next to "his" B-1 just after his introductory flight on the Lancer. At eighty-two years of age, Col. Robert Morgan showed his B-1 crew that he still had the right stuff!

The Memphis Belle C-141C Starlifter, again of the Tennessee Air National Guard. Instead of delivering bombs like her namesake B-17, she delivers critical supplies to areas all over the globe.

During Operation Desert Storm, one A-10 Thunderbolt II was decorated with the famous nose art.

The B-17 owned by David Tallichet. She is painted in the colors of the "Memphis Belle" and is seen here as the "Belle" B-1 Lancer peels away from a historic formation flight and photo shoot over Robins AFB, Georgia, in July 2000.

The only non-military Memphis Belle is operated by Northwest Airlink, a subsidiary of Northwest Airlines. The Bombadier Canadiar Regional Jet makes three daily round trips between Memphis and Charlotte, North Carolina, for the commuter airline. The *"Spirit of Memphis Belle"* is seen here in Iowa after her maiden flight, where she delivered the Memphis Redbirds baseball team to a game there. (With special thanks to Phil Trenary - President of Northwest Airlink!)

14

The 1943 Wyler Documentary

When WWII broke out, no one could predict the importance of bringing the talents of Hollywood to war. Among the contributions of notables like John Ford and William Wyler were scores of producers, cameramen, editors, and technical directors involved in the fight. It would be through their lenses and equipment that the American public would learn what their "boys" were doing far from the safety of their homes.

An unexpected bonus from this marriage of War documentation to film production was the professionalization of the 16mm formatted film. Until the War created the need for massive amounts of film making, 16mm film was considered to be far below the acceptable standards of Hollywood. Even the inventor of the 16mm film, Eastman Kodak, considered it to be an amateur format. By 1935, Kodak was prompted to introduce perhaps the largest innovation to film documentaries for the 20th century—Kodachrome. Designed to compete with the 35mm Technicolor format film, it was an almost instant hit with the public, who was enjoying the ability to create low cost films and still retain the rich deep colors that the film offered. Basically, any 16mm camera could produce beautiful home color movies.

In 1936, the United States Army was intrigued with the benefits of the 16mm format, which eased everything from shooting through distribution because of its smaller equipment sizes. The military had grown used to making training films for years in 35mm, but could easily see the benefits of using the smaller format film in the field in the form of portable projection equipment and the ability to display films in outdoor theatres, at field hospitals, and aboard ships at sea. By the end of the War the U.S. Military was regularly transferring full dramas and training films from 35mm to 16mm for the troops. The unexpected benefit really came into view when it was realized that authentic combat footage could be exposed in the field with much greater ease, then edited and transferred to the 35mm format for distribution to military leaders and the general public throughout America. The result was that, for the first time, strategists could better review the efficiency of combat tactics, and the public's need to be fed current news of the War could be satisfied much easier and with much better pictures. A 16mm camera was simply easier to haul around a battlefield than a huge 35mm

The poster promoting the famous 1943 William Wyler documentary, which was commissioned by Paramount pictures.

The crew of the "Memphis Belle" in a staged pose with William Wyler. He flew at least five actual combat missions aboard the historic bomber and as many as sixteen among other bombers. His film was actually exposed in combat, and to this day remains among the most accurate WWII color battle footage shot in the entire war. Wyler experienced all of the danger and drama along with the flight crews he flew with. On one mission (not on the Belle) his oxygen supply became disconnected and he nearly succumbed to the effects of oxygen starvation. He would have died had a crew member not noticed him passed out on the floor of the plane. Wyler also had to qualify as a gunner, but no one remembers him ever shooting a machine gun—only his 16mm Bell & Howell filmo cameras. Note that Joe Giambrone (crew chief) is up on a ladder painting the 25th bomb symbol on the nose of the Belle.

unit. Millions of feet of 16mm film stock were soon rolling off the production spindles every day.

New standards were being introduced as the film makers gained more experience. Lens design, the brightness of the screen, contact printers, cameras, and projectors were constantly improved. New cameras from Bell & Howell and the Filmo 70 spring-wound 16mm camera followed by GSAP (Gun Sight Aiming Point) cameras, which would find their way into the wings of thousands of Allied fighter planes and even the "Memphis Belle." When Major William Wyler handed the small cameras to the flight crew, he instructed them to shoot the film when they were not shooting the guns!

It is here where the real courage and bravery occurred, when huge Hollywood names like John Huston, Darryl Zanuck, John Ford, George Stevens, and William Wyler made their talents available to the War Department. They volunteered to face the enemy with only their cameras, and often stood on the front lines with their magazines in their Filmo 70 cameras loaded with just 36 seconds of film. There were actors, as well, including Mickey Rooney, Clark Gable,

Major Wyler peers out of the waist window of "Bad Penny," which is being rigged as a camera platform for his movie. Much of the footage was also shot at the 305th Bomb Group at Chelveston. All of the B-17s seen in the 1943 documentary are not from the 91st Bomb Group at Bassingbourn. Also, a bit of deception (because of the war and sensitive censors) is in the film. The briefed target discussed in the wartime film is slated as Wilhelmshaven, which is also stated to be the final target of the "Memphis Belle." Of course, this is not so. Many reasons are given for this, but the most likely reason is that the length of the film required a great bit of shooting, and it would have been impossible to capture all that is seen in the drama in a single mission. The actual final target of the "Belle" was in France. A raid to Germany for the final mission of this plane and crew would certainly play much better in the American theatres back home!

In one of the first shots in the famous 1943 documentary, Morgan discusses the mission with his crew before takeoff.

"Memphis Belle" is staged for takeoff. Visible just below the tail number are the cities that were targeted. They were painted there by the War Department to show Americans where the plane had been. Another example of the positive propaganda value of the Memphis Belle. This image has never before been published.

Ronald Reagan, Charlton Heston, Jimmy Stewart, and more. They all agreed to go into combat instead of making films on quiet and safe sound stages. Anatole Litwak was filming the invasion forces during the landings of North Africa when the landing craft in which he was riding was sunk, taking his priceless footage to the bottom along with it.

For the first time ever the production of film raw stock for non-entertainment uses exceeded Hollywood's! The War Department placed so much importance on the production of films, not only for training but for morale, as well, that an unheard of annual budget of more than fifty million dollars was approved.

This is where Hollywood and the "Memphis Belle" become intertwined. As it was, losses to American Bomber Groups were so acute that the Eighth Air Force instructed Hollywood director Major William Wyler to produce and release a film based on the lives around an operating USAAF air base. Wyler would take it a step further. He wanted a real-life depiction of what aerial combat was like. He asked for and received aerial gunnery training, as it was required for anyone who would be riding along on a bomber mission. If the need arose, the film maker would have to drop his gear and man a weapon. Wyler even took his camera down into the ball turret to capture the sight of the main landing gear of the "Memphis Belle" taking off and landing. It was not allowed, but Wyler quietly stowed himself away in the turret before and after several missions and endured the dangerous takeoffs and landings from below the "Memphis Belle," hoping each time that her wheels would hold.

William Wyler was born in Muelhausen, Germany, and came to America at the age of eighteen to work for his mother's cousin, Mr. Carl Laemmle—the founder of Universal. As a German Jew, Wyler had deep feelings for those being persecuted in his homeland and wanted to do everything he could to beat the Nazis. He had already been successful in Hollywood for directing the popular wartime feature film "Mrs. Miniver."

He was asked to use his talents to create a teaching aid for the up and coming newly formed Bomb Groups. But more importantly, his film would be targeted at the American civilian public. Homefront morale was sagging because of tremendous losses through the precision daylight bombing raids being carried out in late 1942. Military leaders figured that if the average American could feel the importance of what their sons, husbands, brothers, and fa-

Army '487 "Ritzy Blitz" DF-K in position for final engine checks. This shot was made from the radio room overhead hatch in the "Memphis Belle."

B-17s march down the Station 121 perimeter track toward the engine run up area just prior to taking the active runway for a mission.

"Memphis Belle" breaks free of the runway with her wheels spinning at almost one hundred miles per hour. Just visible is the Petty girl that adorns her nose.

Some B-17s of the 91st Bombardment Group (Heavy) on their way to battle.

thers were doing, then the War effort could be enhanced. The B-17 was already the darling of the cameras of the media, and so it was natural that a B-17 Group would be selected.

Of the few Groups in England at that time in the war, only USAAF Station 121 (Bassingbourn) and the 91st Bomb Group offered the accomodations that Wyler was looking for, though much of the finished film includes scenes of the planes of the 305th and 306th Heavy Bombardment Groups, as well. Wyler selected some of the better names that he had known in Hollywood to join him in the fledgling Eighth Air Force Film Unit. Equipment was procured by the truck loads and loaded aboard a ship set for England.

Without really knowing what the finished film was going to look like, Wyler knew that this picture would not be the typical stiff-backed "life on an air base" feature. He wanted to fly in combat. He wanted never before seen images of this new type of warfare from the air. He would mount cameras on different planes through radio-room hatches, waist windows, and throughout the fuselages—anywhere that would give him a shot that had never been recorded before. He himself would ride along in the bombers, seeing over the cameras that he passed out to crewmen.

Wyler and his men made it to England before their 35mm film equipment and began scouting various bases. The search for a bomber that was close to completing the required twenty-five missions began in earnest, and Wyler found several. "Invasion 2nd," "Hell's Angels," and "Memphis Belle" were among several more.

The shooting schedule hit a snag when the liberty ship carrying his cameras, processors, and all of his equipment was sunk enroute to England. It was here where he found that only 16mm equipment would be available to him. He pressed ahead anyway and found that he could capture more film with the smaller cameras, which could be carried throughout the cramped interior of the B-17 more easily than the bulky 35mm gear.

The focus of his lenses began to point toward the plane piloted by Lt. Oscar O'Neill. "Invasion 2nd" appeared to be the bomber that was going to hit the magic 25 missions before any other. Sadly, on 17 April 1943 the bomber and her crew were shot down during a raid to hit targets in Bremen, Germany. All of the crew, including O'Neill, were captured and taken prisoner. Fate was brewing a bit, and this attention toward O'Neill would not be fully realized for a while. Hollywood would, however, catch the O'Neill family. O'Neill had married a British woman, and the two had a daughter. Her name is Jennifer, and she is today a veteran star of the stage and screen!

The combat missions continued for Wyler and his men. On one mission Harold Tannenbaum, a long-time friend and accomplished cameraman, was shot down aboard a B-24 Liberator. The loss affected Wyler deeply. With the loss of Tannenbaum, Eighth Air Force Commanders took a look at Wyler and his climbing combat mission count, then had him grounded. If Wyler were to suffer the same fate as Tannenbaum, then the enemy could exploit the capture or death of William Wyler, especially since he was the one responsible for creating the wartime drama "Mrs. Miniver"—the film that was responsible for creating much empathy among Americans toward the British who were suffering the effects of the Nazis. If Wyler were captured it was likely that his reception among the Nazis would have been very harsh.

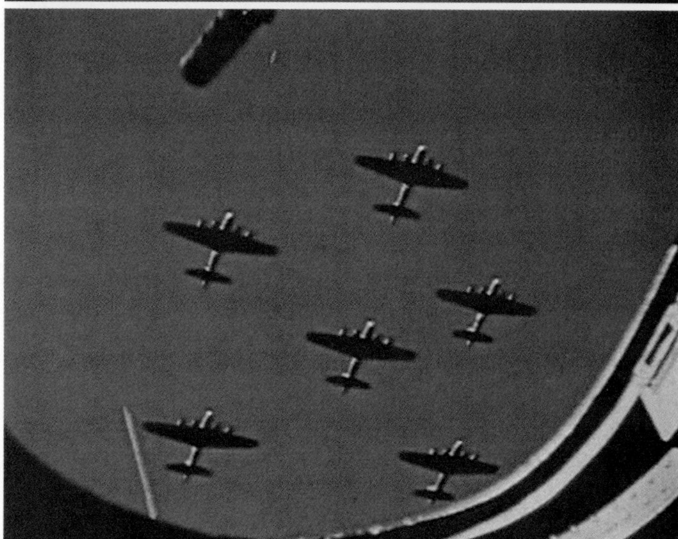

Top: In another very rare image, the "Memphis Belle" (DF-A) is seen in formation with the "Jersey Bounce" on their way to war. Despite the fame of the Belle, almost no pictures exist of this bomber in combat. Careful screening of the finished documenary enabled the author to locate and identify many of the bombers that were captured by Wyler's cameras. Center: Viewed through the radio room hatch in the Belle, six B-17s hold a very tight formation overhead. Bottom: Looking down her left wing, three sisters of the "Memphis Belle" take up a flanking position over the British countryside on their way to hit targets on the continent.

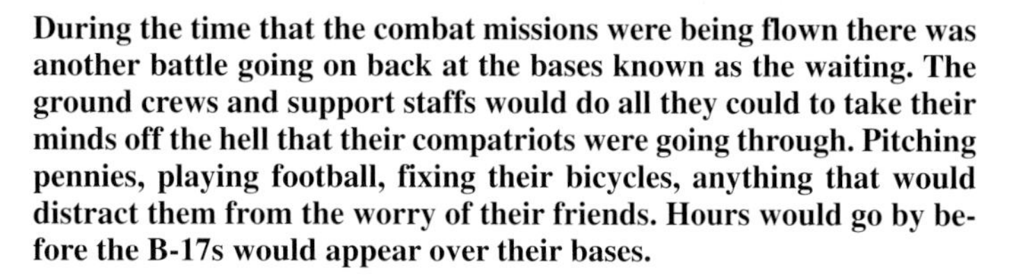

During the time that the combat missions were being flown there was another battle going on back at the bases known as the waiting. The ground crews and support staffs would do all they could to take their minds off the hell that their compatriots were going through. Pitching pennies, playing football, fixing their bicycles, anything that would distract them from the worry of their friends. Hours would go by before the B-17s would appear over their bases.

Right: Morgan and Verinis are seen here behind their oxygen masks, which were required for high-altitude flight during their missions. Back at base, the ground crews would begin converging on the airfield to start the traditional count of returning bombers. With all eyes toward the runway they would see flares shot out from the planes. This told the men on the ground that there were wounded or dead on board, and the tower would give those planes priority to land first.

Even though he had been grounded from combat flight, Wyler had developed so much enthusiasm for the film that he was making aboard the "Memphis Belle" that he characteristically ignored a direct order and flew the very next day for the final mission of the "Memphis Belle."

It is fitting to note here that the film should be considered by the viewer to be wartime propaganda. While none of the scenes were enhanced by photo trickery (the flak and fighters were very real), some innacuracies do exist. It was typical for the War Department to shroud the facts from time to time, and the 1943 film "Memphis Belle" is no exception. The 91st Bomb Group or her four Squadrons are never mentioned. The filming took part on various bases, and the final target is mentioned as being the German Naval installation at Wilhelmshaven. In reality this is, of course, not so. The probable reason for this was simply that it was very likely that the film could fall into enemy hands, especially since it was going to be shown in theatres across America. The War Department was not going to take any chances with this, so Wyler was instructed to create a ficticious raid to settle any concerns about what the enemy might learn by watching the film. Plus, it made the final edit much easier.

One of the largest misconceptions regarding the piece is that it leads the viewer to believe that the "Memphis Belle" was the first Eighth Air Force bomber to reach twenty-five missions. As stated so many times before, this is not true. Again, consider the propaganda value of the work. Leaning on their public relations professionals, it was simply a matter of names that played such a vital part in the selection of the aircraft to be paraded before the American public.

Selecting this plane was simply natural, and it created a media coup that no one could have comprehended. Much to the chagrin of other airmen that made their 25 missions before the Belle, it was very difficult to watch the "Memphis Belle" become America's favorite Flying Fortress.

Taking the perspective of the War Department and the orders which came down from above, no other plane could have or even should have come home first. America needed to see a successful crew—and there were many! But the Belle had already been captured on film for the Wyler documentary, and soon would be seen across America in thousands of movie houses. Her crew, for the most part, was largely intact. No one had died on this plane. She had slugged it out against some of the best Germany had to offer, winning out every time, and perhaps most importantly, this was an unbelievably simple sell to the hungry public. The pilot had a sweetheart! The conference rooms of the upper commands in the Army Air Force must have been roaring with approval. This whole story screamed of "apple pie and baseball," in which the hero does ride (or fly, in this case) off into the sunset. Instead of his girl perched with him on horseback, *she was painted on the nose of the plane!*

Years later, during a visit to the home of William Wyler, Frank Donofrio (Founder of the Memphis Belle Memorial Association) was talking with the director, who proudly showed off his military decorations. There in his living room was his Air Medal that he received for his many combat missions on the Belle. It had a place of higher honor, as it was displayed among his coveted Oscar statues.

"Memphis Belle" Beats Jinx To Lead Fortresses In Raid

NAZIS TRY TO DROP BOMBS ON U.S. FLYING FORTRESSES

BIG BOMBS CASCADE UPON U-BOAT PLANTS

One Belle With Bite

PILOT OF "MEMPHIS BELLE" IS ON THE AIR

Boss of "Memphis Belle" Introduces Fort to King

MEMPHIS BELLE BACK AFTER MAKING MOVIES

Paramount Named To Release Film On 'The Memphis Belle'

NO RETIREMENT NOW FOR MEMPHIS BELLE

Fortress "Memphis Belle" Honored on Eve of Flight Home

Just a few of the 1943 newspaper headlines that appeared in editions all over the United States.

William Wyler was not finished. He went on to produce a second and equally brilliant film about the accomplishments of the famed P-47 Thunderbolt. It has been said that after Wyler took his film to the White House for a private screening with President Roosevelt, a visibly moved FDR offered Wyler a cigarette and accepted a light in return, then said, "This has to be shown right away—everywhere."

That did it. Paramount Pictures jumped on the bandwagon and began cranking out duplication prints, created movie posters and lobby cards, and then took charge of distributing everything all over the country.

The picture was on its way to becoming one of the very best examples of documentary film making the world had ever seen.

Army 41-24617 "Southern Comfort" is seen in the center images on this page. She was one of the B-17s that was never planned to be included in the Wyler documentary. Because of the inflicted damage during a raid on 26 February 1943 she made an impression on Wyler as she taxied in after coming home. This B-17 was assigned to the 364th Bomb Squadron in the 305th Bomb Group at Chelveston—further indication that all of Wyler's film was not shot at Bassingbourn. It cannot be seen in the above images, but a 20mm shell had exploded inside the nose compartment, sending the navigator's head smashing through his chart table. Only his flak helmet saved his life. He was knocked out, but was able to return to his duties in less than 30 minutes. The number three engine was out, and the pilot, Hugh Ashcraft, was unable to feather its prop. With more than four feet of rudder shot away, the plane was difficult to fly back to her base. Repairs saw her ready for action by 31 March, and this would be her final raid. Near Rotterdam, the bomber was bracketed by flak that started an uncontrollable fire on the left wing. After turning back for England, the navigator plotted a return course, allowing them to arrive over friendly soil in time for the flames to begin buckling the metal on the wing. Everyone bailed out and landed safely, except the flight engineer, who sadly slipped out of his parachute harness and fell to his death.

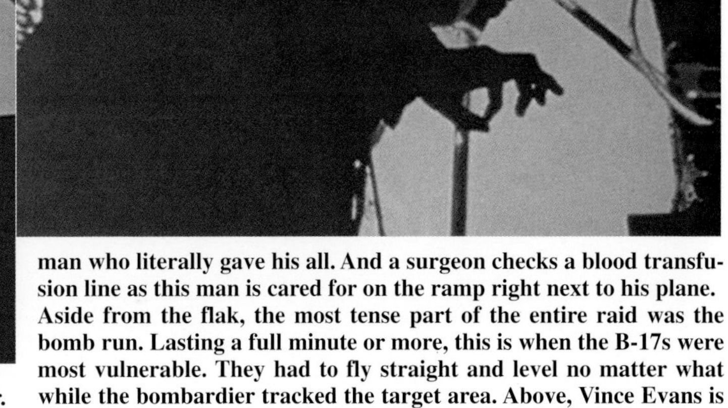

In the images below are scenes of some of the effects of war in the air. A wounded airman is carried on a stretcher to a base hospital. Another badly wounded man is given a puff of a cigarette by his comrades as he is carried from his plane. Officers cover the remains of one man who literally gave his all. And a surgeon checks a blood transfusion line as this man is cared for on the ramp right next to his plane. Aside from the flak, the most tense part of the entire raid was the bomb run. Lasting a full minute or more, this is when the B-17s were most vulnerable. They had to fly straight and level no matter what while the bombardier tracked the target area. Above, Vince Evans is seen hunched over his sight while he flies the "Memphis Belle" on the target run. Moments later he shows the camera that his bombs fell into the "pickle barrel" and destroyed the target.

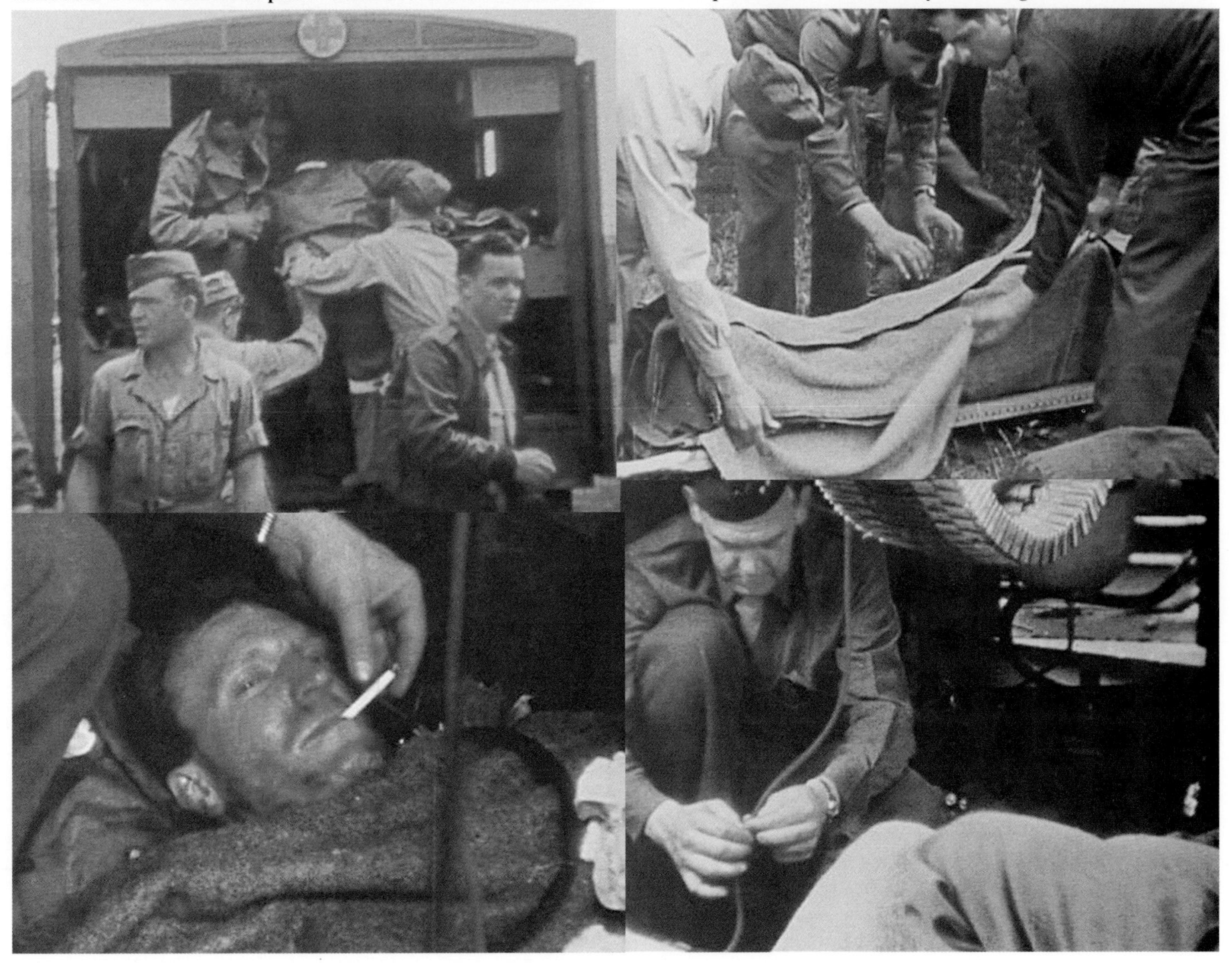

Leighton & Evans relax on the way home after their final mission. Bill Winchell shows his youth, and Tony Nastal peers out over the top of his fifty cal. In the upper left, Bob Morgan is looking out over the left wing at Bassingbourn, and Jim Verinis stares intently at his gauges with number 3 engine just out his window. At the bottom right, Winch and Nastal man the waist guns of the Belle during their last raid.

The "Memphis Belle" is seen here taxiing in from her last combat mission. Note that her inboard engines are cut. Radio operator Bob Hanson and flight engineer Harold Loch can be seen sitting on top of the Belle at the radio room overhead hatch. Ground crews are jubilant and cheer on the crew of the Belle as they slowly roll by.

In the photo in the upper left, Bob Morgan shows a sincere smile and throws a wave to the camera as he taxis the "Memphis Belle" into Bassingbourn. Just below that is a shot looking out of the control tower windows as the Belle slowly rambles to a stop on the perimeter track. Just to the rear of the plane one of Wyler's film cameras is seen on its tripod. It had just captured the image seen at the upper right of Scott Miller and Bill Winchell hanging out of the left waist window celebrating their triumphant return to base. The date was 17 May 1943, and Winchell is making a spinning motion with his finger telling everyone that he had just shot down the eighth and final enemy fighter from that window of the "Memphis Belle."

"Memphis Belle" radio operator is seen smooching terra firma following his final combat mission on the plane. Some believe that this was staged for the Wyler documentary. Hanson promises that this action was a result of his real feelings as he got out of the Belle.

In the final picture at right, Bob Morgan pats his flight engineer Harold Loch on the shoulder, telling him that they had all accomplished a job well done. Most of the crewmembers of the "Memphis Belle" called their pilot "Chief"—a reference that sticks to this day when they gather together for reunions.

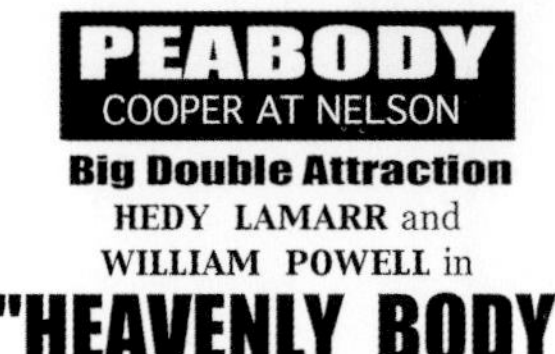

PEABODY
COOPER AT NELSON
Big Double Attraction
HEDY LAMARR and
WILLIAM POWELL in
"HEAVENLY BODY"
—AND—
"Memphis Belle"
(In Technicolor)
Excellent Shorts—12¢ and 35¢

The series of photos shows just a moment in time for the pilot of the "Memphis Belle." Only seconds after emerging from the plane, he is whisked upon the shoulders of his crew and lifted to the girl that helped him through the hell of the past eight months. A smooch, a pat on her bottom, and his smile is only a fraction of the emotions churning within the 24 year old Aircraft Commander. His missions are over!
Aside from large promotional movie posters, film producers included lobby cards, which were distributed to theatres along with prints of the films they were showing. Shown here are three of the seven known lobby cards that were made for the "Memphis Belle" in 1944. The full size measurements were more than 20 inches wide. To the left is a newspaper ad promoting the playing of the film in Memphis.

15

The Belle and the Big Screen

Of the two major film productions covering the "Memphis Belle" and her incredible story, the 1943 documentary is much more accurate from a historical standpoint, even though it must be realized that it took some liberties with realism. That picture was produced for the sole purpose of Wartime propoganda and to boost morale on the homefront during WWII. The 1989 film has taken the brunt of criticisms because of the way in which the crew was portrayed. It must be understood by the viewer that this latter movie was made out of the devotion of a daughter to her father. When Catherine Wyler began shopping around the idea of a WWII bomber story, she had "Memphis Belle" very much on her mind. But to influence a film company to back the huge production costs, some engineered drama would be needed to make the film profitable at theatres. Hence the decision to depart from the real story of the "Memphis Belle" in some areas. Noted WWII author and historian

This poster promoted the 1989 Warner Brothers film co-produced by Wyler's daughter, Catherine.

Wyler's daughter Catherine poses on location in England with her Uncle David's personal B-17, which has been painted like the real "Memphis Belle" for the production. It seems to be an all in the family devotion to remembering the causes of WWII. Catherine's film, like her father's, is considered to be a great treasure to those who wish to feel what it must have been like to fly a combat mission in a B-17.

David Tallichet in the left seat of his B-17. This bomber started Army Air Force life as B-17G #44-83546 and was accepted into the military on 3 April 1945. While she never saw actual combat, the plane served the USAF until 1954 when she was put up for storage at Davis-Monthan AFB near Tucson, AZ. For a time this plane was even based in both Germany and Japan! From 1959 through today four civilian operators claimed ownership to the plane and used the airframe in a variety of roles, from tanker and spraying jobs to fighting forest fires. Tallichet acquired the B-17 in 1986, and when his niece announced her intentions to make her WWII film, Uncle David jumped at the chance to take part. Warner Brothers repainted his bomber similar to the real Belle. Among the differences are the lowered arm on the Petty Girl and the cursive lettering, which differs from the block lettering on the real "Memphis Belle." This B-17, which carries the civil registration N3703G is, as of this printing, the very last B-17 to make a trans-Atlantic crossing!

Roger Freeman was brought on board early in the design phase of the 1989 film to provide his knowledge of bomber stories to the movie. He took great strides to assure that every dramatic moment in the film had actually occurred to some bomber at some point in the war. So the end result of this movie can be summed up by saying that it is a fictionalized account based on hundreds of real bomber stories, and therefore a great film.

The producers of the film never intended to follow the actual story of the "Memphis Belle." Instead, they wanted to incorporate as much drama and experience as possible. As a matter of fact, the film was nearly released under the name "Southern Belle." It was found, however, that several real bombers had carried this name during the War, so there could have been as many as a hundred or more Veterans who would not have been happy to have been portrayed in what could have been thought by some to be a defamatory way. The Veterans of the "Memphis Belle" crew had already read and approved the script, so the decision was made to stick with the name. Plus, only a single "Memphis Belle" bomber was known to exist during WWII.

Before filming began, a massive search was launched for locations and flyable B-17s that could participate.

As the United States is home to the majority of flyable B-17s, America was searched for a suitable location to carry out the filming. It was decided to take the production to England because the surrounding countryside there offered the authenticity needed.

It was then the monumental task of the producers to find as many B-17s as possible to be assembled in the United Kingdom to make the effort worthwhile. Only a single flyable Fort was based in England. Two were brought up from France, and two more made the dangerous flight from America across the Atlantic. One of them belonged to the uncle of the producer. David Tallichet, a pilot with the 100th Bombardment Group (Heavy) out of Thorpe-Abbot during the war, was very well known as a competent warbird operator with extensive experience in WWII heavy bomber operations. He is also the uncle of the producer, Catherine Wyler, and the brother-in-law of the famous Hollywood director William Wyler—and he owned his very own B-17! Uncle David was all too happy to prepare his plane to fly over the ocean to England to take part in the film. He arrived there with a great deal of enthusiasm and provided his expertise and no-nonsense abilities to keep the old planes flying during the rigorous schedule.

While on location in England, actors Matthew Modine and Eric Stoltz are caught taking some time with their real-life counterparts Robert Morgan and radio operator Robert Hanson. The production team felt it was necessary to gather the surviving members of the crew and fly them to England to advise the actors in their work. Modine and Morgan still speak from time to time on special days!

G-BEDF aka "Sally B" evades enemy fire during filming. This proud B-17 joined four others, two from France and two from America, to complete the filming.

Because of the lack of operable Flying Fortresses, huge radio-controlled models were built and used to create the illusion of a hectic WWII airfield. There were five real B-17s and five model B-17s. The scaffolding allowed the film makers to create the take off sequence.
Also, flats were placed strategically in the distance to create even more bombers on the field. The painstaking efforts of the camera director paid off by creating what is a very authentic take off sequence. With a span of more than sixteen feet, the models provided a realism that many did not expect to achieve in the final result. Even though they were just models, they were still handled with great care.
During the crash sequence at the beginning of the film, one of these models was being towed behind a car. The model left the planned track when it became airborne and flew right into the cameraman. It struck him and his equipment, and actually put the poor man in the hospital from the injuries he received. The producer thought the shot was ruined, but when the film was seen, the shot was so good that he refused to keep it from hitting the cutting room floor. It ended up being one of the best shots in the movie.

One of the lighter moments occurred over Lincolnshire during one sequence when the bombers were in the air. Spitfires and a restored Me-109 were aloft with the B-17s over the English countryside. An elderly British farmer took notice when he heard them, looked up to see the "melee," and hurried to call authorities, thinking that he had gone mad and reporting all the while that the damned "Jerries" had started the War again!

One of the more popular discussions among tourists who visit the "Memphis Belle" display in Memphis is the accuracy of that 1989 film. Many ask time and again what really happened on the Belle and was the film true to life. It is with this in mind that I felt it prudent to include explanations of some of the engineered drama that was in that film. This is not an attempt to criticize the effort, but merely to explain what happened and what did not on the "Memphis Belle." From an entertainment standpoint the film has much to offer. Combining the cinematography and the score, the viewer is treated to images that are very realistic. Those who strive to criticize this film often fail to remark that without the necessary scripting (which is what comes under fire most) that the thirty million dollar movie would never have been made, and enthusiasts would have been robbed of the opportunity to share some of the drama and excitement of the B-17 Flying Fortress.

With this in mind, the reader may wish to have this book nearby the next time the film is watched. But keep in mind that every event is verified as having happened to some plane at some point in the War, but did not all happen to the Belle or her crew.

What Really Happened?

Volunteers at the "Memphis Belle" pavilion are asked every day if the 1989 movie was historically accurate. These questions are so frequent that it was decided to include answers to most of them here in this book. It is not the intention of this section to criticize the efforts of the film makers. There were many reasons that surrounded their decisions to make this very fine film. This section was included to give the "Memphis Belle" fan a much better understanding of what really happened aboard the "Memphis Belle."

1. - The film begins on 16 May 1943. There's a party that night and a mission the following day.

It did not happen that way. But there was an anniversary party on 15 May for the 91st Bomb Group. The crew was scheduled to fly their final raid on the 17th. This happened, but on 16 May William Wyler had made arrangements for King George VI and Queen Elizabeth to visit and review the crew of the "Memphis Belle."

2. - Were there dances before missions?

Almost never. If it happened it was the result of a late notice coming from Bomber Command. On the rare instances of this, many of the men who would have to go on the raid resorted to breathing pure oxygen for a while before the mission launch to reduce the effects of the night before.

3. - John Lithgow's character says that the Belle had twenty-four missions without a scratch.

Not true! It is extremely unlikely that any Eighth Air Force bomber flew twenty-four missions without having been scratched. The "Memphis Belle" received major damage several times, and three of her crew received injuries during combat.

4. - Again Lithgow's character comments that the pilot met his girlfriend (for whom the plane was named) while on business in Memphis, Tennessee.

The reason for this inaccuracy was to simplify the explanation for why the plane was named "Memphis Belle." Again, Bob Morgan met Margaret Polk in Walla Walla, Washington, while he was there in his final flight training phase. She was there from Memphis visiting her sister, who was married to one of the doctors assigned to the 91st Bomb Group.

5. - Tate Donovan's character (co-pilot) is disrespectful to Modine (pilot) throughout most of the film.

As far as the "Memphis Belle" is concerned, this did not happen. The real co-pilot of this bomber (Verinis) was actually trained as a first pilot and was assigned to Morgan's crew while he awaited the assignment of his own plane. Jim got his plane after the sixth mission on the Belle and received a B-17 which he named after his home state of Connecticut. The movie also says that the co-pilot was a lifeguard before the War. Jim was active in sports, but not swimming. He was captain of his high school basketball team and won a scholarship to play the game at the University of Connecticut.

6. - Was the Bombardier really in medical school?

No. Vince Evans ran a trucking company in Texas before the War, among other ventures. After the War, he went to Hollywood and became a screenwriter for Humphrey Bogart. He was also close to Ronald Reagan, June Allyson, and Jimmy Stewart. He raced cars, and also owned the famous Andersons Pea Soup Company.

7. - Donovan talks about Modine and called him the "big shot pilot while I'm in the dummy seat."

Not on the Belle. Crew cohesion was good on the "Memphis Belle," and Verinis is often referred to as "the other Belle pilot" to this day.

8. - Lithgow makes a bad speech at the dance; Harry Connick, Jr.'s character saves him by singing Danny Boy.

Nice touch in the film, but John Quinlan—the real tailgunner—was not a singer.

9. - When the "Memphis Belle" crew is seen heading out to their plane, they are riding in a jeep that seems seriously overloaded.

This is reminiscent of the 1943 documentary, mixed here with a great music track, and moves to a very visual series of shots that shows what would be typical pre-mission activities.

10. - The engine start sequence is stirring and very well done.

Good editing compliments the shots of the blowing wheat, which is yet another reminder to us that this film was made by the daughter of the man who made the original in 1943. Make note of the engine vibrations radiating through the actor's bodies. This is a very realistic sequence.

11. - The taxi sequence was another very strong point of the film.

Michael Caton-Jones (director) is successful in bringing the viewer along through this powerful moment. It shows the might of the B-17 in a series of images where he makes five B-17s look like fifty. Don't miss a single shot here.

12. - From throttles forward through the monotone expressions of Gains and Guintoli (we've done this a thousand times before), the yawing fury of looking through the pilots windshield, the gentle back-pressure on the control yoke, and finally the music swell as the bomber breaks free of the runway. Jones has created here one of the best take off sequences captured on film.

13. - The viewer has to endure the bickering between the two waist gunners and the jests of the ball turret gunner.

Again, on the Belle the crew got along well, and some later commented that they did not have time to argue even if they wanted to. While there were jokes and pokes and some fun from time to time, when in combat they were professionals.

14. - The film then takes the viewer to the initial climbout following take off, where a near midair collision almost ends the mission.

This was indeed a constant problem that all crews were well aware of. The British weather did not always accomodate smooth flying conditions, and many collisions in the air did happen. No report, however, was made that this ever happened to "Memphis Belle." This is felt to be an important addition to the movie, as it was a constant concern among the aircrews. During the climbout there were twenty eyes on each plane scanning the sky for friendly planes that were wandering too close for comfort.

15. - On the way to the target Modine warns the crew that the outside temperature is thirty degrees below zero. Then Stoltz's character is seen wrapping a bottle of champagne in a heated flying suit and using a plug-in rheostat to keep it from freezing. Modine also

Actor images courtesy of Warner Brothers / Enigm
copyright 1989
THE
MEMPHIS
BELLE
MEN
Matt Modine
Bob Morgan
Eric Stoltz
Bob Hanson
Tate Donovan
Jim Verinis
Sean Astin
Cecil Scott
Billy Zane
Vince Evans
Courtney Gains
Tony Nastal
D.B. Sweeney
Charles Leighton
Neil Guintoli
Bill Winchell
Lou Reed
Harold Loch
Harry Connick, Jr.
John Quinlan

warns the crew about ice forming in their oxygen masks and causing anoxia.

First point here is that if the temperature outside the bomber is thirty degrees below, then so is the temperature inside the plane. But beyond that, given combat conditions in the film versus real life, it is quite unlikely that any airman would sacrifice his warmth for a bottle of booze no matter how important it was. This did not happen on the "Memphis Belle," but it did play well for the film. As far as the ice forming in the oxygen masks, this occurred quite a bit amongst the flight crews. After some time at altitude the men got into the habit of shortly disconnecting their mask supply hoses and flexing them to break out the ice that formed from the condensation of their breath and then plugging back in. If this was not done with regularity, the hose would become clogged with ice, and the airman would succumb to oxygen starvation. Within minutes this could result in the death of the unfortunate crewman. This did happen to two of the waist gunners aboard the Belle (28 March 1943), but they were rescued by the radioman before they expired.

16. - Then P-51 escorts appear.

For the historian this is unacceptable, as Mustangs were not available at that point of the War for escort duty. In reality, any escorting fighter planes would have been the fantastic British Spitfire or the early P-47 Thunderbolt. The film makers had a very logical reason for using the Mustangs. There were not enough P-47s available for the movie. There are enough Spitfires in England, but as far as those planes were concerned, the producers felt that viewers would have confused the air action with the Battle of Britian, where Spitfires were used extensively. Also, the Mustangs were readily available in large enough numbers, and their owners were willing to participate in the film with enthusiasm.

17. - Donovan says during this part of the flight that he'd give anything to fly a fighter.

This might have been the feeling among some bomber pilots, but not the "Memphis Belle" co-pilot Jim Verinis. He had already been through the fighter plane program and was training to fly the P-40 and the P-51. He actually preferred the much larger B-17. During training he was forced to bail out of one fighter and crash-landed another. That was it for him. He went to his Commanders and told them that he wanted to fly something with "four fans up front," and Jim Verinis was transferred right away to the bomber program.

18. - The bombers were never escorted all the way to the target at this point in the War.

The production does a fine job here showing the feelings of the airmen when the escorts are seen peeling off to head back to England, leaving the bombers to fend for themselves.

19. - Another moment when we see the Wyler influence in the film is when the Group C.O. (David Straitharn) forces Lithgow to read the letters from the families of the deceased airmen. In this scene the viewer is allowed to see some actual film shot by Wyler himself during some of the combat missions that he flew on B-17s.

20. - The "Memphis Belle" ball turret gunner scores a kill on a German fighter.

Cecil Scott (real ball turret gunner) was never credited with shooting down an enemy fighter from the Belle. He shot at many of them and damaged some. And he was responsible for warding off some very intense attacks from the ball.

21. - Giuntoli is shot and wounded in the waist of the "Memphis Belle."

His counterpart Bill Winchell was never hit.

22. - Stoltz is seen during a scene in the radio room bending over to check his camera. This happens just as a piece of flak blows through his table and his logbook. Had he not moved a moment before, he would have been seriously wounded if not killed.

This did happen to the Radio Operator on the "Memphis Belle." Although Robert Hanson does not recall which mission this was, he held on to his logbook, which he still has today.

23. - Donovan says to Modine that he's the only one not doing anything as co-pilot. He wanted to get permission from Modine

Two famous crews! Standing left to right: Navigator Charles Leighton; Co-Pilot James Verinis; Pilot Robert Morgan; Harry Connick, Jr; John Lithgow; Director Michael Caton-Jones; Radio Operator Robert Hanson; Left Waist Gunner Bill Winchell; Right Waist Gunner Cassimer Nastal; Top Turret/Flight Engineer Eugene Adkins; Crew Chief Joe Giambrone; (Kneeling) D.B. Sweeney; Tate Donovan; Matthew Modine; Sean Astin; Eric Stoltz; Neil Guintoli; Courtney Gains; Lou Reed; and Billy Zane.

(Aircraft Commander) to go to the tail of the bomber to shoot the machine guns. The scene shows the co-pilot leaving his seat during the heat of combat to go all the way to the back so he could casually shoot at German fighter planes.

There are many comments over this scene. What was he thinking? Who was going to fly the plane if the pilot were wounded or killed during the dogfight blazing all around them? Verinis never did this, nor did any other co-pilot of the Belle. The co-pilot was critical in the operation of a B-17, especially during combat! Leaving to experiment with machine guns was not the focus of pilots when basic survival was foremost on their minds. If he wanted to play with the guns, all he had to do was turn around and man the top turret about two feet behind him there in the cockpit, instead of moving more than sixty feet to the back of the bomber. Kicking the tailgunner out of this important position on the plane seriously jeopardized the safety of the entire crew, as well as the other bombers in the formation. Co-pilots were not extensively trained as gunners, and unless there was some sort of miracle, it is quite unlikely that he could have been proficient as a tailgunner—one of the most attacked parts of any bomber Combat Wing. The dramatic event which took place after Donovan shot down the German plane was evidently taken from the incident involving the B-17 "All American" described near the beginning of this book. The producers felt that this event was so significant that it needed to be included in this film. Indeed, it was an example of the sturdiness of the B-17, but many viewers wish that the film makers would have chosen another way to portray that horrible midair collision. Of course, that event never happened to the "Memphis Belle."

24. - The Ball Turret on the "Memphis Belle" never jammed.

Before the film's mission launches, some discussion between the Flight Engineer and that gunner prepares the viewer for the inevitable turret problem. This may have indeed occurred on other bombers, but not the Belle. Two things are often discussed about this scene. First, there were redundant manual hand cranks available if electrical power was lost. Many think that this should have been shown. As flight engineer, Diamond would have been responsible for the turret's operation while airborne. Before the mission launched, the ground crew were the people that the gunner would have been talking to. Many veterans have commented about the failure to include these important men in the film.

25. - The Ball Turret is badly hit, and a good bit of it is blown away, leaving Astin hanging out of the bottom of the "Memphis Belle."

Not on the Belle. If there was anything to hold on to, then it would have been nearly impossible to do this in a 200 mph slipstream. If that much of the turret was gone then so was the gunner, or at least many large parts of him. This turret never malfunctioned on the Belle, and Cecil Scott was never injured in it. The Ball was never seriously hit in that manner on this plane. It may have happened to other B-17s, but not the Belle. This scene does have some significance by showing the drama of flying in this important part of a

The majesty of the Flying Fortress is well displayed in this tight shot caught by an alert photographer during the filming in England.

bomber. Much has been documented, and photos show some badly damaged ball turrets on other bombers.

26. - The engine fire and dive.

As depicted in the film, this did occur to the "Memphis Belle." However, it is not clear which exact mission was being flown at the time.

27. - The flight deck is shown being shot up and terribly damaged. There were no reports of any significant combat damage to this area of the "Memphis Belle."

28. - Stoltz is severely injured.

There was never any injury bad enough aboard the Belle that made the crewmen consider throwing another out of the plane on any mission to get medical help from the Germans. This did happen several times aboard other bombers during the War.

29. - The aerial combat scenes.

Most agree that this thirty minute sequence was well done overall. The impression is given that flying these aircraft, at these times and in these conditions was exceptionally difficult.

30. - The landing:

From the time the crew sights the coast of England through the time they land is considered by many to be unrealistic. The crew learns that one wheel will not extend, and Modine (pilot) immediately decides to attempt a wheels-up landing without considering emergency manual extension procedures. The landing gear on the Belle never failed to operate during combat conditions. (The only time the wheels did not work was actually on her very final flight in 1946.) The procedure would have been for the TT/Eng to get into the bomb bay and attempt to crank the mains down. A wheels-up landing was only done as a last resort. However, many B-17s did make successful gear up landings and reported that the B-17 was very good at this maneuver. Because of the exposed main tires in the retracted position, many of these pilots reported significant differential braking control as they slid down the grass.

31. - Falling from the bomb bay. While the flight crew is busy handling the badly damaged Belle, the TT/Flight Engineer finally gets back to the bomb bay to attempt to crank the mains down. He nearly falls from the plane as the bomb doors suddenly pop open beneath him.

This event was probably taken from the story which did happen to a member of the crew of the Belle during a training mission on a different B-17. Navigator Chuck Leighton remembered that during one training sortie over England he had to urinate, and there was a tube in the bomb bay for that purpose. While he was performing this task, the bomb doors did indeed open beneath him as he stood on the catwalk. The blast of wind almost pulled him from the plane, and he scrambled back to the front of the plane to find Bombardier Vince Evans laughing his head off for creating the prank. The more Leighton punched him the harder Evans laughed!

Unless the emergency cable releases on the bomb doors are pulled deactivating the screw drive mechanism, the doors will not blast open like the film shows. All of the stout gearing mechanism would have to fail at once for this to occur. The down cycle of the bomb doors on the B-17 consumed roughly ten full seconds or more. The bomb doors simply do not snap open on a Flying Fortress.

32. - With the landing growing closer and closer, the Belle is seen flying on two engines with three men (1 hanging) in the open bomb bay when a third engine falters. Only one main wheel is down, there are critically injured aboard, and the fuel is dangerously low. A few B-17s were landed on a single engine, but only after they made the landing threshold at the end of the runway. The crew would have been ordered into their crash positions. "Memphis Belle" is seen on short final and over the airfield boundary with all of these problems still existing. The three men in the bomb bay had been there for some time trying to crank the wheel down. Many have

A stacked formation of B-17s over England. With the appearance of these B-17s, many elderly British folks thought the war had started again!

The flightline at Lincolnshire during "Memphis Belle" production. The producers achieved the near impossible by rounding up five flyable Forts. This will likely never happen again.

complained about this predicament, saying that no pilot would ever have risked three of his men in an open bomb bay on final approach with such a badly damaged aircraft. At this moment the stricken main is shown in the full up position. Once over the boundary the landing is merely seconds away. Not even the wheels motor could lower the bad gear in time. There is no way any human could crank the main down in time.

33. - Only a few short seconds before the bomber's wheels touch the runway, the bad gear reaches the down and locked position and the plane heaves back into the air after a very rough, high vertical speed, single engine landing.

The "War" is over for these actors. At the end of the film, the jubilant crew of the "Memphis Belle" is showered with champagne as their skipper soaks them in victory. The emotion of the moment is well done at this part of the film, even though this shot is somewhat inaccurate. For instance. Eric Stoltz (Radio Operator) is seen in the stretcher. No one was ever carried off the Belle with injuries. And in reality, when the crew of this plane finished their missions they lifted their pilot on their shoulders to give the nose art a little kiss. One of the best features of the film was the attention to uniforms. The image above shows just how hard the costume director worked to stay in the period.

Somehow, the three men (1 hanging) in the bomb bay manage to hold on to the plane as the jolts of this rough landing radiate throughout the airframe. Very lucky men indeed!

34. - The Belle rolls to a stop, mission accomplished.

Many viewers really like this scene, as it dramatizes what many bomber crews must have felt as they made their returns aboard damaged bombers through extreme conditions. Take note of the shot where Modine holds the photo of his girlfriend, the Memphis Belle for whom the plane was named. Another nice addition to the film, as Morgan really did place Margaret Polk's picture in the cockpit for all twenty-five missions. (It is still there today!) It would have been nice, though, to see a real picture of Margaret Polk for the film—a nice insider's touch!

35. - Harry Connick, Jr, kisses the ground:

The William Wyler influence is seen again! Only in the 1943 documentary, it is the radio operator Robert Hanson who does this and not the tail gunner.

36. - Modine sprays champagne:

Maybe for other bomber crews, but not the "Memphis Belle." Again, the 1943 film shows the crewmen lifting Robert Morgan on their shoulders to give the nose art a kiss and a little pat on her bottom. Here again many have said that they wish this scene could have been recreated for the 1989 picture.

37. - The closing shot:

Pay attention here. This is a strong visual image of what these bases looked like. A smoking bomber back from its mission, badly damaged, and surrounded by ground crew, officers, pilots, and a good arrangement of ground support vehices. One can almost smell this scene.

What happened and what didn't? Don't be too hard on this film. While the producers took some liberties with engineered drama and the script, these were necessary to sell the film at the box office. And while it does not entirely represent the story of the "Memphis Belle" accurately, it does a great service to those who wish to understand and feel the experiences of the most significant air battles the world has ever seen. The film should be appreciated for this approach. After all, how should four and a half years of combat be condensed into ninety minutes of movie making?

During the filming of "12 O'Clock High," this B-17 just scrapes the roofs off these trucks. Note the stenciling on the vehicles—U.S. Air Force markings from the late 50s and early 60s! Below that are photos from two different movies, but both planes are painted like the "Memphis Belle." In what must be the most unbelievable buzz job ever caught on film, this bomber is literally inches off the ground during the filming of "The War Lover." On the bottom a scene for "The Thousand Plane Raid" is being set. Clearly film makers have always been endeared to the lore of the B-17, especially as it relates to the most well known Fort of them all, the "Memphis Belle."

16

Flight Procedures for the Boeing B-17 Flying Fortress

The Crew

Your assignment to the B-17 airplane means that you are no longer just a pilot. You are now an airplane commander, charged with all the duties and responsibilities of a command post.

You are now flying a 10-man weapon. It is your airplane, and your crew. You are responsible for the safety and the efficiency of the crew at all times—not just when you are flying and fighting, but for the full 24 hours of every day while you are in command.

Your crew is made up of specialists. Each man—whether he is the navigator, bombardier, engineer, radio operator, or one of the gunners—is an expert in his line.

But how well he does his job, and how efficiently he plays his part as a member of your combat team, will depend to a great extent on how well you play your own part as airplane commander.

Get to know each member of your crew as an individual. Know his personal idiosyncrasies, his capabilities, his shortcomings. Take a personal interest in his problems, his ambitions, his need for specific training.

See that your men are properly quartered, clothed, and fed. There will be many times when your airplane and crew are away from the home base, when you may even have to carry your interest to the extent of financing them yourself. Remember always that you are the commanding officer of a miniature army, and that morale is one of the biggest problems for the commander of any army, large or small.

Your success as the airplane commander will depend in a large measure on the respect, confidence, and trust which the crew feels for you. It will depend also on how well you maintain crew discipline.

Your position commands obedience and respect. This does not mean that you have to be stiff-necked, overbearing, or aloof. Such characteristics most certainly will defeat your purpose.

Be friendly, understanding but firm. Know your job and, by the way you perform your duties daily, impress upon the crew that

With her number one engine dead and the prop feathered, this Fort returns to base after a mission. No doubt crew cohesion and training are the reason for many successful returns after harrowing raids on the enemy.

you do know your job. Keep close to your men and let them realize that their interests are uppermost in your mind. Make fair decisions, after due consideration of all the facts involved; but make them in such a way as to impress upon your crew that your decisions are to stick.

Crew discipline is vitally important, but it need not be as difficult a problem as it sounds. Good discipline in air crew breeds comradeship and high morale, and the combination is unbeatable.

You can be a good CO and still be a regular guy. You can command respect from your men, and still be one of them.

"To associate discipline with informality, comradeship, a leveling of rank, and at times a shift in actual command away from the leader, may seem paradoxical," says a brigadier general, formerly a group commander in the VIII Bomber Command. "Certainly it isn't down the military groove. But it is discipline just the same,-and the kind of discipline that brings success in the air."

Train your crew as a team. Keep abreast of their training. It won't be possible for you to follow each man's courses of instruction, but you can keep a close check on his record and progress.

Get to know each man's duties and problems. Know his job and try to devise ways and means of helping him to perform it more efficiently.

Each crewmember naturally feels great pride in the importance of his particular specialty. You can help him to develop his pride to include the manner in which he performs that duty. To do that you must posses and maintain a thorough knowledge of each man's job and the problems he has to deal with in the performance of his duties.

The copilot is the executive officer—your chief assistant, understudy, and strong right arm. He must be familiar with every one of your duties—both as pilot and as airplane commander—to be able to take over and act in your place at any time.

He must be able to fly the airplane in all conditions as well as you would fly it yourself.

He must be extremely efficient in engine operation, and know instinctively what to do to keep the airplane flying smoothly, even though he is not handling the controls.

He must have a thorough knowledge of cruising control data, and know how to apply it at the proper time.

He is also the engineering officer aboard the airplane, and maintains a complete log of performance data.

He must be a qualified instrument pilot.

He must be able to fly good formation in any assigned position, day or night.

He must be qualified to navigate by day or at night, by pilotage, dead reckoning, or by use of radio aids.

He must be proficient in the operation of all radio equipment located in the pilot's compartment.

In formation flying, he must be able to make engine adjustments almost automatically.

He must be prepared to take over on instruments when the formation is climbing through an overcast, thus enabling you to watch the rest of the formation.

Always remember that the copilot is a fully trained, rated pilot just like yourself. He is subordinate to you only by virtue of your position as the airplane commander. The B-17 is a lot of airplane; more airplane than any one pilot can handle alone over a long period of time. Therefore, you have been provided with a second pilot who will share the duties of flight operation.

Treat your co-pilot as a brother pilot. Remember that the more proficient he is as a pilot, the more efficiently he will be able to perform the duties of the vital post he holds as your second in command.

Be sure that he is allowed to do his share of the flying, in the pilot's seat, on takeoffs, landings, and on instruments.

The importance of the copilot is eloquently testified by airplane commanders overseas. There have been many cases where the pilot has been injured or killed in flight and the co-pilot has taken command of both airplane and crew, completed the mission, and returned safely to the home base. Usually the co-pilots who have distinguished themselves under such conditions have been copilots who have respected and trained by the airplane commander as pilots.

Bear in mind that the pilot in the right-hand seat of your airplane is preparing himself for an airplane commander's post too. Allow him every chance to develop his ability and to profit by your experience.

The navigator's job is to direct your flight from departure to destination and return. He must know the exact position of the airplane at all times.

Navigation is the art of determining geographic positions by means of (a) pilotage, (b) dead reckoning, (c) radio, or (d) celestial navigation, or any combination of these 4 methods. By any one combination of methods the navigator determines the position of the airplane in relation to the earth.

Pilotage is the method of determining the airplane's position by visual reference to the ground. The importance of accurate pilotage cannot be over-emphasized. In combat navigation, all bombing targets are approached by pilotage, and in many theatres the route is maintained by pilotage. This requires not merely the vicinity type, but pin-point pilotage. The exact position of the airplane must be not within 5 miles but within 1/4 of a mile.

The navigator does this by constant reference to ground speeds and ETAs established for points ahead, the ground, and to his maps and charts. During the mission, so long as he can maintain visual contact with the ground, the navigator can establish these pin-point positions so that the exact track of the airplane will be known when the mission is completed.

Dead-reckoning is the basis of all other types of navigation. For instance, if the pilot is doing pilotage and computes ETAs for points ahead, he is using dead-reckoning.

Dead-reckoning determines the position of the airplane at any given time by keeping an account of the track and distance flown over the earth's surface from the point of departure or the last known position.

Dead reckoning can be subdivided into two classes:

1. Dead reckoning as a result of a series of known positions obtained by some other means of navigation.

For example, you as a pilot start on a mission from London to Berlin at 25,000 feet. For the first hour, the navigator keeps track by pilotage; at the same time recording the heading and airspeed which you are holding. According to the plan at the end of the first hour the airplane goes above the clouds, thus losing contact with the ground. By means of dead-reckoning from his last pilotage point, the navigator is able to tell the position of the airplane at any time. The first hour's travel has given him the wind prevelant at altitude, and the track and groundspeed being made. By computing track and distance from the last pilotage point, he can always tell the position of the airplane. When your airplane comes out of the clouds near Berlin, the navigator will have a very close approximation of his exact position, and will be able to pick up pilotage points quickly.

2. Dead-reckoning as a result of visual references other than pilotage. When flying over water, desert, or barren land, where no reliable pilotage points are available, accurate DR navigation still can be performed. By means of the drift meter the navigator is able to determine drift, the angle between the heading of the airplane and its track over the ground. The true heading of the airplane is determined by application of compass error to the compass reading. The true heading plus or minus the drift (as read on the drift meter) gives the track of the airplane. At a constant airspeed drift on 2 or more headings will give the navigator data necessary to obtain the wind by use of his computer. Ground speed is computed easily once the wind, heading, and airspeed are known. So, by constant recording of true heading, true airspeed, drift, and groundspeed, the navigator is able to determine accurately the position of the airplane at any given time. For greatest accuracy, the pilot must maintain constant courses and airspeeds. If course or airspeed is changed, notify the navigator so he can record these changes.

Radio

Radio navigation makes use of various radio aids to determine position. The developement of many new radio devices has increased the use of radio in combat zones. However, the ease with which

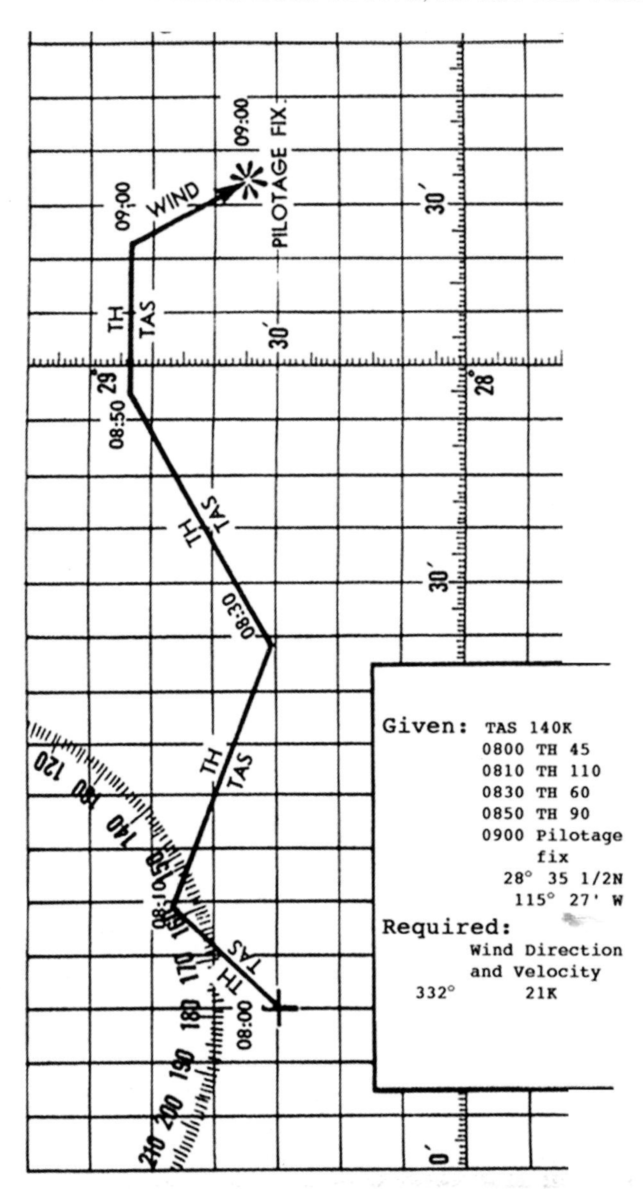

radio aids can be jammed, or bent, limits the use of radio to that of a check on DR and pilotage. The navigator in conjunction with the radio man is responsible for all radio procedures, approaches, etc., that are in effect in the theatre.

Celestial

Celestial navigation is the science of determining position by reference to 2 or more celestial bodies. The navigator uses a sextant, accurate time, and many tables to determine what he calls a line of position. Actually, this line is part of a circle on which the altitude of that particular body is constant for that instant of time. An intersection of 2 or more of these lines gives the navigator a fix. These fixes can be relied on as being accurate within approximately 10 miles. One reason for inaccuracy is the instability of the airplane as it moves through space causing acceleration of the sextant bubble (a level denoting the horizontal). Because of this acceleration, the navigator takes observations over a period of time so that the acceleration error will cancel out to some extent. If the navigator tells the pilot when he wishes to take an observation, extremely careful flying on the part of the pilot during the few minutes it takes to make the observation will result in much greater accuracy. Generally speaking the only celestial navigation used by a combat crew is used during the delivery flight to the theatre. But in all cases celestial navigation is used as a check on dead-reckoning and pilotage except where celestial navigation is the only method available, such as long over-water flights, etc.

Instrument Calibration

Instrument calibration is an important duty of the navigator. All navigation depends directly on the accuracy of his instruments. Correct calibration requires close cooperation and extremely careful flying by the pilot. Instruments to be calibrated include the altimeter, all compasses, airspeed indicators, alignment of the astrocompass, astograph, and drift meter, and check on the navigator's sextant and watch.

Pilot-Navigator Pre-Flight Planning

1. Pilot and navigator must study flight plan of the route to be flown and select alternate airfields.
2. Study the weather with the navigator. Know what weather you are likely to encounter. Decide what action is to be taken. Know the weather conditions at the alternate airfields.
3. Inform your navigator at what altitude and airspeed you wish to fly so that he can prepare his flight plan.

G model B-17s of the 401st Squadron / 91st Bombardment Group (Heavy) sit canted on their Bassingbourn taxiway during their pre-takeoff mag checks. The first plane shown here (LL-G) went missing in action during a raid to Peenemunde on 25 August 1944. All nine of her crew were killed.

4. Learn what type of navigation the navigator intends to use; pilotage, dead-reckoning, radio, celestial, or a combination of all methods.
5. Determine check points; plan to make radio fixes.
6. Work out an effective communication method with your navigator to be used in flight.
7. Synchronize your watch with your navigator's.

Pilot-Navigator In Flight

1. Constant course - For accurate navigation, the pilot must fly a constant course. The navigator has many computations and entries to make in his log. Constantly changing course makes his job more difficult. A good navigator is supposed to be able to follow the pilot, but he cannot be taking compass readings all the time.
2. Constant airspeed must be held as nearly as possible. This is as important to the navigator as is constant course in determining position.
3. Precision flying by the pilot greatly affects the accuracy of the navigator's instrument readings, particularly celestial readings. A slight error in instrument reading can cause considerable error in determining positions. You can help the navigator by providing as steady a platform as possible from which he can take readings. The navigator should inform you when he intends to take readings so that the airplane can be leveled off and be flown as smoothly as possible, preferably by using the automatic pilot. Do not allow yur navigator to be disturbed while he is taking celestial readings.
4. Notify the navigator of any change in flight, such as change in altitude, course, or airspeed. If change in flight plan is to be made, consult the navigator. Talk over the proposed change so he can plan the flight and advise you about it.
5. If there is doubt about the position of the airplane, pilot and navigator should get together, refer to the navigator's flight log, talk the problem over and decide together the best course of action to take.
6. Check your compasses at intervals with those of the navigators, noting any deviation.
7. Require your navigator to give position reports at intervals.
8. You are ultimately responsible for getting the airplane to its destination. Therefore, it is your duty to know your position at all times.
9. Encourage your navigator to use as many navigation methods as possible as a means of double-checking.

Pre-Flight Critique

After every flight get together with the navigator and discuss the flight and compare notes. Go over the navigators log. If there have been serious navigational errors, discuss them with the navigator and determine their cause. If the navigator has been at fault, caution him that it is his job to see that the same mistake does not occur again. If the error has been caused by faulty instruments, see that they are corrected before another navigation mission is attempted. If your flying has contributed to inaccuracy in navigation, try to fly a better course next time.

Miscellaneous Duties

The navigator's primary duty is navigating your airplane with a high degree of accuracy. But as a member of the team, he must also have a general knowledge of the entire operation of the airplane.

He has a .50-cal. machine gun at his station, and he must be able to use it skillfully and to service it in emergencies.

He must be familiar with the oxygen system, know how to operate the turrets, radio equipment, and fuel transfer system.

He must know the location of all fuses and spare fuses, lights and spare lights, affecting navigation.

He must be familiar with emergency procedures, such as the maual operation of landing gear, bomb bay doors, and flaps, and the proper procedures for crash-landings, ditchings, bailout, etc.

The Bombardier

Accurate and effective bombing is the ultimate purpose of your entire airplane and crew. Every other function is preperatory to hitting and destroying the target.

That is your bombardier's job. The success or failure of the mission depends upon what he accomplishes in that short interval of the bombing run.

When the bombardier takes over the airplane for the run on the target, he is in absolute command. He will tell you what he wants done and until he tells you "Bombs Away", his word is law.

A great deal, therefore, depends on the understanding between bombardier and pilot. You expect your bombardier to know his job when he takes over. He expects you to understand the problems involved in his job, and to give him full cooperation. Teamwork between pilot and bombardier is essential.

Under any given set of conditions - groundspeed, altitude, direction, etc.-there is only one point in space where a bomb may be released from the airplane to hit a pre-determined object on the ground.

There are many things with which a bombardier must be thoroughly familiar in order to release his bombs at the right point to hit this predetermined target.

He must know and understand his bombsight, what it does, and how it does it.

He must thoroughly understand the operation and upkeep of his bombing instruments and equipment.

He must know that his racks, switches, controls, releases, doors, linkage, etc., are in first-class operating condition.

He must understand the automatic pilot as it pertains to bombing.

He must know how to set it up, make any adjustments, and minor repairs while in flight.

He must know how to operate all gun positions in the airplane.

He must know how to load and clear simple stoppages and jams of machine guns while in flight.

He must be able to load and fuse his own bombs.

He must understand the destructive power of bombs and must know the vulnerable spots on various types of targets.

He must understand the bombing problem, bombing probabilities, and bombing errors, etc.

He must be thoroughly versed in target identification, and in aircraft identification.

The bombardier should be familiar with all duties of the crew and should be able to assist the navigator in case the navigator becomes incapacitated.

For the bombardier to be able to do his job, the pilot of the aircraft must place the aircraft in the proper position to arrive at a point on a circle about the target from which the bombs can be released to hit the target.

Consider the following conditions which affect the bomb dropped from an airplane: -

1. Altitude: Controlled by the pilot. Determines the length of time the bomb is sustained in flight and affected by atmospheric conditions, thus affecting the range (forward travel of the bomb) and deflection (distance the bombs drifts in a crosswind with respect to the airplane's ground track).
2. True airspeed: Controlled by the pilot. The measure of the speed of the airplane through the air. It is this speed which is imparted to the bomb and which gives the bomb its initial forward velocity and, therefore, affects the trail of the bomb, or the distance the bomb lags behind the airplane at the instant of impact.
3. Bomb Ballistics: Size, shape and density of the bomb, which determines its air resistance. Bombardier uses bomb ballistics tables to account for type of bomb.
4. Trail: Horizontal distance the bomb is behind the airplane at the instant of impact. This value, obtained from bombing tables, is set in the sight by the bombardier. Trail is affected by altitude, airspeed, bomb ballistics and air density, the first 2 factors being controlled by the pilot.
5. Actual time of fall: Length of time the bomb is sustained in air from instant of release to instant of impact. Affected by altitude, type of bomb and air density. Pilot controls altitude to obtain a definite actual time of fall.
6. Groundspeed: The speed of the airplane in relation to the earth's surface. Groundspeed affects the range of the bomb and varies with the airspeed, controlled by the pilot. Bombardier enters groundspeed into the bombsight through synchronization on the target. During this process the pilot must maintain the correct altitude and constant airspeed.
7. Drift: Determined by the direction and velocity of the wind, which determines the distance the bomb will travel downwind from the airplane from the instant the bomb is released to the instant of im-

pact. Drift is set on the bombsight by the bombardier during the process of synchronization and setting up course.

These conditions indicate that the pilot plays an important part in determining the proper point of release of the bomb. Moreover, throughout the course of the run, as explained below, there are certain preliminaries and techniques which the pilot must understand to insure accuracy and minimum loss of time.

Prior to takeoff the pilot must ascertain that the airplanes flight instruments have been checked and found accurate. These are the altimeter, airspeed indicator, free air temperature gauge, and all gyro instruments. These instruments must be used to determine accurately, the airplane's attitude.

When forced to leave the plane while over enemy territory, the sworn oath to protect the secret bombsight was very much on the mind of the bombardier. There were several options available to him when it came time to destroy the device. He could throw it out of the plane, that is, if they were high enough to ensure destruction when it hit the ground, and if he could manage to dismount the sight and heft its more than forty pounds toward an exit in the gyrating plane. He could wait until a crash landing, then burn it up with a timed thermite charge. Or, like the honor bestowed upon a dying calvaryman's horse, draw his .45 automatic sidearm and place the regulation three rounds through it. The first through its lenses and optics, and then two more through the rate-end mechanism—the heart of the bombsight.

The Bombardier's Oath

Mindful of the secret trust about to placed in me by my Commander in Chief, the President of the United States, by whose direction I have been chosen for bombardier training...and mindful of the fact that I am about to become guardian of one of my country's most priceless military assets, the American bombsight...I do here, in the presence of Almighty God, swear by the bombardier's Code of Honor to keep inviolate the secrecy of any and all confidential information revealed to me, and further to uphold the honor and integrity of the Army Air Forces, if need be, with my life itself.

The pilot's Preliminaries

The autopilot and PDI should be checked for proper operation. It is very important that PDI and autopilot function perfectly in the air; otherwise, it will be impossible for the bombardier to set up an accurate course on the bombing run. The pilot should thoroughly familiarize himself with the function of both the C-1 autopilot and PDI.

If the run is to be made on the autopilot, the pilot must carefully adjust the autopilot before reaching the target area. The autopilot must be adjusted under the same conditions that will exist on the bombing run over the target. For this reason the following factors should be taken into consideration and duplicated for initial adjustment.

1. Speed, altitude, and power settings at which run is to be made.
2. Airplane trimmed at this speed to fly hands off with bomb bay doors open.

The same condition will exist during the actual run, except that changes in load will occur before reaching the target area because of gas consumption. The pilot will continue making adjustments to correct for this by disengaging the autopilot elevator control and re-trimming the airplane, then re-engaging and adjusting the autopilot trim of the elevator.

Setting up the Autopilot

One of the most important items in setting up the autopilot for bomb approach is to adjust the turn compensation knobs so that a turn made by the bombardier will be coordinated and at constant altitude. Failure to make this adjustment will involve difficulty and delay for the bombardier in establishing an accurate course during the run - with the possibility that the bombardier may not be able to establish a proper course in time, the result being considerably large deflection errors in point of impact.

Uncoordinated turns by the autopilot on the run cause erratic lateral motion of the course hair of the bombsight when sighting on a target. The bombardier in setting up course must eliminate any lateral motion of the fore-and-aft hair in relation to the target before he has the proper course set up. Therefore, any erratic motion of the course hair requiresan additional correction by the bombardier which would not be necessary if autopilot was adjusted to make coordinated turns.

Use of the PDI: The same is true if PDI is used on the bomb run. Again coordinated smooth turns by the pilot become an essential part of the bomb run. In addition to added course corrections necessitated by uncoordinated turns, skidding and slipping introduce small changes in airspeed affecting synchronization of the bombsight in the target. To help the pilot flying the run on PDI, the airplane should be trimmed to fly practically hands off.

Assume that you are approaching the target area with autopilot properly adjusted. Before reaching the initial point (beginning of bomb run) there is evasive action to be considered. Many different types of evasive tactics are employed, but from experience it has been recommended that the method of evasive action be left up to the bombardier, since the entire aircraft patterns fully visible to the bombardier in the nose.

Evasive action: Changes in altitude necessary for evasive action can be coordinated with the bombardier's changes in direction at specific intervals. This procedure is helpful to the bombardier since he must select the initial point at which he will direct the airplane onto the briefed heading for the beginning of the bomb run.

Should the pilot be flying the evasive action on PDI, (at the direction of the bombardier) he must know the exact position of the initial point for beginning the run, so that he can fly the airplane to that point and be on the briefed heading. Otherwise there is the possibility of beginning to run too soon, which increases the airplanes vulnerability, or beginning the run too late, which will affect the accuracy of the bombing. For best results the approach should be planned so the airplane arrives at the initial point on the briefed heading and at the assigned bombing altitude and airspeed.

At this point the bombardier and pilot as a team should exert an extra effort to solve the problem at hand. It is now the bombardiers responsibility to take over the direction of flight, and give directions to the pilot for the operations to follow. The pilot must be able to follow the bombardiers directions with accuracy and minimum loss of time, since the longest possible bomb run seldom exceeds 3 minutes. Wavering and indecision at this moment are disastrous to the success of any mission, and during the crucial portion of the run, flak and fighter opposition must be ignored if bombs are to hit the target. The pilot and bombardier should keep each other informed of anything which may affect the successful completion of the run.

Holding a level: Either before or during the run. the bombardier will ask the pilot for a level. This means that the pilot must accurately level his airplane with his instruments (ignoring the PDI). There should be no acceleration of the airplane in any direction, such as an increase or decrease in airspeed, skidding or slipping, gaining or losing altitude.

For the level the pilot should keep a close check on his instruments, not by feel or watching the horizon. Any acceleration of the airplane during this moment will affect the bubbles (through centrifugal force) on the bombsight gyro, and the bombardier will not be able to establish an accurate level.

For example, assume that an acceleration occurred during the moment that bombardier was accomplishing a level on the gyro. A small increase in airspeed or a small skid, hardly perceptible, is sufficient to shift the gyro liquid bubble 1° or more. An erroneous tilt of 1° on the gyro will cause an error of approximately 440 feet in the point of impact of a bomb dropped from 20,000 feet, the direction of error depending on the direction of tilt of gyro caused by the erroneous bubble reading.

Holding altitude and airspeed: As the bombardier proceeds to set up his course (synchronize), it is absolutely essential that the pilot maintain the selected altitude and airspeed within the closest possible limits. For every additional 100 feet above the assumed 20,000-foot bombing altitude, the bombing error will increase approximately 30 feet, the direction of error being over or "long". For erroneous airspeed which creates difficulty in synchronization on the target, the bombing error will be approximately 170 feet for a 10 mph change in airspeed. Assuming the airspeed was 10 mph in excess, from 20,000 feet, the bomb impact would short 170 feet.

The pilot's responsibility to provide a level and to maintain selected altitude and airspeed within the closest limits cannot be over-emphasized.

If the pilot is using PDI (at the direction of the bombardier) instead of autopilot, he must be thoroughly familiar with the corrections demanded by the bombardier. Too large a correction or too small a correction, too soon or too late, is as bad as no correction at all. Only through prodigious practice flying with the PDI can the pilot become proficient to a point where he can actually perform a coordinated turn, the amount and speed necessary to balance the bombardier's signal from the bombsight.

Erratic airspeeds, varying altitudes, and poorly coordinated turns make the job of establishing course and synchronizing doubly difficult for both pilot and bombardier, because of the added necessary corrections required. The resulting bomb impact will be far from satisfactory.

After releasing the bombs the pilot or bombardier may continue evasive action - usually the pilot so that the bombardier may man his guns.

Three important areas of the "Memphis Belle." At the left is the plane's autopilot, which mostly held altitude as well as heading. The PDI gauge is seen in the photo above. This aided the pilot during the bomb run. Some engine instruments are seen to the right. These are positioned in front of the co-pilot. The flight instruments are in front of the pilot on the panel.

The intervalometer panel of the "Memphis Belle." It is mounted on the left side of the bombardier's station and gave him critical information, including the status of the bombs in the bay racks, as well as airspeed and altitude.

The pilot using the turn control may continue to fly the airplane on autopilot in a position to be engaged by merely flipping the lock switches. This would provide potential control of the airplane in case of an emergency.

Reducing circular error: One of the greatest assets towards reducing the circular error of a bombing squadron lies in the pilot's ability to adjust the autopilot properly, fly the PDI, and maintain the designated altitude and airspeeds during the bombing run. Reducing the circular error of a bombing squadron reduces the total number of aircraft required to destroy a particular target. For this reason both pilot and bombardier should work together until they have developed a complete understanding and confidence in each other.

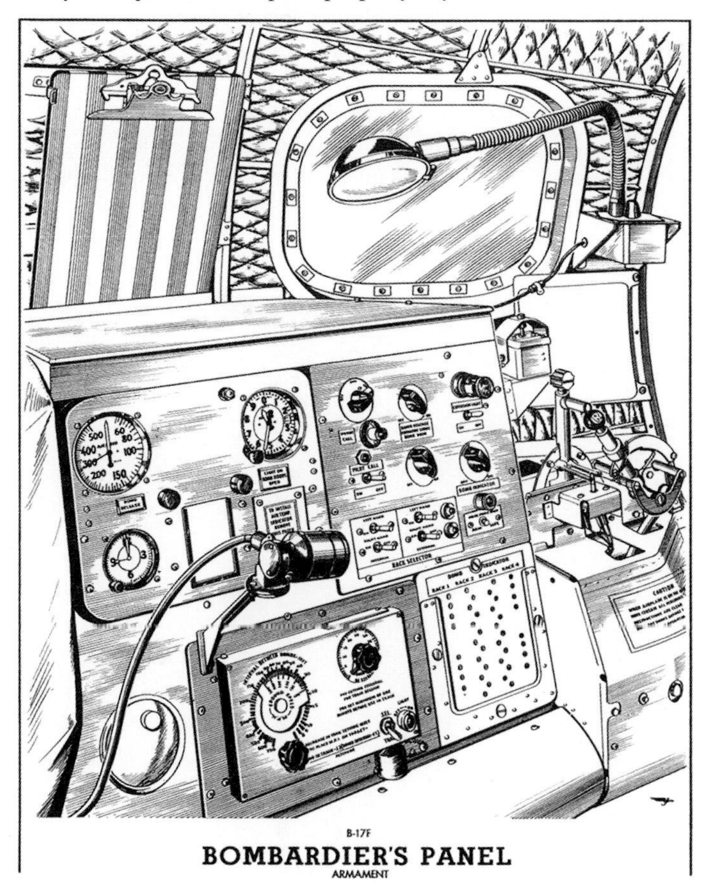

There is a lot of radio equipment in today's B-17's. There is one man in particular who is supposed to know all there is to know about this equipment. Sometimes he does, but often he doesn't. And when the radio operators defecincies do not become apparent until the crew is in the combat zone, it is then too late. Too often the lives of pilots and crew are lost because the radio operator has accepted his responsibility indifferently.

Radio is a subject that cannot be learned in a day. It cannot be mastered in 6 weeks, but sufficient knowledge can be imparted to the radioman during his period of training in the United States if he is willing to study. It is imperative that yu check your radio operators ability to handle his job before taking him overseas as part of your crew. To do this you may have to check the various departments to find any weakness in the radio operator's training and proficiency, and to aid the instructors in overcoming such weaknesses.

Training in various phases of the heavy bomber program is designed to fit each member of the crew for the handling of his jobs. The radio operator will be required to:

1. Render position reports every 30 minutes.
2. Assist the navigator in taking fixes.
3. Keep the liason and command sets properly tuned and in good operating order.
4. Understand from an operational point of view:
    (a) instrument landing
    (b) IFF
    (c) VHF

and other navigational aids equipment in the airplane.

5. Maintain a log.

In addition to being a radio operator, the radio man is also a gunner. During periods of combat he will be required to leave his watch at the radio and take up his guns. He is often required to

learn photography. Some of the best pictures taken in the Southwest Pacific were taken by radio operators. The radio operator who cannot perform his job properly may be the weakest member of your crew-and the crew is no stronger than its weakest member.

Top Turret / Flight Engineer

Size up the man who is to be your engineer. This man is supposed to know more about the airplane you are to fly than any other member of the crew.

He has been trained in the Air Force's specialized technical schools. Probably he has served some time as a crew chief. Nevertheless there may be some inevitable blank spots in his training which you, as a pilot and airplane commander, may be able to fill in.

Think back on your own training. In many courses of instruction you had a lot of things thrown at you from right and left. You had to concentrate on how to fly; and where your equipment was concerned you learned to rely more and more on the enlisted personnel, particularly the crew chief and flight engineer, to advise you about things that were not taught to you because of the lack of time and the arrangement of the training program.

Both pilot and engineer have a responsibility to work closely together to supplement and fill in the blank spots in each other's education.

To be a qualified combat engineer a man must know his airplane, his engines, and his armament equipment thoroughly. This is a big responsibility; the lives of the entire crew, the safety of the equipment, the success of the mission depend upon it squarely.

He must work closely with the co-pilot, checking engine operation, fuel consumption, and the operation of all equipment.

He must be able to work with the bombardier, and know how to cock, lock and load the bomb racks. It is up to you, the airplane commander, to see that he is familiar with these duties, and, if he is hazy concerning them, to have the bombardier give him special help and instruction.

He must thoroughly familiar with the armament equipment, and know how to strip, clean and re-assemble the guns.

He should have a general knowledge of radio equipment, and be able to assist in tuning transmitters and receivers.

Your engineer should be your chief source of information concerning the airplane. He should know more about the equipment than any other crewmember-yourself included.

You, in turn, are his source of information concerning flying. Bear this in mind in all your discussions with your engineer. The more complete you can make his knowledge of the reasons behind every function of the equipment, the more valuable he will be as a member of the crew. Who knows? Someday that little bit of extra knowledge in the engineer's mind may save the day in some emergency.

Generally, in emergencies, the engineer will be the man to whom you turn first. Build up his pride, his confidence, his knowledge. Know him personally; check on the extent of his knowledge. Make him a man upon whom you can rely.

Gunners

The B-17 is a most effective gun platform, but its effectiveness can be either applied or defeated by the way the gunners in your crew perform their duties in action.

Your gunners belong to one of two distinct categories: turret gunners and flexible gunners.

The power turret gunners require many mental and physical qualities similar to what we know as inherent flying ability, since the operation of the power turret and gunsight are much like that of airplane flight operation.

While the flexible gunners do not require the same delicate touch as the turret gunner, they must have a fine sense of timing and be familiar with the rudiments of exterior ballistics.

All gunners should be familiar with the coverage area of all gun positions, and be prepared to bring the proper gun to bear as the conditions may warrant.

They should be experts in aircraft identification. Where the Sperry turret is used, failure to set the target dimension dial properly on the K-type sight will result in mis-calculation of range.

They must be thoroughly familiar with the Browning aircraft machine gun. They should know how to maintain the guns, how to clear jams and stoppages, and how to harmonize the sights with the guns.

While participating in training flights, the gunners should be operating their turrets constantly, tracking with the flexible guns even when actual firing is not practical. Other airplanes flying in the vicinity offer excellent tracking targets, as do automobiles, houses, and other ground objects during low altitude flights.

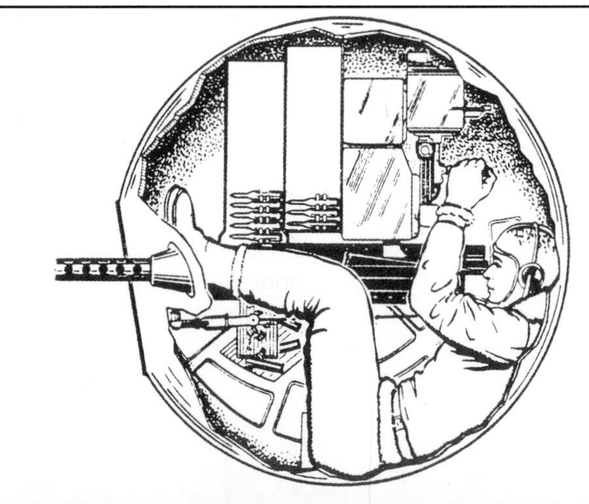

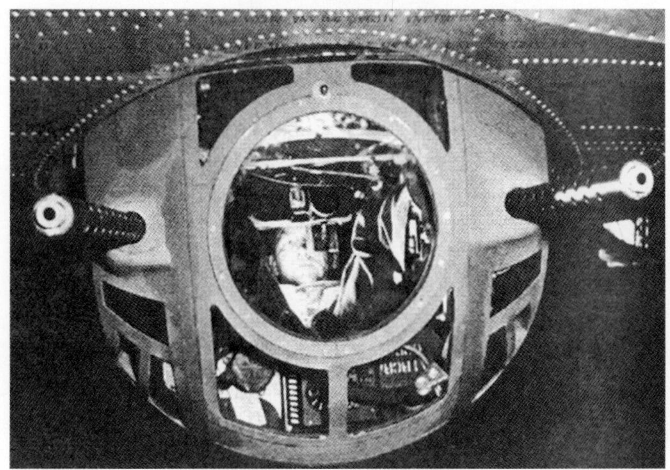

The Sperry Company of Long Island, New York, designed the famous ball turret during the late 1930s. The company actually contracted with a firm named Briggs to build the unit through the War, but it retained the name Sperry. The type A-2 half-inch thick armor-plated sphere was only 44 inches in diameter and weighed in at 850 pounds. Framework inside the bomber supported the turret, which hung below the center of the plane. The guns were fixed to the turret and moved through azimuth and elevation as a whole through electro-hydraulic actuators. The ball could rotate through 90 degrees of elevation and 360 degrees of azimuth. The rate of spin in azimuth was 45 degrees per second. In the event of power failure, the gunner could move the turret through manual handcranks—albiet not very quickly. Turret rotation was controlled by hand levers and foot pedals. Enemy fighters were sighted through a Sperry computing gunsight which the gunner would use his left eye to peer through. Naturally, the gunner was usually selected primarily because of his small physical stature. As can be seen, there is almost no room for the gunner to carry his parachute along in the ball with him. In the event of a bail-out, the gunner would have to get back into the plane, which was likely gyrating, perhaps even upside down, find his parachute, clip it on, and then get out of the bomber!

The importance of teamwork cannot be over-emphasized. One poorly trained gunner, or one man not on the alert, can be the weak link as a result of which the entire crew may be lost.

Keep the interest of your gunners alive at all times. Any form of competition among the gunners themselves should stimulate interest to a high degree.

Finally each gunner should fire the guns at all stations to familiarize himself with other man's position and to insure knowledge of operation in the event of an emergency.

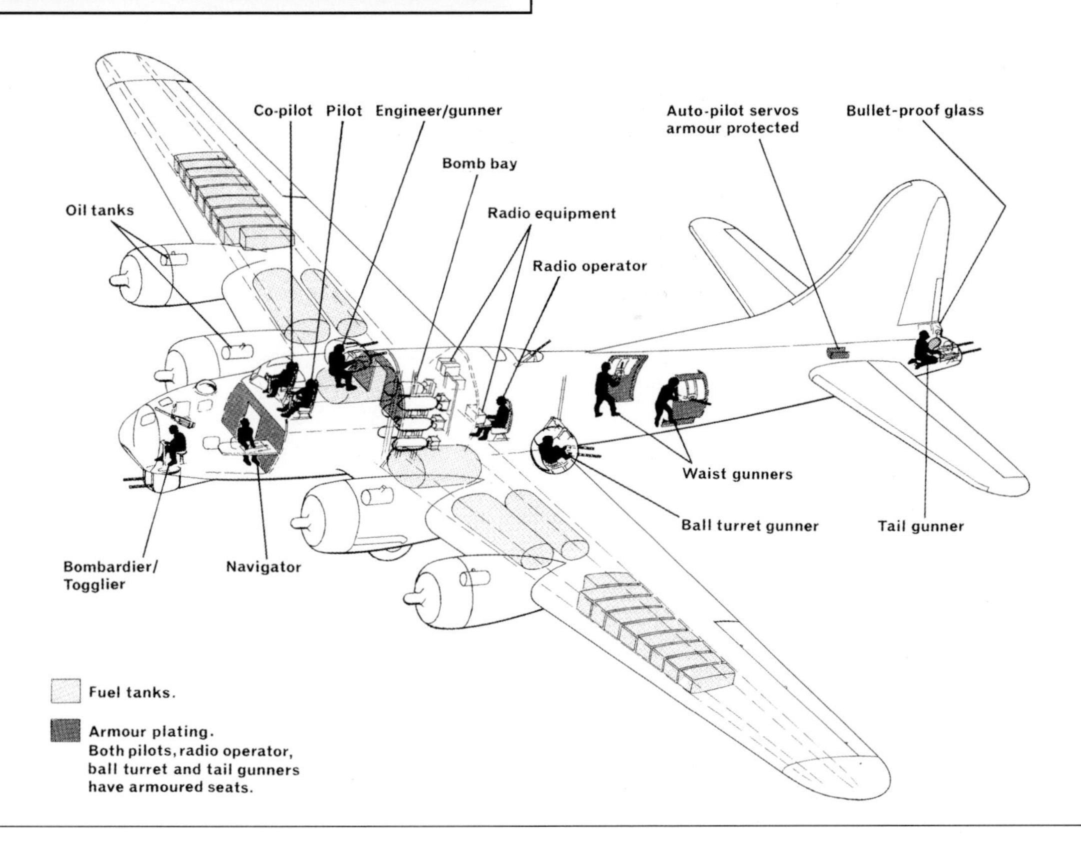

Flying the Fortress

Before the student ever takes to the left or right seat of the B-17, he must understand what a Flying Fortress is. Very simply, this is a bombardment type airplane. Basically, a large dump truck with wings that has a very intricate yet beautifully simple arrangement of innards that allows the ship to live and breathe.

Her nerves are made up of a network of wire harnesses that serve the intercom, turrets, lights, heaters, radios, and many of the flight instruments. At the same time the B-17 needs blood to survive, and she has several types. Grade 100/130 high-octane aviation fuel to feed her four hungry engines. More than 140 gallons of engine oil to lubricate them. Hydraulic fluid to operate her brakes and cowl flaps. Glycol to be slung from her propeller hubs down the blades to free accumulating ice that would rob precious thrust.

Engine turbo-superchargers help the powerplants lift her more than 70,000 pounds into the air. Then they help the engines breathe properly when flying six miles high. There are the heating, vacuum, oxygen, and de-icing systems that must be understood. She protects herself with the sting of thirteen heavy-barreled fifty-caliber machine guns, and can deliver an offensive blow whenever and just about wherever it is needed.

This is a four-engined, midwing monoplane almost 75 feet long, and to get her inside, a door 105 feet wide by twenty feet high is needed. With a crew of ten men and a maximum bomb load of 12,000 pounds, the B-17 is pulled through the sky by her 4,800 horses inside the four nine-cylinder engines out on her massive wings.

So complex is this machine that, at first, the military was not very interested. There was not much confidence in a plane that was thought to be way too much for her crew to ever handle.

After careful study is made of the ship's electrical, hydraulic, fuel, vacuum, oxygen, oil, heating, de-icing, and various other systems and networks aboard the Flying Fortress, and after completing some basic flight training in both primary and advanced trainers, it is time to begin learning how to fly what was, in her day, the most advanced bomber ever built.

From an engineer's point of view, the works of the B-17 are as meticulous as a fine Swiss watch. The redundancy of her critical systems was and still is a marvel of engineering. Many do not believe even today that none of the flight control surfaces were boosted in any way. All axes of flight are manipulated with nothing more than the muscles in the arms and legs of her pilots. Even the engine controls are of cable and pulley technology.

Seemingly archaic by today's cockpits, the automatic pilot did a fine job of flying the B-17. Even the marvel of automatic engine controls that would take over and assume pre-determined settings should the controls and/or cables be shot away or damaged in flight. In cases such as this the stricken engine or engines, as the case may be, would automatically set the throttle wide open; superchargers would set at 65% power; the intercoolers would go immediately to a setting of cold; and props to 1,850 rpm.

There are simply too many things to check and re-check before any flight aboard a plane such as the B-17 is made. Checklists became standard aboard military aircraft following the loss of the Boeing model 299 (one of the first B-17s) at Wright field, Ohio, in the late 1930s. A control surface lock was left engaged during the takeoff and the plane crashed.

Things like fastening a harness, plugging in a headset, adjusting the seat and rudder pedals, and plugging in the oxygen mask to the proper regulator were easy to remember and perform. But with

The B-17F cockpit mock-up shown here is actually a series of photographs that have been enlarged and mounted on panels at appropriate points so that the whole set-up looks for all the world like a cockpit model. The panels were designed to assist instructors at the Hobbs Army Airfield, New Mexico, in teaching pilots the operation and flying characteristics of the aircraft. This method of instruction was designed and originally used at Luke field, Arizona. It was widely recognized and even suggested by many that it be made a standardized training procedure. This teaching method was supplemented by other mock-ups, charts, diagrams, training films, film strips, and other training aids.

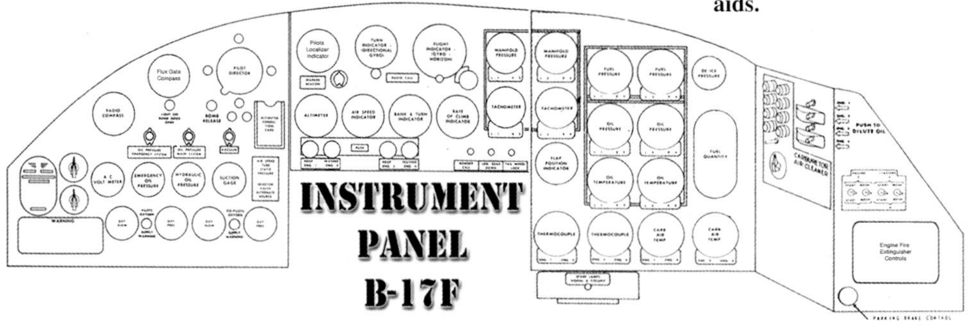

Instrument panels varied slightly among the various models of the B-17.

the thought of a combat mission before them, or perhaps just the fallibility of the human mind, the items that needed to be gone over could easily be skipped or forgotten. The checklist was a critical part of each mission. Failure to check all systems could and did result in catastrophe for many crews.

With the list on his lap and well before engine start, the pilot would track his thumb down the sheet and go over every item. Some pilots made a ritual of this task, and actually performed it twice or more.

The exchange between the pilot and copilot after reaching the flight deck of the B-17 typically went by a precise routine. Of course, there has been much that has happened since the time when a young sergeant awoke the crew hours earlier. The aircraft had been on the hardstand receiving hundreds of checks by the ground crew. Ammunition trucks delivered the serviced machine guns and fuel had been topped off. The crew had dressed and eaten their breakfast after briefing. Some would skip the morning meal, not wanting to have to relieve themselves later during the raid.

The crew arrived at their plane and began the common checks of their equipment. All unnecessary items, such as anything that could mark them as American would be left back at the base. They had their silk escape maps, foreign currency, and their fake ID pictures with them. They checked every item in their stations. Most often it was for the mutual protection of a fellow crewman or a friend on another plane that would fly alongside them. No one wanted to carry the guilt that maybe the loss of a friend or comrade might be their fault. A somewhat fatalistic approach to missions often found these airmen caring more about the lives of their compatriots than their own.

The officers of the crew completed their exterior walk around the bomber with the crew chief right alongside them. It was his job to assure the airplane commander that the bomber was more than ready to fly. Reports that had been made out from previous missions listed problems and combat damage. All the crews at one time or another would work throughout the entire night patching holes and fixing fuel lines, often cannibalizing other war weary planes for parts.

Several hours had passed from the time these men rolled out of bed until the time they were ready to make their way to the flight deck. While waiting for the signal to start engines, the all-important check list made its way into the lap of the pilot, and he would begin to scroll down carefully, noting each item with a verbal query.

Pilot: "Parking Brake." The co-pilot depresses the foot pedals and releases the knob near the fire extinguisher controls on the extreme right side of the instrument panel before he sets it again to assure that the brakes are set hard.

OFFICIAL A.A.F. PILOT'S CHECK LIST

B-17F AND B-17G AIRPLANES

For detailed instructions see Pilot's Handbook AN 01-20EF-1 or AN 01-20EG-1 in data case.

PILOT

BEFORE STARTING

1. Pilot's Pre-flight—Complete.
2. Form 1A, Form F, Weight and Balance—Checked.
3. Controls and Seats—Checked—Checked.
4. Fuel Transfer Valves and Switch—Off.
5. Intercoolers—Cold.
6. Gyros—Uncaged.
7. Fuel Shut-off Switches—Open.
8. Gear Switch—Neutral.
9. Cowl Flaps—Open Right—Open Left—Locked.
10. Turbos—Off.
11. Mixture Control—Checked.
12. Throttles—Closed.
13. High RPM—Checked.
14. Auto Pilot—Off.
15. De-icers and Anti-icers Wing and Prop.—Off.
16. Cabin Heat—Off.
17. Generators—Off.

STARTING ENGINES

1. Alarm Bell—Checked.
2. Wheel Chocks—In Place.
3. Fire Guard and Call Clear—Left—Right.
4. Master Switches—On.
5. Battery Switches and Inverters—On and Checked.
6. Parking Brakes—Hydraulic Check—On—Checked.
7. Booster Pumps — Pressure — On and Checked.
8. Carburetor Filters—Open.
9. Fuel Quantity—Gallons per tank.
10. Start Engines
    a. Fire Extinguisher Engine Selector—Checked.
    b. Prime—As Necessary.
    c. Energize
    d. Mesh
    e. Both Magnetos ON after one revolution.
11. Flight Indicator and Vacuum Pressures—Checked.
12. Radio—On.
13. Check Instruments—Checked.
14. Crew Report.
15. Radio Call and Altimeter—Set.

ENGINE RUN UP

1. Brakes—Locked.
2. Trim Tabs—Set.
3. Exercise Turbos (Hydraulic Regulators Only) and Props.
4. Check Generators—Checked and Off.
5. Run Up Engines.

AFTER LANDING

1. Hydraulic Pressure—OK.
2. Cowl Flaps—Open and Locked.
3. Turbos—Off.
4. Booster Pumps—Off.
5. Wing Flaps—Up.
6. Tail Wheel—Unlocked.
7. Generators—Off.

END OF MISSION

1. Dilute Engine Oil—When Necessary.
2. Engines—Cut.
3. Radio—On ramp.
4. Switches—Off.
5. Chocks.
6. Controls—Locked.
7. Form 1.

GO AROUND

1. High RPM and Power — High RPM.
2. Wing Flaps—Coming Up.
3. Power Reduction.
4. Wheel Check—OK Right—OK Left.

RUNNING TAKE OFF

1. Wing Flaps—Coming Up.
2. Power.
3. Wheel Check—OK Right—OK Left.

CO-PILOT

BEFORE TAKE OFF

1. Tail Wheel—Locked.
2. Gyro—Set.
3. Generators—On.

AFTER TAKE OFF

1. Wheels—Pilot's Signal.
2. Power Reduction.
3. Cowl Flaps.
4. Wheel Check—OK Right. OK Left.

BEFORE LANDING

1. Radio Call Altimeter—Set.
2. Crew Positions—OK.
3. Ball Turret—Stowed.
4. Auto Pilot—Off.
5. Booster Pumps—On.
6. Mixture Controls—Auto Rich.
7. Intercooler—Set.
8. Carburetor Filters—Open.
9. Wing De-icers—Off.
10. Cabin Heat—Off.
11. Landing Gear
    a. Visual—Down right
    Down left
    Tail wheel Down, Antenna In
    b. Light—OK
    c. Switch—Neutral
    d. Manual Check.
12. Hydraulic Pressure—OK. Valves closed.
13. RPM 2300—Set.
14. Turbos—Set.
15. Flaps 1/3 — 1/3 Down.

FINAL APPROACH

1. Flaps—Pilot's Signal.
2. High RPM—Pilot's Signal.

SUBSEQUENT TAKE OFF

1. Trim Tabs—Set.
2. Wing Flaps—Up.
3. Cowl Flaps—Open Right, Open Left.
4. High RPM—Checked.
5. Fuel—Gallons per tank.
6. Booster Pumps—On.
7. Turbos—Set.
8. Flight Controls—Unlocked.
9. Radio Call.

SUBSEQUENT LANDING

1. Landing Gear
    a. Visual—Down Right
    Down Left
    Tail Wheel Down
    b. Light—On.
    c. Switch—Neutral.
    d. Manual Check.
2. Hydraulic Pressure—OK.
3. RPM 2300—Set.
4. Turbo Controls—Set.
5. Wing Flaps 1/3 — 1/3 Down.
6. Radio Call.

FINAL APPROACH

1. Flaps—Pilot's Signal.
2. High RPM—Pilot's Signal.

30 January 1945

This list supersedes pilot's check lists of previous dates.

(Items not underlined co-pilot answers.)

Co-pilot: "Parking brake set - hydraulic pressure is 700 pounds."

Pilot: "Release flight control locks." On the floor of the flight deck between the two pilot seats are levers that release the pins that lock the elevators and the rudder. Then without pause he instinctively reaches for the aileron lock and removes the pin. The red banner attached to it is a caution which stands out so as not to be forgotten. There are clips on the yoke column to stow the pin. This is a good place to stow a pin as important as this. It is good to be reminded at a glance that this pin has been removed, freeing the ailerons.

Co-pilot: "Flight controls free." The windows of the cockpit have been slid back to allow the pilots to see to the rear of the bomber as the yoke is moved through its entire range of motion and rudder pedals are shifted from full left to full right rudder deflection. At the same time, the ailerons are visually scanned to see that when the wheel is turned hard left, the outer trailing edge of that wing moves up as it should. Then again the wheel is turned over hard all the way to the right and the left aileron moves down. At the same time the copilot is looking at his aileron which should be, and is, in the full up position. During this check, the flight crew is paying attention to any strange sounds or stiffness, basically anything they would not like in the range of movement. As the yoke is shoved all the way forward, the men gaze at the rear of the plane to assure that the elevators on the tail have gone all the way down. The opposite is checked as well, and as the wheel is pulled back into the laps of the men in the cockpit, the elevators reach their full up positions.

Pilot: "Fuel Transfer valves off, transfer valves neutral."

CoPilot: "Transfer valves and switches all off."

This is done so that when electrical power from the engine start comes on line, the fuel transfer pumps will not begin sending fuel from one wing into the other. If that were to happen, then there would be an overflow in the tanks that are already full.

Pilot: "Fuel shut-off switches open?"

Co-Pilot: "Fuel shut-off switches open." The purpose of these switches (one for each engine) is to give the pilots immediate control of the flow of fuel in the event of an engine fire.

Pilot: "Cowl flaps open?"

Co-Pilot: "Cowl flaps open and locked - left and right." Opening these flaps, which surround the nacelle just aft of each engine, will prevent the engine from overheating during the start and taxi. Also, in the event of an engine fire, the foam which will be sprayed into the affected engine will be drawn in between the engine's cylinders and baffle plates, thus saturating the fire.

Pilot: "Turbo controls off?"

CoPilot: "Turbo controls off." The sophisticated supercharger is never used during engine start. A backfire would more than likely damage the unit, which was very difficult to replace.

Pilot: "Mixture idle cut off."

Co-Pilot: "Mixture controls in engine off."

Pilot: "Prop controls full up, high rpm."

Co-Pilot: "Okay, Props up."

Pilot: "Throttles closed and set." At this time the pilot pulls the iron gate throttles on the central pedestal all the way back. Then for the engine that is to be started first, that lever is advanced, or "cracked," about an inch. This will give around 1,000 rpm once the engine fires.

Pilot: "Auto pilot off."

CoPilot: "Auto off." In the event that the auto pilot is left on, an accident could occur during takeoff, as the device would be trying to fly the bomber. The auto pilot could be easily overcome physically, but no one would want to be in a fight with it during the takeoff roll.

Pilot: "Carburetor air filters on." A yellow indicator lamp glows the moment the co-pilot throws the switch on the right side of his instrument panel.

Pilot: "Intercoolers."

CoPilot: "Intercoolers cold." Once the superchargers are running, they will be provided with cool air ensuring that they will not overheat. Once at altitude, the sliding intercooler levers will be moved into the "hot" position to prevent carburetor icing.

Pilot: "Propeller anti-ice switches?"

CoPilot: "Prop de-icers off."

Pilot: "Cabin heat off."

CoPilot: "Cabin heat off."

Pilot: "Generators off."

CoPilot: "Okay, generators are off." The generators, which are driven by the engine, are left off until the engines are running. The plane will then be on its own internal power and will no longer require the 24 volts provided by the auxillary power unit cart, which has been providing power for the entire start sequence.

Pilot: "Uncage flight indicator and gyro."

CoPilot: "Gyros okay."

Pilot: "Landing gear switch."

CoPilot: "Gear down and locked - safetied." If the switch for the landing gear was in the "up" position, then the instant the electrical system came alive, the mains would retract while the plane was sitting in its hardstand.

It is at this time that the crews aboard the bombers have a moment or two to think about the upcoming raid. Some of them say a prayer. The cold plane is armed and ready, and the crews are suited and in their takeoff positions. The wind can be heard coming through the open spaces in this aluminum warrior. The pilots check their watches for the "hack" time to start engines. All around the base, the aircrews prepare for what is ahead. Some are nervous, some are excited. Without warning a bomber across the field has started his engines. He is a minute or so early, but no one minds, and as if cued, many others begin their engine start rituals. Soon the field is swamped in the din of "round" sound. The pilot cranes his head out his window and hollers down to the crew chief.

Pilot: "Clear Chief?"

Crew Chief: "Clear! - God Bless!"

The pilot's left hand reaches immediately for three battery switches, and at the same time he uses his right hand to move the master bar ignition switch up and into its "on" position. The pilot gives the battery level gauges a quick glance, even though he knows that they are good because his crew chief has been working on the plane all night.

Pilot: "Fuel booster pump on," as he reaches for the switch. This will bring the pump on line, providing the engine with a mere 8 pounds of pressure in the fuel line feeding the carburetors. Also the fuel pump will prevent any possibility of a vapor lock in the fuel lines once the bomber reaches altitude.

He then reaches out his window and holds up one finger, making a spinning motion. This tells all those on the ground that he is about to start engine number one. A couple of the men on the ground crew position themselves near that engine with fire extinguishers, watching closely for any sign of a fire.

Pilot: "Start one!" The co-pilot, using his left hand, pushes the start switch down, holding it there for a little more than 15 seconds. From out on the left wing, the sound of the starter motor can be heard building its speed. While continuing to hold that switch, he moves other fingers from the same hand to the "mesh" switch and engages it. The starter clutch is pushed into the engine's flywheel, instantly turning the 1,000 pound 13 foot propeller. With his right hand on the engine primer, the copilot is giving a few steady strokes that feed atomized fuel into the top five cylinders of the engine. The pilot gazes out at the slowly spinning prop and carefully counts 9 blades. This will tell him that the starter has rotated the engine 3 times. With his hand on the magneto switch for the number one engine, he throws the mags to "both" and the propeller immediately flies into a barely visible arc.

The copilot releases the start and mesh switches, as well as the hand primer. The engine is a little rough and shakes the entire aircraft while billowing huge clouds of oil smoke, which blow out toward the rear of the plane from the propwash. The pilot moves the mixture control on the center pedestal into "auto rich" and makes very slight throttle adjustments, seeking 1,000 rpm on the tachometer. The co-pilot and flight engineer are staring intently at the right side of the instrument panel, watching the gauges for that engine. The oil pressure, cylinder head temperature, manifold pressure, and fuel flow all arc into the green while the bucking engine begins to smooth out. The process is repeated for the remaining three engines. Many times an engine had a developed personality. Some were reluctant to start, while others gave much more than expected. If all goes well, there will be no engine problems with anything in this early phase of the mission.

With all four fans turning, a last minute check is made of the vacuum system, which runs some of the critical flight instruments in the cockpit. Next the pilot's attention is turned toward the radios mounted over his head between the two windows in the roof of the flight deck. The co-pilot is going over a required 11 point check list intended to confirm proper operation of the instruments before him. The pilot has turned on the transmitter and receiver of the command set. "Voice" is selected on the filter box, and a knob on his headset jackbox, which is mounted on the fuselage wall at his left thigh, is turned to "command."

Pilot: "Crew check." He calls each of his ten men by their assigned station and waits for them to acknowledge. Until all the men have responded accordingly to the query of the pilot, he could not know if they were all at their respective positions, and that the plane's interphone was operating properly.

The ground crew stands by outside the bomber waiting for a signal from the pilot to remove the wheel chocks. This was usually done with both hands held together in fists with thumbs pointed outwards. The men on the ground would take special care to approach the main wheels from below and behind the wings to avoid the invisible spinning mass of the propellers. It was very easy to become casual about this procedure, and this behavior could and did result in the loss of life to the unwary man who walked right into the propeller's arc.

When the man on the ground returned to the front of the plane the pilot knew that the chocks had been moved from beneath the wheels and the plane was free to taxi.

A code word is broadcast from the tower's short-range radio to all the B-17s on the field. The mission is on, and from all over the field there are engines coming up on their power settings as the huge dinosaur like beasts begin to move out of their hardstands and to the taxi perimeter strip. While the pilot negotiates the narrow taxiways, he pays particular attention to the wing tips—he cannot see over the front of the plane. A wide turn could easily see the plane off the pavement and in the soft grass, and a plane could become bogged down and stop the convoy of B-17s on their way to the assigned active runway. The inboard engines are turning at a slow 700 rpm, and the outboards are used to turn the plane with just a touch of braking. This practice is much easier on the brakes and

The well-known Boeing "Totem Pole" logo which graced all of the control yokes of their fine products. These were often among the first items taken by souvenier hunters, even before the "Memphis Belle" left military service.

tires. Care is taken to not pivot the B-17 on one of the tires. This could scrub a flat spot into the low-pressure tire and ruin it very quickly. The pilots watch the waist codes on the moving planes as they go by to catch their position in the take off order. In the briefing, the planes had been scheduled a very prescribed order. This would make the assembly of the formation much less confusing after the launch. The separation between the planes as they proceed toward the far end of the field is kept at a couple of hundred feet, because it is very hard to stop a B-17 quickly.

The flight engineer has taken up a position in the cockpit and stands just between the pilots. From time to time he will even go up into the top turret and swing it around to scan the progress of the slowly moving bombers. From here, he can warn the pilot if the planes begin to stop and what others around them are doing. This can be a rather delicate ballet of sorts, and accidents are frequent. The tail gunner is back in his position, advising the pilots how the dance behind them is progressing.

As the bombers finally make their way to the end of the field, they all marshall near the edge of the active runway and gracefully angle their planes at roughly 45° for their engine run-ups and magneto checks. This is done so that the resulting propwash from the advancing engines will not affect the planes now situated only several yards away and behind them. Here is where a flight crew wants to find out if the engines are not running properly. On the minds of the men are B-17s that have had tragic take offs which have resulted in a crash when an engine or two was not running just right. And a B-17 full of fuel, bombs, and men always made an unbelievable fire when it hit the ground. Almost twenty minutes have passed from the time these lumbering beasts left their roosts across the field.

Trim tabs are set to zero, and the flaps are in the fully retracted position. Brakes are set. Engine run-up includes a test of the magnetos, which provide spark plug ignition to each engine. Throttle levers are advanced one at a time, and while watching the manifold pressure rise to 28 inches, the mags are moved from the "both" position to the "left" position, and then "right." While this is going on the three men in the cockpit closely monitor the engines' tachometer, noting the drop in engine speed, which is usually between 50 to 75 rpm. After returning the magneto switch to the "both" position, the prop rpm lever on the central control pedestal is moved through its range from low rpm to high rpm. Pilots will look out on the wings and at the prop arc, because as this is done the angle of attack of the blades changes, and this change can be visually noted. It is also audibly noted, because this action creates an unbelievable and unforgettable roar. Also, a glance at the tachometer again will note a drop of 300 to 400 rpm.

With the final pre-take off checks complete and assurances that the entire crew is ready and in position, the plane idles and waits for their turn on the active runway. The crewmen in the back do not know how long this will take. Their only cue to the take-off is the reduction of airframe vibrations as the B-17 comes up on power and the long, sometimes two mile roll begins. Along with this, of course, is the unmistakable song of their four Wright Cyclone engines beating out their hymn of synchronized fury as they begin to accelerate the mass down the runway.

Each turbo-supercharger is checked for proper operation, and the pilots complete their before take-off rituals. A scan of the other planes in the formation and their progress is made. The planes in the front of the assembly begin rolling down the runway at intervals of 45 to 60 seconds, and soon it is time for this plane to move forward.

The turn onto the active is not complete until the plane has tracked straight forward for several yards. This assures the pilots that the tail wheel is in position to be locked for the runway roll. The next pieces of equipment to be checked are done so in fast order, and the take-off is only moments away.

The list on their laps directs them to set the gyros and turn on the generators and the fuel booster pumps before setting the fuel mixture to "auto-rich." The props are moved into their high rpm settings, and a thumbs up from the pilot signifies that they are ready to go flying. On cue from the aircraft director's truck, which is parked in the grass just off the left wing of the plane, the pilot will take the traditional palm-up grip in the center of the throttle quadrant and advance all four engines to their stops. The manifold pressure gauge needle swings through 46 inches as the pilots asks the engines for no less than 100 percent and all 4,800 horses to make their take-off speed. Slight twists of the wrists manipulate the power demands from the engines, keeping the bomber in a straight track down the center line. The rudder comes alive at about 60 miles per hour, and the effect is noted in the feet of the pilots. A touch of back pressure on the yoke lifts the tail wheel from the concrete just a little bit so as to keep a nose high stance on the plane.

While the co-pilot takes over the throttles, he and the flight engineer note the critical engine instruments, watching various temperatures, fuel flow, pressure, even an overspeeding prop and velocity while the pilot guides the aircraft down the runway. Around 90 to 100 mph is achieved, the end of the runway is fast approaching, and the B-17 unsticks itself from the ground. Throughout the take-off the men are at the ready to get to the proper switches and controls should there be an emergency.

The ritual continues, and the well-rehearsed crew immediately works in synch. The pilot is careful not to pitch this heavy plane too high, and the co-pilot will tap the brakes to stop the rotating tires just before throwing the switch to bring the wheels up. Until they hear from the gunners in the back, they are not satisfied that all wheels are stowed. The pilots will look out at each wing to check the main wheels, and the gunners will report the status of the tail wheel. This is a little redundant, because the three indicator lamps on the instrument panel will show that all three wheels are up.

Instrument panel manufacturer's plate for the Boeing B-17

Somewhere around a mile or so from the airfield the pilot reduces the throttles to achieve a climbing speed of 150 mph. Engine rpm will show between 2,300 to 2,500 rpm at 38 inches of manifold pressure. Most of the time this is done by adjusting the pitch of the propellers. In just a little while the pilots will begin to turn the plane in an appropriate direction to begin the first stages of the assembly of the formation. This is done while the airplane climbs at a rate of only 300 feet each minute. Pilots often found themselves on instruments at this part of the mission because of the typical English weather. The planes would ascend into the gray skies and quickly lose all references to the ground and other bombers in the sky.

When the planes had passed through a couple of thousand feet the fuel booster pumps were turned off. If the engine-driven pump failed below this height, then the fuel booster pump continued to feed the affected engine while the pilot tried to get back to the field for a safe landing. If this were to occur at a reasonable height, then more time was available to the pilots to remedy the situation.

From the time of engine start through the landing hours away, constant monitoring of all flight and especially engine instruments kept the flight deck crew very busy. The strain on the engines of a heavy bomber was terrific, and there were many things that could go wrong. Knowing just how, when, or why to say crack the cowl flaps open a bit could mean the difference between getting back to base or not. The finesse of operating these sturdy engines properly and efficiently was only learned through time. It was also crucial to know every emergency procedure to plan for the inevitable engine problem. In the early G and all F model B-17s, the supercharger controls had to be excercised constantly to keep the regulator lines from freezing and creating sluggish operation or even failure. A good pilot had the habit of looking at the critical temperature gauges every few seconds.

While the bomber climbed through 10,000 feet, the crew was ordered by the pilot to go on oxygen. Masks were cinched tightly to the faces of the men, hoses were connected to supply stations, and their respective regulators were checked for efficiency. It was critical that the correct mixture of air and oxygen was fed to each man, and the regulator was built for this task. It fed the proper mixture depending on how high the plane was. The higher the plane, the thinner the oxygen, and the "on demand regulator" compensated for the changes in altitude. Crewmen got into the habit of unclogging their hoses of accumulated ice every twenty minutes or so. Condensation from their warm breath built up inside the hose and could close off the flow of oxygen. The affected crewman would not realize this until it was too late and he passed out, and death could follow quickly.

The gunners would busy themselves with the task of preparing their Browning fifty caliber machine guns. Ammunition would already be pulled down the feed chutes from their boxes and up to the weapon. The bullets were pulled up into the top of the breech and then seated before the top cover of the gun would be slapped shut. The charging handle on the right side of the machine gun was pulled back and the weapon was ready to be test fired. This would wait, though, until the formation was assembled and headed out over the English Channel. The gunners took great care to know the operation of the gun and were considered specialists. A thick coating of grease covered the weapon, because at high altitude the temperature of more than fifty degrees below zero could make the gun freeze up. Nothing was worse to a gunner than getting to altitude and facing the enemy with a weapon that when the trigger was pulled, instead of the expected high speed stuttering blasts, only sporadic and slow rates of fire were available. The bullet achieved the same velocity and just the rate of fire was affected—that is, if the gun did not freeze completely.

The ball turret gunner prepared to enter his tiny cocoon suspended below the B-17. By hand-cranking, the ball was de-elevated so that the twin fifties were pointed straight down. This exposed the entry hatch to the inside of the plane. Here the gunner was treated to a view of the ground thousands of feet below. Stepping down into the unit, he then connected his oxygen, interphone, and electrically heated suit. Another crewman often helped him secure the partially armored hatch and dog off its two closure pins. He was now free to grab the handles that controlled both azimuth and elevation and bring the turret into a position that had him sitting somewhat comfortably. He was also without a parachute—it did not fit into the turret with him. Should a bail-out be necessary, he woud require roughly a minute's time to maneuver the turret into position, throw the pins that were positioned over each shoulder, disconnect his suit, interphone, and oxygen supply, and stand up into the plane. This could be nearly impossible to do in a spinning plane with high "G" loadings. Once inside the B-17 he would have to locate his parachute and clip it on before he could effect his escape through the nearest available opening.

The ball turret gunner's position on the plane is often mistakenly referred to as the most dangerous crew station on a Flying Fortress. This is not true. The positions on a B-17 that resulted in the highest injury and casualty rates were those in the extreme nose and tail. Technically speaking, those that flew in the ball turret were probably the safest men on the crew. They certainly had, however, the most notorious station on the plane. From below the plane it was found that not only was this man a protective gunner, he was also a necessary extra set of eyes, often reporting damage to the pilots that they could never possibly see.

At this point the assembly of the formation was well underway. This could be a confusing time, which was often amplified by poor weather. Pilots would watch intently for different colored flares fired from the lead planes that would mark their place in the formation. On the ground, "buncher" and "splasher" radio stations broadcast pre-determined beacons to the planes, which were straining to climb for altitude as they headed out over the water from England. This while the formation was being made into the strength of an entire combat wing of B-17s and B-24s. Aboard the bombers, navigators and radio operators watched the needles of the automatic direction finder/radio beacon equipment to make assurances that the combat wing was headed in the right direction. These frequencies and beacons were queried for authenticity because the enemy had the practice of broadcasting false beacons to the American planes in the hopes of drawing them into a fight in their territory and on their terms.

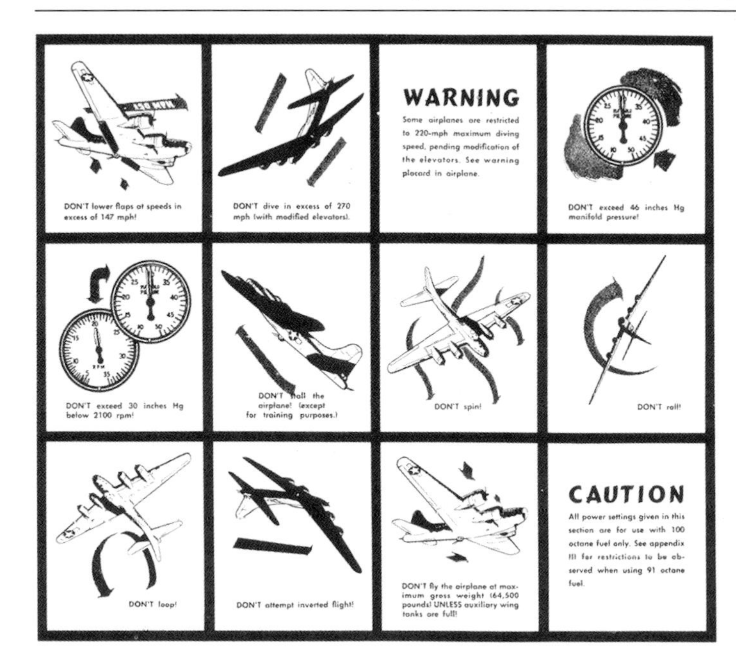

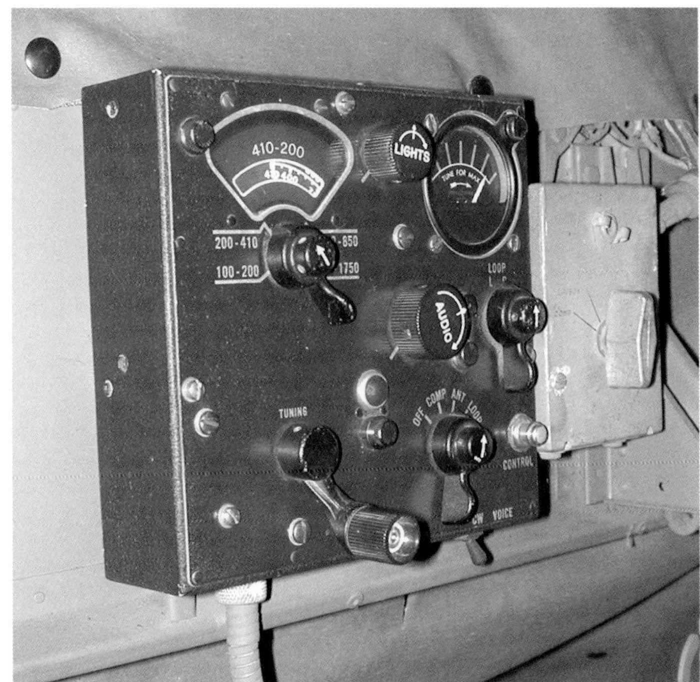

One of the radio beacon control boxes in the "Memphis Belle." The other is mounted in the ceiling of the cocpkit. The loop antenna for this unit lives inside the large football shaped housing mounted just below the nose of the plane. The operator would literally turn that loop with this control unit, enabling him to have a better idea of heading and distance from ground stations known as "bunchers" and "splashers." This was a critical piece of equipment that took training to become qualified to use proficiently. In this installation, the unit is mounted within the bombardier/navigator compartment on the bulkhead just at the pilot's feet.

As the planes headed out over the water, the gunners test-fired their guns—but only enough to see that they were operating correctly. They did not have large amounts of ammunition, and they knew that they had to conserve their bullets.

The pilots alternate flying the bomber to ease the strain and tension of flying the tight box formations. This often grew very tiring, and switching control of the plane from time to time reduced these effects. Plus it gave the co-pilot valuable time at the controls. In all likelihood, he would soon be transferred off this crew and given the command of his own ship.

Squadrons assembled into groups, which then joined with other groups to become combat wings, which then echeloned together to become combat divisions. When the "Memphis Belle" was built in July 1942, the USAAF reported a total of 44 B-17s in the European Theatre of Operations. That number rose to 108 B-17s by the time "Memphis Belle" flew her first operational raid on 7 November 1942. When the Belle was finished flying combat in May 1943 there were 599 operating Flying Fortresses in the United Kingdom. The highest number of operational B-17s in the Mighty Eighth Air Force was during the month of March 1945 when no less than 2,367 B-17s were flying raids. Given these numbers, it is easy to appreciate just what an assembled Combat Division of Flying Fortresses and Liberators—along with their escorting fighters—must have looked like as they left England to bomb targets in the Rhineland.

Radio silence between planes was critical, because everyone knew the Germans were listening. Signals from plane to plane were handled by everything from vari-colored flares to special lights installed in the leading edge wing roots, as well as the tail of each Fortress. The pilots were concerned with the status of the formation as they headed into each and every raid. There were the inevitable events when one or more bombers returned to base early with difficulty. This ranged from mechanical trouble to combat damage. There were many times when the groups were unclear as to what altitude they would fly at, and where they were to assemble in the formation. This resulted in pilots that felt they were pushed out of their positions. So it almost never happened that the bombers showed up over the target in the manner that the formation was laid out well before the raid launched. As airplanes left the formation, others would move into their spot to fill the vacancy. This confusion was always enhanced by the rigors of high-altitude flying, combat damage to the plane, injuries or death among crewmen, weather, and even navigational errors.

The gunners were always on the lookout for the moment the enemy fighters would arrive. Hours were spent looking into the sun, as well as scanning the blue sky all around. The gunners would report the status of the formation to the pilots, keeping them aware of the bombers all around them. The severe cold, the monotony of looking for fighters, the fallability of the human mind—and for that matter the plane, never left their minds. And then there was the flak.

Boeing B-17 Engineering Schematics

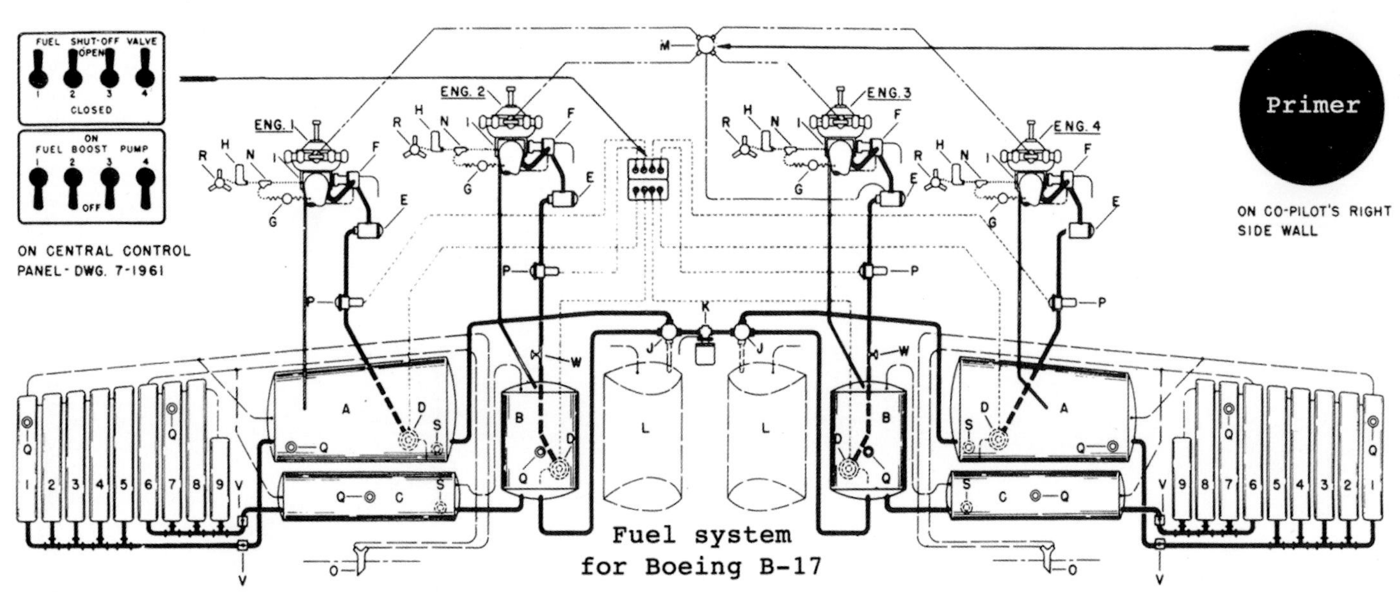

Fuel system for Boeing B-17

The fuel tanks and associated fuel lines aboard the B-17 are laid out in this diagram to show both the location and the fuel supply routing from tanks to tanks and from tanks to engines. Fuel consumption was always very carefully monitored, and the transfer of fuel between various tanks was frequently done with great care. Weight distribution was a large factor in the transfer of fuel. The plane's center of gravity would be affected, and the pilot had to be prepared for this. There were no long-range tanks ever installed in the outer wing sections of the "Memphis Belle" as the diagram above shows, but they were included in the building of most of the later models of the B-17. The transfer of fuel was the job of the flight engineer, who had his transfer valves located near the floor at the rear of the flight deck.

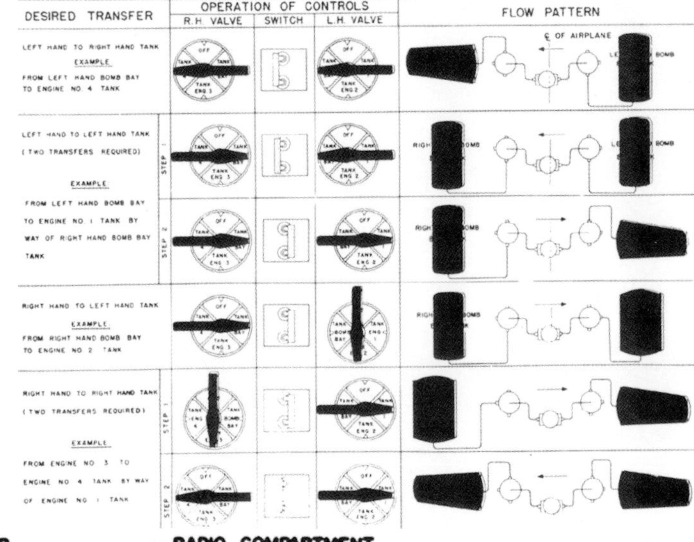

DESIRED TRANSFER		OPERATION OF CONTROLS			FLOW PATTERN
		R.H. VALVE	SWITCH	L.H. VALVE	
LEFT HAND TO RIGHT HAND TANK EXAMPLE FROM LEFT HAND BOMB BAY TO ENGINE NO. 4 TANK					℄ OF AIRPLANE
LEFT HAND TO LEFT HAND TANK (TWO TRANSFERS REQUIRED) EXAMPLE FROM LEFT HAND BOMB BAY TO ENGINE NO. 1 TANK BY WAY OF RIGHT HAND BOMB BAY TANK	STEP 1				
	STEP 2				
RIGHT HAND TO LEFT HAND TANK EXAMPLE FROM RIGHT HAND BOMB BAY TO ENGINE NO. 2 TANK					
RIGHT HAND TO RIGHT HAND TANK (TWO TRANSFERS REQUIRED) EXAMPLE FROM ENGINE NO. 3 TO ENGINE NO. 4 TANK BY WAY OF ENGINE NO. 1 TANK	STEP 1				
	STEP 2				

The oxygen schematic is shown below. This triple-redundant system supplied breatheable air at altitude through its network of tanks, lines, and regulators. Twenty large low-pressure bottles contained oxygen, which was distributed to the crew from various lo-

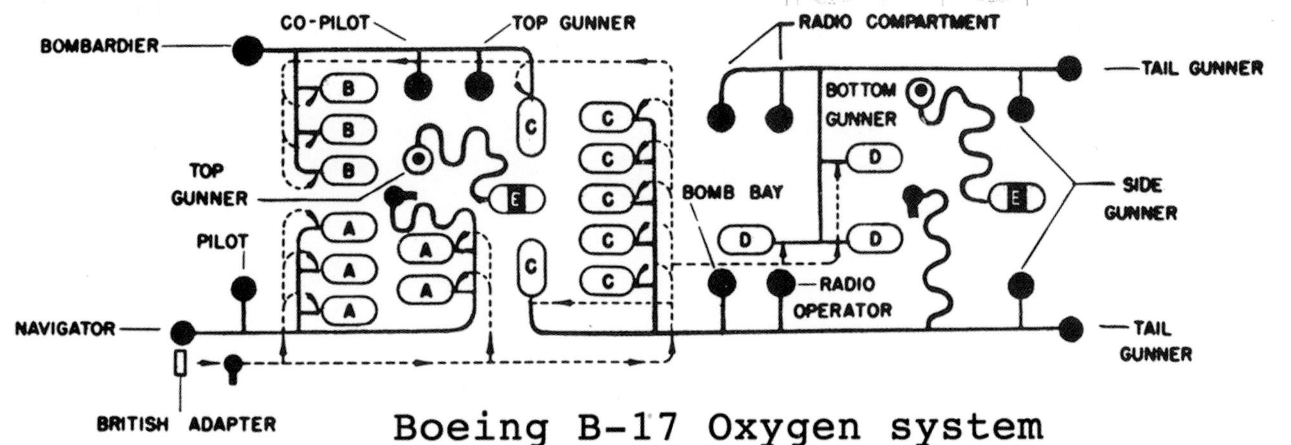

Boeing B-17 Oxygen system

Ⓐ- CYLINDERS AT LEFT SIDE OF COCKPIT
Ⓑ- CYLINDERS AT RIGHT SIDE OF COCKPIT
Ⓒ- CYLINDERS UNDER PILOT'S FLOOR
Ⓓ - CYLINDERS UNDER RADIO COMPARTMENT FLOOR
Ⓔ - TURRET CYLINDERS
● REGULATOR, TYPE A-12
◔ REGULATOR, TYPE A-9A
▬ DISTRIBUTION LINE
FILLER SYSTEM
FLEXIBLE FILLER SYSTEM

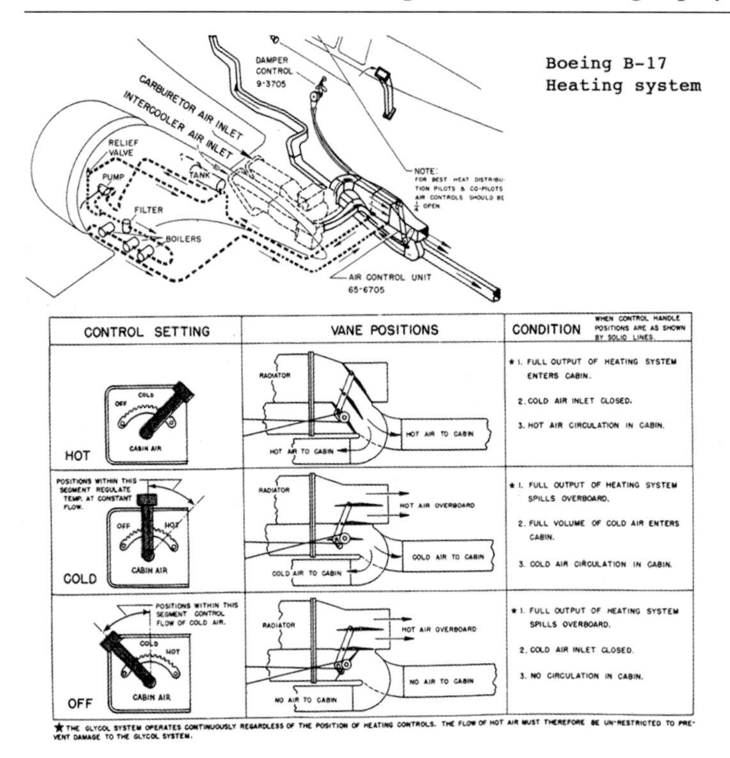

CONTROL SETTING	VANE POSITIONS	CONDITION WHEN CONTROL HANDLE POSITIONS ARE AS SHOWN BY SOLID LINES.
HOT	RADIATOR; HOT AIR TO CABIN	★ 1. FULL OUTPUT OF HEATING SYSTEM ENTERS CABIN. 2. COLD AIR INLET CLOSED. 3. HOT AIR CIRCULATION IN CABIN.
COLD (POSITIONS WITHIN THIS SEGMENT REGULATE TEMP. AT CONSTANT FLOW.)	RADIATOR; HOT AIR OVERBOARD; COLD AIR TO CABIN	★ 1. FULL OUTPUT OF HEATING SYSTEM SPILLS OVERBOARD. 2. FULL VOLUME OF COLD AIR ENTERS CABIN. 3. COLD AIR CIRCULATION IN CABIN.
OFF (POSITIONS WITHIN THIS SEGMENT CONTROL FLOW OF COLD AIR.)	RADIATOR; HOT AIR OVERBOARD; NO AIR TO CABIN	★ 1. FULL OUTPUT OF HEATING SYSTEM SPILLS OVERBOARD. 2. COLD AIR INLET CLOSED. 3. NO CIRCULATION IN CABIN.

★ THE GLYCOL SYSTEM OPERATES CONTINUOUSLY REGARDLESS OF THE POSITION OF HEATING CONTROLS. THE FLOW OF HOT AIR MUST THEREFORE BE UN-RESTRICTED TO PREVENT DAMAGE TO THE GLYCOL SYSTEM.

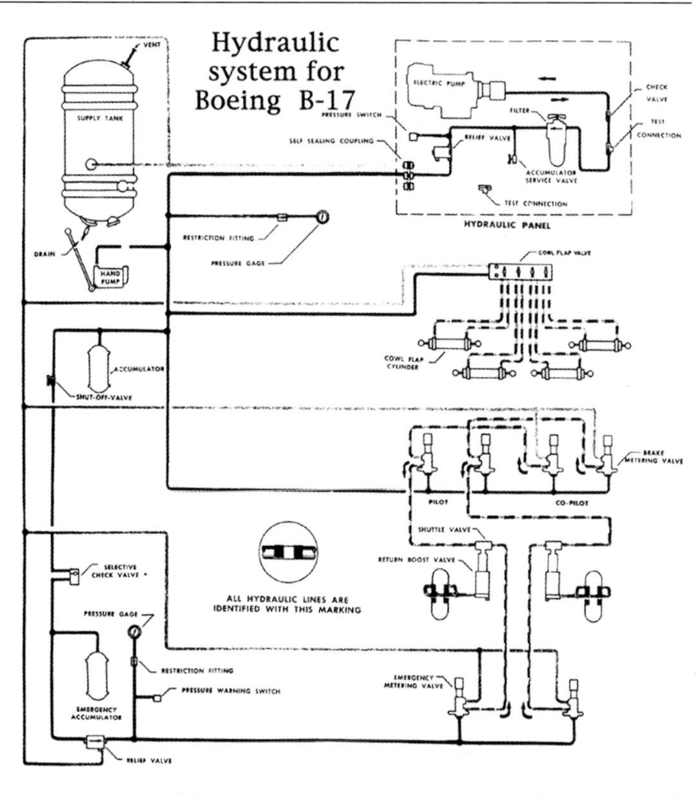

cations throughout the plane. All regulators were not the same at each crew station. There were even replenishment points where portable walk-around bottles could be refilled by members who needed to move thoughout the plane in flight. The heating system could be described as marginal at best. The B-17 was not pressurized and was lightly insulated. Heat from the engines was ducted to the nose, flight deck, and radio room of the bomber. This provided little measurable difference between the outside air temperature and the temperature inside the plane.

The diagram in the upper right shows the important hydraulic schematic. The fluid resevoir was located at the rear of the flight deck and was mounted on the bulkhead just to the rear of the top turret assembly. The components that were operated with this fluid were the cowl flaps and brakes. A hand-pump was situated on the bulkhead of the cockpit at the right thigh of the co-pilot. He would use this pump to pressurize the system and open the cowl flaps prior to engine start.

Wing de-icer boots were fitted to the leading edges of the main wings, as well as the horizontal stabilizers and vertical stabilizer. These were inflated on demand by the vacuum system, which also operated several critical cockpit instruments. This included the important turn and bank indicator. The boots themselves were thought to be somewhat troublesome in combat, for if they were damaged they could peel off in flight, damaging the aircraft or others nearby in the formation.

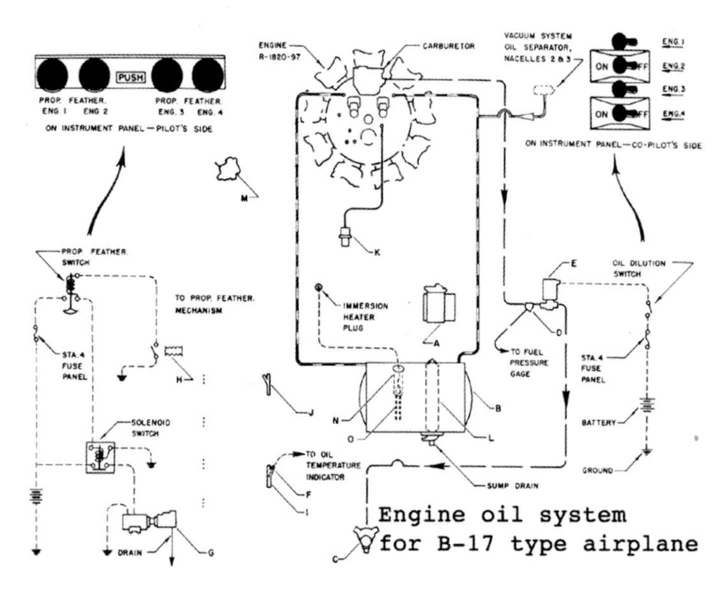

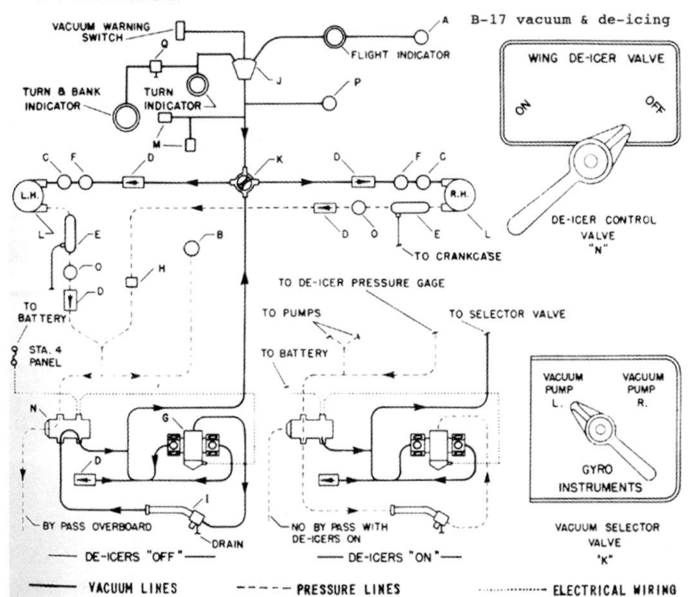

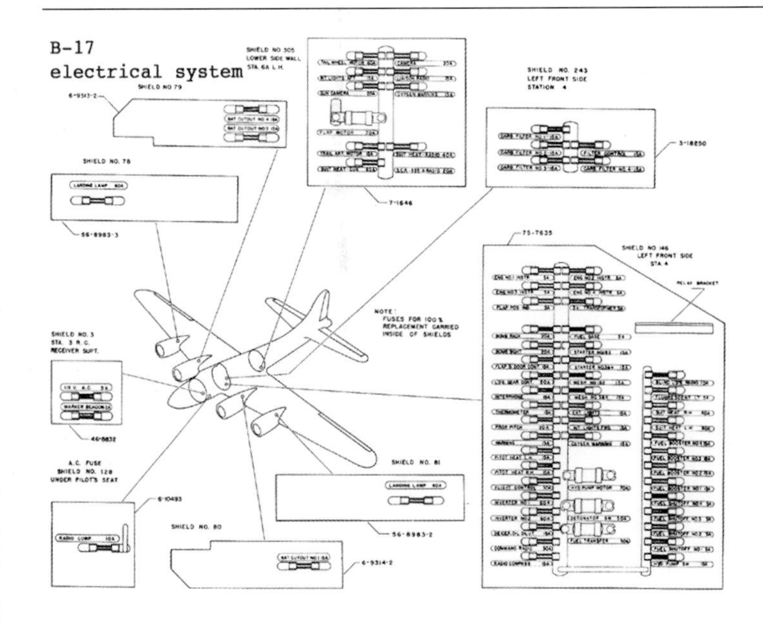
B-17 electrical system

The engine oil system was of critical importance. Almost 147 gallons of oil were provided in four tanks for engine operation. The tech order omission of a critical stand pipe in the tanks resulted in some combat tragedies. When the oil in the tanks was lost (in combat) there was no longer enough left in the system reserve to feather a propeller. A windmilling prop was the result, and often these wrenched free of the plane, causing terrible damage. This was rectified by the ordered reinstallation of these important stand pipes.

The electrical system in the B-17 is rather extensive, and its network runs, of course, throughout the entire aircraft. It services everything wingtip lights, flap motors, landing gear, gun turrets, and cockpit instruments, as well as the radios and intercoms. The main fuse panel is located at the rear of the flight deck on the bulkhead behind the top turret assembly. This completed the area occupied by the flight engineer/top turret gunner, who aside from gunning was available to give his attention to the flight instrument panel, fuel transfer controls, hydraulic equipment, and main electrical panel. Most of the equipment in the Flying Fortress operates on a 24 volt DC supply. Three batteries for the supply of power before engine start are located within the leading edge of the main wing roots. Large electric inverters are situated beneath the floor of the co-pilot's seat. These provide alternating current for various systems.

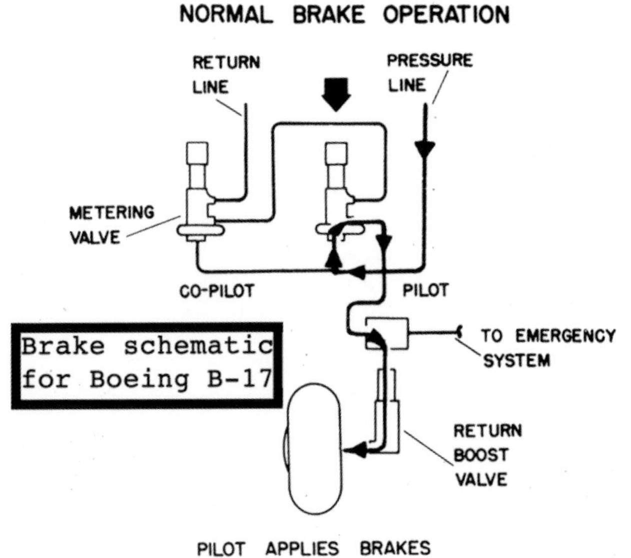

Brake schematic for Boeing B-17

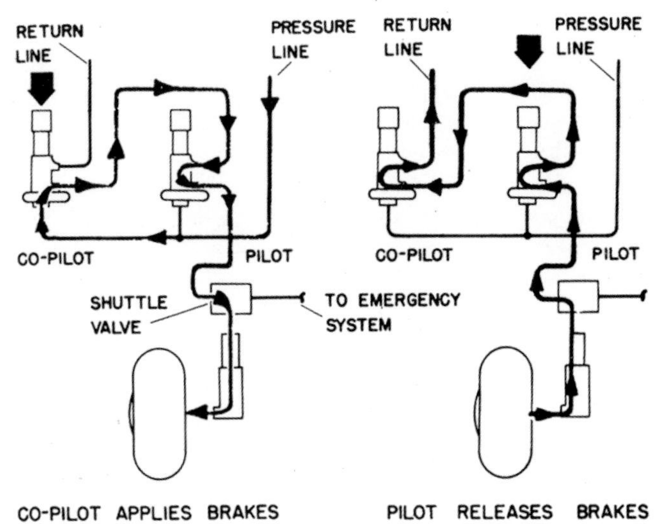

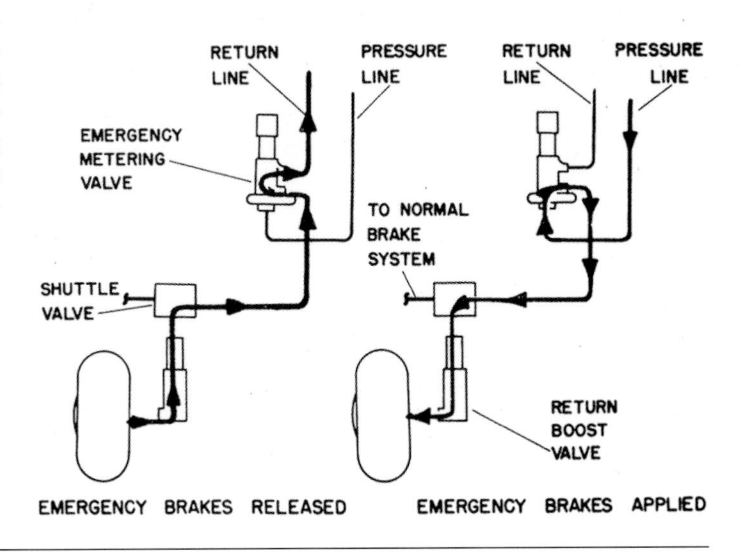

Fliegerabwherkanone

The enemy you could not shoot back at! This was often referred to as flak, and was as damaging emotionally as its explosive destructive power. The word described both the gun as well as the round it fired. Typically, high value areas such as aircraft factories, oil refineries, and ball bearing plants were ringed with these cannons, which could fire a barrage into the bomber stream five miles above. The Germans designed a multi-purpose weapon with a rifled 88mm barrel mounted on a stabilized cruciform. The gun could be configured for general artillery, anti-tank, and of course, anti-aircraft roles, with the latter being its most effective adaptation.

There were three systems of fire control; direct line-of-sight; indirect fire; and anti-aircraft. Some targets of the Eighth Air Force had hundreds of these cannons protecting them. Their main objective was to ward off attacks from the heavy Allied bombers. The idea was to fill the sky with shrapnel that would perforate the thin skinned airplanes flying overhead. A round would be fired into the general direction of the bomber stream, and its fused explosive head would detonate at altitude, sending blossoming steel fragments in every direction. The range and speed of the American formations was calculated, and hundreds of guns on the ground would erupt, sending these shells into the skies. This was a very effective weapon in the German arsenal that resulted in the loss of thousands of Allied planes.

American bombardiers were often able to see the muzzle flashes through their bombsights. And the other men aboard the B-17s and B-24s would sit and wait at their stations for the blast that had their number on it. There was absolutely nothing they could do but sit there and "wet your pants," as one crewman put it.

The German tactics had the Luftwaffe fighter planes attacking the bombers on the way into the target. After slicing through the formations, they would stand off some distance from the target area and let the flak gunners on the ground would take over, trying to destroy every American airplane they could. When the bombers left the target area the Allied planes would be out of the range of the guns, but back in the sights of the German fighter planes. Allied bomber crews were subject to attack from the moment they crossed over the enemy coast all the way into and away from the target, and back out across the English Channel. This often lasted between an hour for a coastal raid through eight hours for deep mission penetrations.

The general feeling among Allied bomber crews was that there was nothing worse than the flak. And the flak was always there—on every mission. It was dreaded even before engine start. Many men could not take its impact on their minds and found that they could not fly missions because of the toll it took on their nerves.

There were hundreds of tasks that had to be performed on a regular basis, and several tasks that had to be made up as the crews went along. Some bomber crews found themselves in very unique situations and adapted to the emergency at hand. Some flight engineers made repairs to critical systems in the air. Pilots were forced to find new ways to fly a B-17 after it had been severely damaged during a raid. Sometimes they were trying to fly on a single engine while trying desperately to restart another or more after they had been shot out. Commanders had to make decisions when targets were covered by clouds, and also what to do with their crew if they were unable to make it home and a bail out was inevitable.

Every man on the crew was a specialist in their job, and they often cross-trained so that they could fill in if necessary. Eyes always scanned the skies for anything from enemy fighters to flak, and even an unwary Allied plane that would wander too close during formation flight. Allied cover fighter planes were sometimes shot at when their forgetful pilots accidentally pointed their noses at the bomber streams.

The crew was exhausted when the mission was over. They had been hard at work for some eight to ten hours. They had fought in extreme conditions and even rendered first-aid to their comrades. There are plenty of tales from these men where they endured hours of flying back aboard the plane with their friends' bodies on board.

Now that it was time to head back to the base, the crews would take a moment to reflect on what they had just been through. Most of them were counting another mission checked off and thought of the number left to go. But they learned to never let their guard down. Danger was still everywhere, and some found themselves falling victim to relaxation. As the formation loosened up, the opportunity for midair collisions arose. It was of paramount importance to con-

German flak gunners open up on a formation of bombers overhead. Some Allied targets had several hundred of these guns protecting their factories and towns.

tinue to keep the formation stacked tightly. Just as important was the continued monitoring of the engines. They had been hammering out their drone for hours and were often damaged. A wrong input to the tired powerplants or failure to give them attention could result in the loss of that engine on the way home.

Even as the formation approached the coast of England, precise attention was needed to fly back into the country. Entering a no-fly zone would bring the quick attention of British flak gunners. London was a no-fly zone, as some navigators sadly found out when they were fired upon.

The formation had begun to let down from altitude. As they passed through 10,000 feet or so frozen airframes began to thaw, but it was still very cold and many of the men had frostbite. Sometimes this was so bad that fingers had turned black. About this time the pilot would tell the crew that they could go off oxygen. Some of the crew used this time to eat a snack or have some coffee from a thermos, while others headed to the bomb bay relief tube to urinate—that is, if they felt like going through the task of locating their "equipment" through the layers of flying gear. The gunners unloaded and make their machine guns safe. The radio operator continued to monitor ground transmissions, and the ball turret gunner exposed his hatch to the inside of the Fortress while he began to climb up and out of the ball. It would take some time for him to straighten up because he was sometimes inside the 44 inch sphere for seven hours or more.

As the bomber descends, the pilots turn off the fuel booster pumps and turn on the carburetor air filters. The formation crosses the coast of England and familiar landmarks are seen, as both turrets are moved to their stowed positions. The top turret is azimuthed so that the guns are pointed straight back towards the vertical stabilizer. The ball turret is pointed the same way, and special care is taken to assure that the weapons are elevated to the horizontal. If they were left in the lowered elevation, then the barrels would strike the ground when the B-17 landed.

The combat wing begins to break up as various groups point toward their bases. The descent is planned so that the planes are only flying at several thousand feet when their airfield is spotted. Bombers who suffered combat damage, crew injuries, or fatalities were given priority to land first. The pre-landing checklists make their way onto the laps of the pilots, and the landing ballet begins.

The pilots begin to stack their planes so that an orderly procession of B-17s will fly along the edge of the airfield parallel to the assigned runway for landing. On the ground, all the members of the group are busy counting the returning planes and hoping that there are the same number that had taken off hours earlier. The planes reach a point at around 1,000 feet and begin individual turns to the base leg of the approach. This is done to assure that there will be sufficient spacing between the planes as they touch. As they peel off, one by one, several seconds elapse before the next bomber for the base leg approach phase of the landing makes his turn.

The pilot calls "crew positions" over the interphone, and an acknowledgement comes back from a member of the crew that has checked that the men on the plane had moved to their positions in the radio room or the floor near the ball turret. The next item on the checklist is "auto-pilot off, fuel booster pumps on." The fuel mixture is moved to auto-rich, and the intercooler shutter levers are then moved to their off positions at the co-pilot's right leg. Then the co-pilot reaches in front of him and switches the four carburetor air filters to the on position.

The pilot reaches for the throttle quadrant, and with his palm up grip retards the levers some to slow the B-17. When the airspeed falls below 147 miles per hour, the pilot will ask the co-pilot for a "notch" of flaps. This has an even greater slowing effect on the plane, and power is cautiously applied to keep the aircraft from becoming too slow and headed for a stall.

The pilot calls for "gear down," and the co-pilot reaches for the switch located near the right of the central pedestal, moves the switch to the "down" position, and looks for the gear down indicator lamps to illuminate. As the screw rods at the mains turn, the huge main tires begin to lower and the drag on the airframe increases. With the flaps being moved to their full down 30 degree setting a little more power is added to avert a stall. With the flaps and wheels all down the B-17 remains a very predictable and stable platform. The turn to final approach has already been made, and the B-17 continues its descent—the altitude of the plane being controlled mainly now by the throttles. The pilots have a clear view of the runway through their windscreens and continue to monitor the track of the plane as they maintain glide scope and glide path. The switch for the landing gear is confirmed to be moved into its neutral detent, and the wheels are scanned visually to assure they are down and locked.

The pilot asks for "high rpm," and the co-pilot responds by moving the prop pitch control levers to a point where they register around 2,000 rpm while they look for about 38 inches of manifold pressure. If they bolter this landing and have to go around again, the plane will need enough power to get back into the air. These settings will allow a margin of safety.

The central control console of the "Memphis Belle" shows the magneto switches on the left, booster pump switches, cowl flap controls, and flap and gear switches, as well as turbo-supercharger and fuel mixture levers. Note that the levers for the number one engine turbo and the number four engine mixture settings are missing. They were among the items taken by souvenier hunters over her long neglected years of sitting unprotected.

When the plane nears the runway threshold the pilot asks for "props, high rpm" and "full flaps" as the airspeed needle degrades through 100 miles per hour. The bomber is now situated just a touch nose high, and the picture of the runway is never as clear as the pilots of this tail-dragger would like. But the idea is to stall the plane just inches above the concrete—much in the same manner that a bird lands on a fence post. A three-point landing is the desired posture, and a good deal of finesse is needed to achieve this, even in a B-17 that is not battle-damaged.

The airspeed falls through 90 mph, and the degrading air pressure above and below the wings forces the plane to surrender to earth's gravity. A noticable jolt assures the crew that the B-17 has returned to terra-firma and is again out of her natural element in the sky. Careful attention is paid to braking, and these are not applied until the plane slows sufficiently. Hard braking in a heavy bomber will assure that the linings are burned out very quickly, and even a fire could result. The co-pilot opens the cowl flaps during the rollout phase of the landing. This slows the plane even more because of the induced drag, and also assures that the engines won't overheat during the taxi to the hardstand.

The tail wheel is unlocked so the plane can be turned from the runway as the turbos, fuel pumps, and generators are all turned off. Then the flaps are raised while the plane turns down the taxi strip. As earlier in the day while they were moving out for the take off, asymmetrical power settings are used on the outboard engines for turning control. The fuel mixture levers for engines two and three are moved to "idle cut-off," and the magnetos for those engines are turned off. The propellers whisper into silence but are not feathered.

Turning into the hardstand, the flight crew sees their ground crew standing by. One has his arms in the air in front of and on the left wingtip so the pilot can see him. Two more have the wheel chocks in their hands, and as the Fortress' brakes emit a final squeal, the plane stops and her wheels are chocked. Pilots are careful not to apply the parking brake, because the linings could weld themselves to the brake drums. This will be done later when they sufficiently cool off. Engines one and four are briefly run up to clear the spark plugs of any carbon build-up that may have occurred during the taxi, and then their fuel mixture levers are moved down into "idle cut-off." "Magnetos one and four off." "All electric switches and master ignition bar off."

All flight control surfaces are moved to their neutral positions, and the pin with the red flag that had been stowed on the control column is unclipped from its holder and put into place.

What was only moments before a roaring beast now sits silently in her roost, and the voices of the men on the ground are a little unusual. The flight crew has to acclimate their ears to the absence of the din of four Wright Cyclone engines. As ever, the men are careful to not walk beneath an engine at this time. The concern is not that they might walk into a spinning propeller, but that scalding oil would drip onto them. It has been said that a radial engine only stops leaking when its out of oil. Light wisps of smoke waft from around the engines and nacelles, and an ever present ticking sound comes from the cooling vanes of each engine.

The B-17 has come home to her base and is ready for a rest. Her loving ground crew will immediately begin patching and repairing. They will not sleep until their baby is ready for another raid. They'll work hard—it will probably be tomorrow.

Above, Opposite: July 26, 2000, Warner-Robins Air Force Base, Georgia—116th Strategic Bomb Wing, the ultimate heritage formation. B-1 #86-0133, dedicated as "Memphis Belle," takes up the position on the starboard wing of the Tallichet movie "Memphis Belle" B-17. This is the first time the two very different types of bombers had ever flown together. While the B-17 flies her missions from air show to air show throughout the United States, her counterpart takes up her role as enforcer of freedom flying for America all over the world. The photo was taken from an F-15 Eagle at an altitude of 9,500 feet. To remain in formation, the B-1 slowed and has her leading edge wing slats deployed while the B-17 had her throttles all the way to the stops. The nose art on the B-1 was painted by noted airbrush artists Dru Blair and Mickey Harris. The B-17 nose art was done by Ron Kaplan. The formation flight was made possible by Col. Tom Lynn and C/Msgt. Glenn Parker of the 116th Bomb Wing.

Left: This is the battery/generator control box of the "Memphis Belle. It is situated at the pilot's left leg on the fuselage wall of the cockpit. Note on the upper left of the panel is a passing light switch. This illuminates a red lamp that is located just outboard of the number one engine in the same housing as the landing light in the wing's leading edge. It was used to signal other planes in the formation that the bomber was moving in the flight. This was done to maintain radio silence during the mission.

The crew of the "Memphis Belle," seen here in a series of poses in front of their charge. The photo in the upper left was snapped in 1943 by the War Department. Below that is a picture from a gathering of the crew during the 1960s. On the right are five of the original combat crew of the "Memphis Belle" gathering in 1987 in Memphis. They are seen at the nose of the C-141 "Memphis Belle," which is operated by the 155 A.S. of the Tennessee Air National Guard. And below that the men are seen in front of the very plane that made them famous. Left to right: Harold Loch; Jim Verinis; Bob Morgan; Tony Nastal; and Bob Hanson.

In what must be one of his proudest moments, the author is seen here standing behind the greatest man in "Memphis Belle" history. Eighty-six year old Frank G. Donofrio is often referred to as the man who saved the Belle for Memphis. Frank formed the non-profit Memphis Belle Memorial Association, Inc., in 1967. Partially in tribute to his brother, who was killed during the invasion of Iwo Jima, Frank began his career with the famous plane to enshrine and memorialize the veterans of WWII. He tells his friends today that some day he will get himself a job that pays him money! Frank has tirelessly carried the cause of this great airplane for more than 33 years, and is often seen to this day at the pavilion, answering the questions of tourists during their visits. (Photo by: S/Sgt Angela Stafford, USAF photographer, Washington, D.C. - Pentagon)

These WWII veterans, most of them former airmen, have spent years volunteering their services at the Memphis Belle pavilion in downtown Memphis, TN. Visitors and tourists get to hear much more than just the story of the Belle first hand from some of these very important men. - Chip Long

The pilot and the author pose for a picture at the front of the Belle. Perkins has been a close friend to Bob Morgan for nearly a decade, and has had many chances to hear Morgan's special stories from the Colonel himself. Right: The business end of the cockpit in the "Memphis Belle." The iron gate throttle quadrant of the B-17 was quite unique and a brilliant design. The arrangement of each lever enabled the pilot, the co-pilot, or even the flight engineer to quickly answer the powerplant demands of each 1,200 horsepower Wright Cyclone engine. To be one of the fortunate to sit in the cockpit of this special plane today becomes a trip into the past. A time filled with bravery, honor, duty, and courage. Everything from bullets and bursting flak to bandages. A love story that beat the passage of time in an old plane with round engines.

The "Memphis Belle" during her very last days in England. The bomber stands out against the British countryside despite her mottled camouflage. With her combat days behind her, the plane was pointed in the direction of American aviation legend and fame. Of the forty-seven or so B-17 airframes left in all the world, "Memphis Belle" today stands in honor of all 12,731 built. Her missions today no longer take place in the freezing danger of Nazi controlled airspace, but instead on the banks of the Mississippi River in Memphis, Tennessee. Here she "flies" on in tribute to all those who paid the ultimate sacrifice by giving their lives to defend freedom. The "Memphis Belle" is recognized by many as one of the definitive bombers of the Twentieth Century—and some historians even say that this plane is ranked among the top five American aircraft ever built. In these minds, she flies in an honored formation with the Wright Flyer, Spirit of St. Louis, Enola Gay, and General Chuck Yeager's Belle X-1. The "Memphis Belle"—just one airplane of one Squadron, of one Group, of one Air Force, of fifteen Air Forces. Remembering all those brave young men who are to be Forever Honored - Never Forgotten.

Thank you Mickey Harris and Dru Blair for your hard work and dedication to the "Memphis Belle" War Memorial Foundation here in Memphis, TN. For those readers who wish to purchase one of these limited edition prints, contact Dru on the world wide web at www.drublair.com. The purchase of this print will help to build the museum that will enshrine this great aircraft.

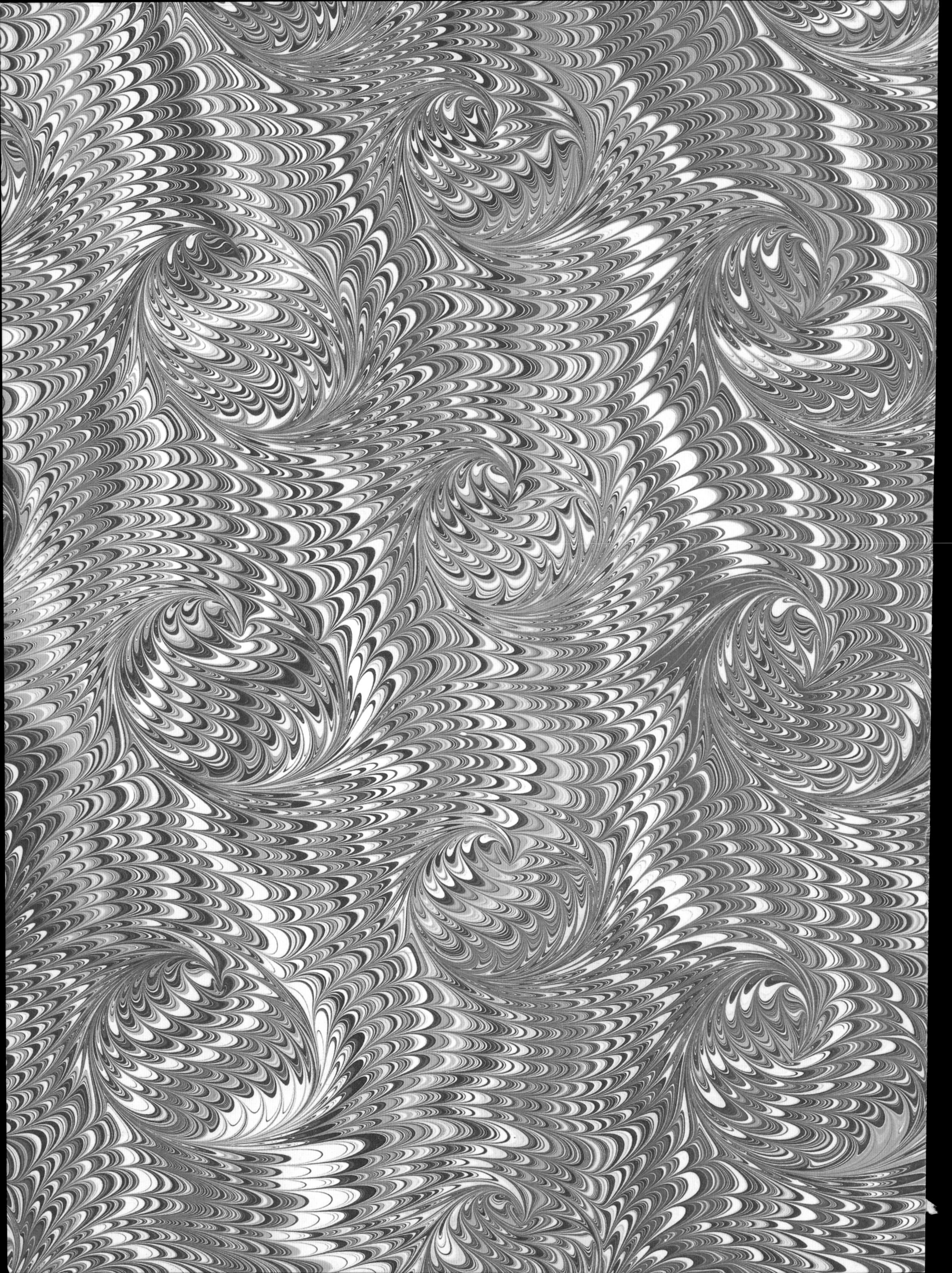